Environment and Climate Change in Asia:
Ecological Footprints and Green Prospects

Victor R Savage and Lye Lin-Heng

NUS
National University
of Singapore

Environment and Climate Change in Asia:
Ecological Footprints and Green Prospects

Victor R Savage and Lye Lin-Heng

PRENTICE HALL

Singapore   London   New York   Toronto   Sydney   Tokyo   Madrid
Mexico City   Munich   Paris   Capetown   Hong Kong   Montreal

Published in 2011 by
Pearson Education South Asia Pte Ltd
23/25 First Lok Yang Road, Jurong
Singapore 629733

Acquisitions Editor: *Angela Yeo*
Project Editor: *Neo Yi Ling*
Prepress Executive: *Kimberly Yap*

Pearson Asia Pacific offices: *Bangkok, Beijing, Hong Kong, Jakarta, Kuala Lumpur, Manila, New Delhi, Seoul, Singapore, Taipei, Tokyo*

Printed in Singapore

4   3   2
14   13   12

ISBN   978-981-06-9014-4

PRENTICE HALL   www.pearsoned-asia.com

This book is dedicated to all the endangered organisms on Earth. May we listen to the voices of Nature, find wisdom in indigenous knowledge, nurture leaders with political will and moral courage, and may each of us take on the responsibility of ensuring the sustainability of our beautiful world.

The MEM programme would like to thank Shell Companies in Singapore for generously sponsoring the publication of this book.

CONTENTS

PREFACE

This book marks a significant milestone in the history of the Master of Science in Environmental Management (MEM) Programme in the National University of Singapore (NUS). This book is only one of the celebratory events to mark the MEM's 10th anniversary (2001–2010).

We celebrate the anniversary for three reasons.

Firstly, there are few postgraduate programmes in NUS that are as multi-disciplinary in pedagogical practice. The MEM was initiated with the academic inputs of seven Faculties: Arts and Social Sciences, NUS Business School, Engineering, Law, Medicine, Science, and the School of Design and Management (SDE) which hosts the Programme. It also has teaching staff from the Lee Kuan Yew School of Public Policy (which was not yet in existence at the programme's inception in 2001), and the expertise of professors from the Yale School of Forestry and Environmental Studies, with which the programme has a Memorandum of Understanding since its launch in 2001. Few postgraduate programmes, locally or overseas, have the distinction of being able to run modules so successfully based on the teaching inputs from varied Faculties and disciplines. Students are thus fortunate to be exposed to a wide-range of disciplines, from public health to green technologies, environmental law to economics, and environmental planning to industrial ecology. The success of the Programme is best measured by the number of foreign students it attracts annually – 40 per cent of the students are from the Asia-Pacific region and beyond.

Secondly, the MEM is a rewarding academic experience for students by exposing them to the linkages between 'town and gown'. Not only are certain modules taught by people involved in the private sector and non-government organizations (NGOs), there are also the fortnightly seminars on Friday evenings which tap the expertise of industry experts and government officials on their views in environmental management. These seminars may take the form of public forums. They form part of the MEM students' curriculum, designed to broaden their perspectives on environmental issues and effective management.

Thirdly, it is heartening to state that this MEM Programme was not a top-down directive from the University's Administration, but rather a bottom-up initiative from environmentally-interested faculty members. It was these passionate faculty members across Faculties and disciplines who pulled together their interests and helped establish, with their respective Dean's endorsement, the MEM programme. Their environmental interest has continued to sustain and support it over the last 10 years. Indeed, many of these faculty members are also involved in the newly-initiated (AY2011–2012) multi-disciplinary undergraduate Bachelor of Environmental Studies (BES) programme.

Given the cascading impacts of climate change affecting environments and organisms, including human beings, no programme can be more pertinent in understanding the science of Nature and the need to effectively understand the socio-economic, political and cultural perspectives in managing environmental sustainability. The contributions in this book are as varied and multi-disciplinary as is required to comprehend environmental management. I am sure there are many lessons and insights from varied disciplinary and practitioners' vantage points for both the research and applied environmental benefits of students, corporate personnel and public administrators. Most of all, I hope this book serves as a focus for continuing debate and dialogue among interested stakeholders in ensuring a sustainable world.

I applaud all the administrators, Faculty deans and faculty members for their support of the MEM Programme. In particular, I thank various institutions and corporations (Shell Group of Companies in Singapore, Tan Chay Bing Educational Fund and the National University of Singapore Society) for sponsoring bursaries and scholarships for our MEM students. Specifically, I thank the Shell Group of Companies in Singapore for their unwavering financial support for the programme; this book is one such outcome of their corporate endorsement of the Programme.

Professor Tommy Koh,
Chairman of the International Advisory Committee,
MEM Programme,
and Singapore's Ambassador at Large

ABOUT THE AUTHORS

Marian CHERTOW

Marian Chertow is a professor of industrial environmental management at the Yale University School of Forestry and Environmental Studies. Her research and teaching focus on industrial ecology, business/environment issues, waste management, and environmental technology innovation. She is also appointed at the Yale School of Management and the National University of Singapore where she is an adjunct member of the faculty of the M.Sc in Environmental Management program, teaching the course on Business and the Environment. Professor Chertow serves on the National Advisory Council for Environmental Policy and Technology (NACEPT) that advises US EPA and is incoming President of the International Society of Industrial Ecology (2013–2015).

CHOU Loke-Ming

CHOU Loke-Ming is Professor in the Department of Biological Sciences, National University of Singapore. His research interests include coral reef ecology and restoration, and integrated coastal management. He serves on the Science & Technical Advisory Committee of the Global Coral Reef Monitoring Network (International Coral Reef Initiative) and is a member of the Joint Group of Experts on the Scientific Aspects of Marine Environmental Protection. He has provided consultancy services in the field of marine environment management to UNEP, FAO, World Bank, WorldFish Center and many national agencies. He is a member of the MEM's Programme Management Committee, representing the Faculty of Science, and teaches the module "Environmental Science".

Esther Sekyoung CHOI

Esther Choi graduated in 2011 from the master's program in environmental management at the School of Forestry and Environmental Studies at Yale University. She holds a bachelor's degree in environmental sciences from the University of California, Berkeley. Her research interests include economic and social incentives to influence consumers' behavior and challenges associated with industrialization and urbanization around the world. Esther is currently working as a researcher at the Ecological Economics Research Laboratory at Seoul National University.

Richard T CORLETT

Richard T Corlett is a Professor in the Department of Biological Sciences at NUS. His major research interests include terrestrial ecology and biodiversity conservation in tropical East Asia, plant-animal interactions, urban ecology, invasive species, and the impacts of climate change. He is the author of several books, including *The Ecology of Tropical East Asia*, published in 2009 by Oxford University Press, and *Tropical Rain Forests: An Ecological and Biogeographical Comparison*, co-authored with Richard Primack, published by Wiley in 2011.

Asanga GUNAWANSA

Dr. Asanga Gunawansa holds a Ph.D. in law from the National University of Singapore (NUS) and a LL.M in International Economic Law from the University of Warwick. He is an Attorney-at-Law of the Supreme Court of Sri Lanka and has over 15 yrs of experience as a legal counsel. Dr. Gunawansa is currently attached to the School of Design and Environment of NUS. His current teaching and research areas include: Construction Law, Arbitration, Legal Aspects of Project Financing, Public-Private Partnerships, and International Environmental Law. Dr. Gunawansa is also a Faculty Research Associate of the Institute of Water Policy, Lee Kuan Yew School of Public Policy, and an Associate Member of the Executive Committee of the Law Faculty's Asia Pacific Centre for Environmental Law (APCEL). He is a member of the MEM's Programme Management Committee, representing the School of Design and Environment.

Carsten M HÜTTCHE

Carsten M Hüttche is the founder and director of environmental consultancy firm Environmental Professionals (Enviro Pro), which he started in 1998 in Singapore. Since then, he has expanded his business activities to include green infrastructure projects for Asia's water resources sector. He is a green entrepreneur aiming to push the envelope for more natural environments and green designs in Asia's fast growing cities. In 2006, his firm and partners were awarded Singapore's "Innovator of the Year" award.

He holds a Masters Degree in Biology from the Freie Universität Berlin, Germany and held a position as a Senior Adjunct Lecturer for the School of Design and Environment, National University Singapore, between 2008 and 2010, teaching in the MEM programme.

Carsten is a regular senior consultant for International Funding Institutions such as the Asian Development Bank, World Bank and bi-lateral development agencies. He specializes in environmental safeguards and climate change adaptation.

HO, Hua Chew

Dr Ho Hua Chew is currently a member of the Executive Committee (EXCO) and also the Vice-Chairman of the Conservation Committee in the Nature Society (Singapore). He helps to co-ordinate the conservation activities & projects of the Society, such as the formulation of conservation proposals, feedbacks to government land-use & development plans, biodiversity surveys, etc. He has been doing conservation work for the Nature Society for more than a decade, in the course of which he was involved with the formulation of the conservation plan for Sungei Buloh and the Master Plan for the Conservation of Nature in Singapore. His main fields of expertise are biodiversity conservation and environmental ethics. He also lectures part-time on Environmental Ethics as well as on Biodiversity Conservation at tertiary institutions.

David KOH

Professor David Koh is Director of the Centre for Environmental and Occupational Health Research in the Saw Swee Hock School of Public Health in the NUS. He is an editorial board member of several journals, including Occupational and Environmental Medicine (UK), Occupational Medicine (UK), Journal of Occupational Health (Japan) Australasian Journal of Dermatology, and a member of the International Advisory Panel of the East West Centre, USA. His research interests are in occupational and environmental health. He is a member of the MEM's Programme Management Committee, representing the Medical School.

Harn Wei KUA

Dr Kua Harn Wei graduated with a Master's degree in theoretical physics from the National University of Singapore, before joining the Department of Building, School of Design and Environment, in 1999 as a full-time teaching assistant. After graduating with Masters degrees in Civil & Environmental Engineering, and Technology & Policy, from the Massachusetts Institute of Technology (MIT), he went on to earn his Ph.D. from the MIT Building Technology Program in 2006. His research interest is in creating integrated science-based sustainability policy models (related to energy and resource management) for climate change mitigation/adaptation in the built environment. Integrated policies are those that consider the complex interactions among different sustainability-related issues and prescribe multidimensional yet coordinated measures to address the problems arising from these interactions. Dr. Kua uses a range of concepts and methodologies in his analyses of problems and formulations of solutions, including industrial ecology, case studies and action research. He is a member of the MEM's Programme Management Committee, representing the School of Design and Environment, and co-teaches the course on Business and the Environment.

Keith LEE

Keith Lee is currently a Ph.D. student in the Department of City and Regional Planning at the University of California, Berkeley, where his research interests include industrial ecology, sustainable urban development and the influence of planning policy and urban consumer behavior on urban metabolism in the Asian context. He graduated in 2011 from the Yale School of Forestry and Environmental Studies with a Master of Environmental Management and also holds a Bachelor of Arts in economics from the University of Chicago.

LYE Lin-Heng

Lye Lin Heng is Associate Professor at the Faculty of Law, NUS, and Chair of the University's MSc (Env Mgt) [MEM] programme's Management Committee, as well as its representative for the Law School. She is also Deputy Director of the Law Faculty's Asia-Pacific Centre for Environmental Law (APCEL). Her teaching and research interests lie in Property Law and

Environmental Law. She is Visiting Associate Professor at the Yale School of Forestry and Environmental Studies, and also teaches at Sydney University's Law School. She is a member of the IUCN Academy of Environmental Law's Board of Governors. She is also a member of the Singapore Ministry of Environment and Water Resources' Committee on Clean Drinking Water Standards, and the Strata Titles Board, Singapore.

George OFORI

George Ofori is a Professor at the Department of Building, National University of Singapore. He is a Fellow of the Chartered Institute of Building, Royal Institution of Chartered Surveyors, and Society of Project Managers (Singapore). He is Director of the M.Sc. (Environmental Management) programme and a Co-Director of the Centre for Project Management and Construction Law. His research is on construction industry development, international project management, sustainability in construction, and leadership in construction. He has been a consultant to several international agencies, and governments on various aspects of construction industry development.

Nicholas A ROBINSON

Nicholas A Robinson is University Professor for the Environment at Pace University, and holds positions as the Kerlin Distinguished Professor of Environmental Law at Pace Law School, and as adjunct professor at the Yale University School of Forestry & Environmental Studies. He specializes in international and comparative environmental law, edited the proceedings of the 1992 UN Conference on Environment & Development, and is co-author of *Capacity Building in Environmental Law in the Asia and Pacific Region* (ADB). He chaired the Commission on Environmental Law of the International Union for the Conservation of Nature and Natural Resources (IUCN, 1996-2004), and received the Elizabeth Haub Prize in Environmental law from l'Université libre de Bruxelles (1992). He is a member of the Advisory Committee for the MEM programme as well as the Law Faculty's Asia-Pacific Centre for Environmental Law.

Victor R SAVAGE

Associate Professor Victor R Savage holds a joint position in Geography and the NUS Environmental Studies Programme (BES). His research interest is mainly on Singapore and the Southeast Asian region – historical and cultural landscapes, sustainable environments, environmental education, sustainable urban development and cross-cultural issues. A/P Savage is the current Editor of the Singapore Journal of Tropical Geography and an International Editorial Board member of the Sustainability Science (Japan) and Geografiska Annaler (Sweden). He is Deputy Chair of the MEM's Programme Management Committee and represents the Faculty of Arts and Social Sciences.

Judy SNG

Dr Judy Sng is an Assistant Professor in the Saw Swee Hock School of Public Health, NUS, and Associate Program Director of the National Preventive Medicine Residency program in Singapore. She has received several research and academic awards, among them the Society of Medicine Gold Medal in Occupational Medicine, NUS (2007), the Young Asian Scientist Award, Asian Conference on Occupational Health (2008) and the Getrude T Huberty Warren Memorial Award, UCLA (2010). Her research and teaching interests are in occupational and environmental medicine and epidemiology.

Hugh TW TAN

Hugh TW Tan is an Associate Professor at the Department of Biological Sciences and Deputy Director of the Raffles Museum of Biodiversity Research at the National University of Singapore. His research centres mainly on Singapore and Southeast Asia conservation biology, urban ecology, urban agriculture, and the horticulture of Singapore's native plants. He is editor of *Nature in Singapore* and Raffles Museum Books. He edited and co-wrote the recently published *Singapore Biodiversity: An Encyclopaedia of the Natural Environment*.

Dodo J THAMPAPILLAI

Dr Jesuthasen ("Dodo") Thampapillai is an economist and Associate Professor at the Lee Kuan Yew School of Public Policy, National University of Singapore. He also holds a Personal Chair in Environmental Economics at Macquarie University. Dodo's current research focus is on Macroeconomics and the Environment and the revised edition of his text *Environmental Economics: Concepts Methods and Policies* (Oxford University Press 2002, 2006). He teaches the course "Environmental Economics" in the MEM programme.

Darren CJ YEO

Dr Yeo is an Assistant Professor in the Department of Biological Sciences, NUS. He is also a Research Associate of the Raffles Museum of Biodiversity Research, NUS, and Research Affiliate of the Tropical Marine Science Institute, NUS. His research interests include invasion biology, freshwater ecology, and systematics and biogeography of freshwater decapods crustaceans. He is currently an Associate Editor of the Raffles Bulletin of Zoology. Teaching areas include Ecology of Aquatic Environments, Freshwater Biology and Biodiversity, and Invasion Biology.

Asia's 'Sustainability' Quests:
An Introductory Overview

Victor R SAVAGE, LYE Lin-Heng and George OFORI

> We are now twelve years into a unique century, the first century in the
> 35 million centuries (3.5 billion years) of life on Earth in which
> one species can jeopardize the planet's future.
>
> *–Holmes Rolston III* (2012: 1)
> American environmental philosopher

I. Motivation and Occasion

Having published two volumes of the research of our Masters in Environmental
Management (MEM) students over the last decade (2001–2011) (see *Sustainability
Matters: Environmental Management in Asia,* World Scientific, Lye *et. al.,* 2010;
and also, *Sustainability Matters: Challenges and Opportunities in Environmental
Management in Asia,* Pearson Custom Publishing, Lye *et al.,* 2011), the Programme's
Management Committee felt it was fitting for faculty members who teach and
supervise students' research work (Dissertations and Study Reports) to pen their
views on relevant aspects of the environment from their disciplinary perspectives,
personal vantage points or research interests. These contributors are, by no means,
exhaustive of the faculty involved in teaching the MEM programme, but reflect the
wide-ranging disciplinary engagement in MEM; from Science to Building, from
the Social Sciences to Law, and from Medicine to the Humanities. We also have
contributions from adjunct faculty as well as non-faculty members – people who
might be seen as applied practitioners of environmental management contributing
their insights from a non-government organization (NGO) or industry vantage
point. In addition, we are glad to have the inputs of American environmental
lawyer, Nicholas Robinson, a member of the advisory committee and former
external examiner of the MEM Programme. Given the wide-ranging nature of
environmental issues and the current climate change interventions, one can expect
that a book of this nature would be multidisciplinary and interdisciplinary in scope
and theme.

The tenth anniversary (2001–2011) of the MEM programme is cause for academic
celebration. Since its inception in 2001, the Programme has grown and stabilized
over the last decade not only because of administrative support but also due to
the dedication of environmentally-conscious and passionate faculty members who
are personally engaged in environmental issues, both international and local. This
interdisciplinary and multidisciplinary dialogue and commitment amongst faculty
members are testimony to the world we live in, where we all want a sustainable
world for all creations, including humankind. Indeed, the National University
of Singapore (NUS) can be proud that this multidisciplinary programme has
developed over the years with a large number of foreign students (about 40 per

developed over the years with a large number of foreign students (about 40 per cent annually), public-sector endorsement and private-sector financial support in the form of bursaries and scholarships. The MEM programme is indeed fortunate to have a growing body of corporations and foundations contributing to the welfare of students, outreach programmes and academic vitality. We are glad to have support from environmental-active corporations. In our time, nothing can be more important and immediate than dealing with environment and climate change challenges because they have such profound impacts on ecosystems and, in turn, human life, health and property.

II.   Synthesis and Overview

Environmental issues sprung into the public consciousness since Rachel Carson's (1962) path-breaking book, *Silent Spring*, and this public profile was subsequently heightened with the 1987 Brundtland Report (World Commission on Environment and Development, 1987), which gave the global community the concept of 'sustainable development'. Indeed, politicians and governments have bandied the term 'sustainable development' around to demonstrate their commitment to being 'green' and environmentally friendly. Both George Ofori and Asanga Gunawansa critically review the concept of 'sustainable development' and provide different criticisms and interpretations as well as interventions of its concepts and the term 'practical usage'. In most cases, however, the use of the term 'sustainable development' has merely served as environmental lip service and 'green' hypocrisy. Governments and corporations have, unfortunately, pursued less-than-friendly environmental policies and programmes. Ho Hua Chew, in his chapter on conserving biodiversity in Singapore, demonstrates that, even in a model state like Singapore, the authorities also have their struggles with regard to their commitment to 'green' causes and environment-friendly living.

Environmental concerns ironically took on a new dimension and high public profile with the increasing problems arising from climate change. Ever since climate change was brought to the public's attention by the American climatologist James Hansen in the late 1970s, the issue has become internationally legitimized with the Kyoto Protocol. Since then, many books and several United Nations (UN) conferences (in Bali, 2007; Copenhagen, 2009; Cancun, 2010; and Durban, 2011) have been held to discuss the global response to climate change and global warming (Flannery, 2005; Stern, 2007, 2009; DiMento & Doughman, 2007; and Linden, 2007). The difference between environmental impacts and climate change is that the latter is more deadly and more impactful; as hazards are more frequent and more damaging to environments than have ever been experienced by communities. The impact of climate change, which climate scientists would prefer to call 'weather changes', is relentlessly unfolding its venom and destruction almost every week in Asia. However, Nicholas Robinson notes in his chapter that carbon dioxide emission increases over the last 420,000 years based on Antarctic ice cores have never been higher than the current rates. This points to Victor Savage's endorsement of the view that the 'Antropocene', or 'Anthropozoikum', has arrived and become an 'integrated part of the Earth System'. Such weather changes hint at human-induced climate change outcomes. The link between current environmental global disasters and natural hazards to climate change, as well as the link between such disasters to human causes, is now finally getting public endorsement by the global community of scientists. A draft document of the report of the United Nations Inter-Governmental Panel on Climate Change (IPCC), to be tabled in Kampala in November 2011, has come to the conclusion that man-made climate change has boosted the frequency or intensity of heat waves, wildfires, floods and cyclones. More importantly, such disasters are likely to multiply (*The Straits Times,* 2 November 2011: A19).

In 2011, one sees massive floods in Pakistan, Thailand, Cambodia and Malaysia; endless typhoons and cyclones hitting Bangladesh, the Philippines and Vietnam; prolonged droughts in parts of China, India and Indonesia; and rising sea levels that threaten the Mekong Delta and the islands of Indonesia and the Philippines. In their articles, both Carsten Hüttche and Victor Savage delineate some of the impacts of climate change in the Asia-Pacific region, arguing that small and poor Asia-Pacific countries have little ability to mitigate global warming, hence having to find ways to adapt to climate change at both national and local levels. They cannot be passive observers but must become active participants in global forums to ensure that their climate change plights are heard.

While the main theme of this volume of essays is environment and climate change, the regional amplification and setting is Asia – a massive continental region with vastly different communities, cultures, states, urban systems, ecosystems and bio-regions. Every political entity in Asia, however, shares the same problems with environment and climate change, and seeks to find a common pathway to minimize environmental damage, avert human impacts and uncover green solutions. Hence this book provides a whole array of case studies at various scalar spatial levels ranging from the local level (villages along the Kampar River in Sumatra), cities (Singapore), national state levels (China, Timor Leste, India, the Solomon Islands, Indonesia), regional representations (Association of the Southeast Asian Nations, Asia) and global interventions.

Unfortunately, while there might be a science to physical environment and climate change, there is no science for evaluating social and cultural sustainability, political development or economic progress. Responses to environmental and climate change should be customized for each society's and community's cultural and social ballasts, levels of economic and political development, as well as the ecosystem parameters it resides in. There is thus no 'one size fits all' remedy for any community or state facing environmental problems. This makes the subject of ensuring national goals of environmental sustainability and quality of life immensely challenging for Asian governments. In large countries like China, India and Indonesia, there is as much ecosystem diversity as there is cultural and social heterogeneity. In short, every sector (government, private corporations, educational institutes or NGOs) needs to constantly engage in dialogue with other sectors to find common-pool solutions to the variable problems faced by countries. This book of interdisciplinary and multidisciplinary essays is but one contribution to this dialogue. The academic interventions reflect a variety of disciplinary and interdisciplinary environmental perspectives: biological science, building, architecture and real estate, medicine, law, economics, geography, ethics, ecology and industrial ecology, including the viewpoints of nature lovers, bird watchers and environmental consultants.

The themes in this book can be organized in many ways. However, we would like to offer our own perspective of what central issues they have in common. Firstly, all the chapters clearly note that environment and climate change outcomes reflect human interventions and impacts. In short, environment and climate change cannot be seen as a product of natural phenomena and elements *per se* – they are taking place because of human activities, negligence and greed as well as environmental commodification, economic motivations and political mismanagement of natural resources. In the essays by Chou Loke-Ming on Southeast Asia's coral reefs and by Darren Yeo, Richard Corlett and Hugh Tan on biodiversity issues in Singapore, the evidence is clear that human activities are directly and indirectly undermining natural ecosystems, already leading to the extinction of aquatic and terrestrial species alike.

Clearly, while there are alternative views on the current environmental crisis that the regions face, it seems evident from the various papers that human beings cannot be absolved of liability in creating the environmental mess we are currently trying to solve. Chou Loke-Ming's intervention on coral reef deterioration in Southeast Asia demonstrates the anthropogenic causes (global climate change, local overfishing and marine pollution) arising from developmental initiatives and population pressures in coastal areas. The identification of these causes cannot be over-emphasized, as, without identifying the real causes of problems, one can never prescribe the right solutions. Hence grounded academic research such as David Koh and Judy Sng's work on the pollution of the Kampar River in Sumatra is pertinent in addressing real-world environmental problems. Their chapter on villagers residing along the Kampar River in Sumatra is evidence of environmental impacts on human health and diseases.

Secondly, all the papers have tried to discuss the issue of 'sustainability', whether it is environmental sustainability, sustainable development, sustainable buildings, sustainable cities or generally sustainable societies. It is difficult to operationalize the concept of 'sustainability' since it covers a wide range of subject matters. But using various disciplinary vantage points, the papers have tried to engage ideas of sustainability in specific areas. It ranges from Ho Hua Chew's questioning of whether the current economic (capitalistic) trajectory of Singapore's developmental agenda can, in fact, bring forward a more sustainable way of living in the city-state, to Lye Lin-Heng's perspective on sustaining cities by developing various infrastructure (legal, administrative, environmental), land use laws and other benchmarks such as the ISO 14000 certifications. Quite the opposite of Hua Chew's views, Asanga Gunawansa is a firm advocate of the compatibility of economic growth and environmental sustainability when he states the way to go is to realize that 'we cannot revert to living in caves, and it will be difficult to convince masses in the developing countries to give up their dreams of economic development. We condemn the poor to aspire to less prosperity'.

Thirdly, given that cities now hold 50 per cent of the world's human population and that 44 per cent of Asia's population live in urban areas, one cannot deny the relevance of finding the right environmental nexus for sustaining cities. Renowned environmental lawyer Nicholas Robinson puts it succinctly that cities are 'already artificial experiments in sustainability'. While city governments deal with the challenges of managing natural ecosystems for water supplies for survival, they also have to manage an artificial landscape (refuse, sewerage, pollution) that requires sustainable management. Two chapters, by Kua Harn-Wei and George Ofori squarely address the urban hardware (infrastructure and buildings) and the challenges of keeping cities 'green'. Both papers focus on issues of sustainable building materials. In Harn-Wei's paper on the cement industry in China, he argues that much needs to be done to ensure the sustainability of the industry as cement is a major resource in laying down the urban concrete landscape. The prescription is for a more integrated approach for the environment, economy and employment. In the case of George Ofori, the new pathways to eco-friendly cities lie in developing green buildings. This is a complex issue that involves a matrix of eight drivers: financial incentives, building regulations, client awareness, client demands, planning policies, taxes and levies, investments, as well as labelling and measurements. Among his suggestions for underscoring green buildings is the need for strong leadership, the development of human resources and greater regional collaboration on green building programmes.

Fourthly, 'sustainability' underscores the complex human-nature relationships in societies and that untangling the variables and finding solutions to issues are never an easy direct cause-effect relationship. Hence several authors reflect on trying to find an ecological loop that minimizes waste and hence imitates the lessons of nature in creating a sustaining system. Marian Chertow, Esther Choi and Keith Lee's chapter deals squarely with an industrial ecology approach by using the material flow analysis (MFA) tool to analyze Singapore's 'urban metabolism'. Basing their findings on 2000, 2004 and 2008 statistics, the authors show that only 56 per cent of Singapore's waste was recycled in 2008; that Singapore imports considerable sand from the region (Malaysia, Indonesia, Cambodia and Vietnam) to support its economic growth, to the environmental detriment of its sand exporters; and that Singapore's reliance on foreign food production has led to increasing 'virtual water' imports. The authors demonstrate that there are considerable fluctuations in material imports and exports that affect 'Domestic Material Consumption inversely as compared to global economic activity'. This MFA study underscores the ongoing debates on the scale of the ecological footprint that cities leave on other ecosystems in their consumption habits. In another vein, Nicholas Robinson argues for the need to operationalize the Chinese model of the 'circular economy' to create a more sustainable economic model for dealing with waste and recycling. Given that cities stick out like sore thumbs in the natural environment, the discussion of the concept of eco-cities by Nicholas

Robinson and Victor Savage is meant to convey ecosystem closure and ensure more sustainable cities, attuned to the rhythms of nature rather than just human schedules and time-keeping.

Fifthly, any book discussing a pertinent theme such as eco-development with sustainability objectives is bound to consider issues of development versus nature trade-offs, economic versus environmental assessments and the whole debate about land-use options. According to Dodo Thampapillai, one needs to consider the 'minimum threshold level of ecosystem support' in considering the trade-off between, say, the issue of water conservation versus that of housing development. Ho Hua Chew's argument is more bio-centric – that one has to accept the view that pristine-nature areas are irreplaceable and hence no economic trade-offs can be made. Robinson views the nature development issue in more anthropocentric terms in relation to sustaining, restoring and maintaining the 'resilience of the natural systems' where human life exists. Both Carsten Hüttche and Asanga Gunawansa take a more developmental perspective to sustainable development. Hüttche places priority on 'poverty' alleviation when he states 'addressing poverty in ecologically sustainable ways will reduce vulnerability to climate change by increasing the ability of both social and environmental systems to adapt to climate change'. However, Gunawansa argues that there is no trade-off between a 'healthy environment' and 'healthy growth' of the economy because both could lead to an 'improved environment, together with economic and social development'. Clearly, embarking on the green economy can be a new catalyst for economic development for many countries in Asia. George Ofori, for example, notes that the retrofitting of Singapore's 210 million square metres of built-up space to meet the government's Green Mark standards is estimated to be some SGD300 billion'.

Managing environments is thus a multidimensional exercise that involves formal educational inputs, various stakeholders and the reliance on traditional knowledge and culture. It is, as Gunawansa advocates, the 'collective responsibility' amongst government, the corporate sector and the general public in each state to cultivate 'social responsibility'. This collection of essays has a plethora of solutions, suggestions and samples for getting our 'eco-logic' correct. It begins with well-tried examples like Carsten Hüttche's critical review of the applications of Environmental Impact Assessments in Asia-Pacific countries (China, Southeast Asian countries) to more current methods for risk assessment with regard to climate change impacts and ends with essays on environmental management. Lye Lin-Heng makes a strong call for education and ethics, advocating endorsement of the Earth Charter as well as environmental education, right to information, access to justice and public participation. Nicholas Robinson sheds light on the challenges of creating environmental educational programmes at the tertiary level to equip new-generation environmental managers. He argues that the new breed of environmental managers certainly require multidisciplinary knowledge to tackle sustainability challenges. These include using a whole gamut of expertise in new technologies, analyzing and applying public policy rules and incentives, analyzing financial costs and benefits, and managing organizational changes. In a broader context, Savage notes that sustainable management requires not only

Western developments in modern science but also a grounded understanding of Asian traditional, cultural conventions of 'relational nature' and the 'relational self', as referred to in Buddhism.

III. Conclusion

The twin themes in this book are: to uncover the human-nature relationships of communities at various scales (countries, cities, villages); and to discuss the options for the right pathways between state development and environmental and climate change. Needless to say, the natural hazards arising from climate change are undermining developmental programmes and disrupting the livelihoods of communities in Asia. The devastating floods in Thailand in October–November 2011 which has claimed over 500 lives is a poignant reminder that there is no easy human solution to containing the wrath of nature. Bangladesh and the Philippines, both poor countries, are also the most hazard-prone countries in Asia, which provides enormous challenges for their respective governments to move development projects forward. Unfortunately, despite the growing evidence of natural hazards taking place in Asia and around the world, there remain people who are skeptics about climate change. It seems as though the prognosis by the United Nations International Panel on Climate Change (IPCC) has been conveniently ignored by certain sectors of political leaders, government administrators and corporate titans. The reason is that accepting these environmental realities mean accepting Al Gore's view of 'the inconvenient truth' – oil and car companies do not want to hear that they are culprits of CO_2 emissions as it questions their industrial productivity and profit margins. In this light, one of the modest purposes of this book is to bring awareness of these pressing environmental challenges to the informed public, government administrators and the corporate world. There are many interpretations to the problems confronting countries, cities and communities, but rather than shying away from the problems, we all need to engage in dialogue and debate, so that clearer perceptions, rational perspectives and pragmatic views may prevail.

Most Asian countries have neither the financial muscle nor the technological and scientific expertise to deal with the mitigation of climate change. Indeed the research institutes that have these scientific and technological abilities are located in developed western countries (the US and Europe). Hence, what the British economist Nicholas Stern (2009) advocates, is for developing countries to concentrate on finding adaptive mechanisms to handle climate and environmental change. Environmental adaptation is not new to many Asian communities – it has been going on for centuries. There is tremendous folk wisdom and science that the 'little peoples' in Asia can tap on in confronting environmental challenges. This is best encapsulated in what Daniel Goleman (2009:43) defines as 'ecological intelligence' – "an understanding of organisms and their ecosystems, and *intelligence* connnotes the capacity to learn from experience and deal effectively with our environment". Therefore, in advocating environmental dialogue, one needs to ensure that the dialogue is not only among academics, intellectuals, and the literati, but also among folk peasantry, tribal communities, entrepreneurs, administrators, religious laity and the informed public. We cannot replace the knowledge and ecological intelligence of peasant and tribal communities who have developed intimate knowledge of their ecosystem they reside in through accumulated experience. Our conceptions of nature should not be based merely on the tyranny of modern science – one should welcome alternative points of view as well.

The authors in this book do not pretend to have the answers and prescriptions to all the life-threatening problems that Asian communities confront. But in grounded research we have also learnt and shared how other village, tribal and urban communities handle quotidian environmental challenges. At the end of the day, we want to ensure a better quality of life for all communities, a more effective and efficient management of ecosystems, and a more sustainable world. Time is ticking away for decisive action. We hope that in reading this book, you may be able to see different vantage points, critically evaluate our views and adopt ideas for a more sustainable environment.

References

Carson R (1962) *Silent Spring.* The Riverside Press, Cambridge.

Dimento JFC, Doughman P (2007) (eds) *Climate Change: What It Means for Us, Our Children, and Our Grandchildren.* The MIT Press, Cambridge, Massachusetts.

Flannery T (2005) *The Weather Makers: How Man Is Changing the Climate and What It Means for Life on Earth.* Grove Press, New York.

Goleman, D (2009) *Ecological Intelligence: Knowing the Hidden Impacts of What We Buy.* Allen Lane, London.

Linden E (2007) *The Winds of Change: Climate, Weather, and the Destruction of Civilization.* Simon & Schuster, New York.

Lye LH, Chou LM, Chia A, Asanga G, Kua HW (eds) (2011) *Sustainability Matters: Challenges and Opportunities in Environmental Management in Asia.* Pearson Custom Publishing, Singapore.

Lye LH, Ofori G, Malone-Lee LC, Savage VR, Ting YP (eds) (2010) *Sustainability Matters: Environmental Management in Asia.* World Scientific, Singapore.

Rolston H III (2012) *A New Environmental Ethics: The Next Millennium for Life on Earth.* Routledge, New York & London.

Stern N (2009) *A Blueprint for a Safer Planet: How to Manage Climate Change and Create a New Era of Progress and Prosperity.* The Bodley Head, London.

Stern N (2007) *The Economics of Climate Change: The Stern Review.* Cambridge University Press, Cambridge.

The Straits Times (2011) 'UN Report ties climate change to extreme weather' *The Straits Times*, 2 November 2011: A19.

World Commission on Environment and Development *Our Common Future.* Oxford University Press, Oxford.

The Asian Century: Environmental Impacts and Implications

Chapter 1: Whither the Sustainability and Development Twain
Meet: Implications and Challenges for Asian States
Victor R SAVAGE

Chapter 2: Understanding Sustainable Development and the
Challenges Faced by Developing Countries
Asanga GUNAWANSA

CHAPTER ONE

Whither the Sustainability and Development Twain Meet: Implications and Challenges for Asian States

Victor R SAVAGE
Department of Geography, National University of Singapore

"Given that most Asian states are too poor, economically powerless and lacking in expertise to initiate climate change mitigation processes, they need to engage the developed countries to put in motion concrete steps to slow down global warming. Asian states cannot take a back seat position on climate change. They need to be actively involved to ensure that the international community acts responsibly and effectively in halting global warming. If not, developing Asian states will remain the losers in the unrelenting impacts of climate change in the latter half of the 21st century."

– Victor R SAVAGE

Whither the Sustainability and Development Twain Meet: Implications and Challenges for Asian States

Victor R SAVAGE[1]

Department of Geography, National University of Singapore

Abstract

The increasing impacts of climate change on communities and countries in Asia have become more apparent over the last two decades and are likely to continue into the following decades of the 21st century. These unfolding natural hazards come at a time when the world's financial and economic problems continue to develop. Climate change is a global problem that requires global solutions, but the international meetings at Kyoto, Bali, Copenhagen and Cancun have not produced encouraging results. Asian states do not have the scientific knowledge, the research and development ballast or the financial capabilities to tackle climate change mitigation issues. All most Asian states can do is to find adaptive mechanisms at the local level to reduce the impacts of climate change.

Key words: adaptive mechanisms, climate change, floods, global warming, mitigation strategies, nature, prolonged drought, relational self, sea level rise

I. Introduction

If we are successful in finding a sustainable way of living in the twenty-first century, then perhaps the principles we develop will become the guiding principles of a truly sustainable global civilization.

— *Flannery* (2009: 105)

As we come closer to the vortex of one of probably the worst global economic crises in history in 2011, there is a growing sense of foreboding, despair and resignation amongst people around the world, and we may never see the light at the end of the tunnel. Given the massive loss globally of USD26 trillion (Laszlo, 2009: 17) during the financial crisis in 2007–2008, one can understand the grave trepidation felt about the impact of the Euro crisis on international trade, jobs, inflation and food insecurity. Ancient Indian historians and philosophers would view this economic 'depression' as a cyclic process in history and accept that we are in a down cycle, a trough in history after experiencing the golden years. The Javanese would see this period as *zaman edan*, a crazy, dark period in

---

[1] Department of Geography, Faculty of Arts and Social Sciences, National University of Singapore. 1 Arts Link, AS2 #04-34, Singapore 117570.

which order would be upset, 'the greedy and fools would thrive, the lazy would be rewarded and people would no longer know manners or shame' (Mulder, 1994: 43). Does this description of our time sound familiar?

Yet, ironically, measured in other human terms and given a longer historical trajectory, we might be seen as enjoying a golden age in human history. From the viewpoint of *Homo sapien* history, we have arrived in the 21st century with the most positive balance sheet: we have the largest population to date (over seven billion), who are in better health, with a longer lifespan (67 years). We are more literate and wealthy, and live with greater comforts than those at any other time in human history. It is on this very optimistic note that the Danish political scientist, Lomborg (2001), shared his rather sceptical environmental views that the state of the world environment was much better than had been presented by environmental non-government organizations (NGOs) and environmentalists like Brown (2001) and the global green NGOs: Greenpeace, World Wide Fund for Nature, and the Worldwatch Institute. Environmentalists might disagree with Lomborg's perspective, but there is evidence that human beings are clearly better off though the planet might be in deteriorating health. Lomborg's 'measurement of the real state of the world' was an update of an earlier equally optimistic report card assessment of the 'state of humanity' by Julian Simon (1995). In his words: 'The decrease in the death rate, and the attendant increase in life expectancy – more than doubling – during the last two centuries in the richer countries, and in the 20th century in the poorer countries, is the most stupendous feat in human history' (Simon, 1995: 26).

There is no doubt that Lomborg (2001) and Simon (1995) are correct about the enormous strides human societies have made towards what the French historian Fernand Braudel (1995) views in his holistic and transnational perspective of human civilization. Holistically speaking, as a species, we progressed, albeit unequally. While 30 per cent of the world's population consume 70 per cent of the world's resources, the other 70 per cent of the world's population live in poverty, malnutrition, starvation, disease and without proper housing, clean water and sanitation. Escobar's (1995: 212) book, *Encountering Development*, puts the global inequality more starkly: 'The industrialized countries with 26 per cent of the population account for 78 per cent of world production of goods and services, 81 per cent of energy consumption, 70 per cent of chemical fertilizers and 87 per cent of world armaments. One US resident spends as much energy as seven Mexicans, 55 Indians, 168 Tanzanians and 900 Nepalis'. In Southeast Asia, Cambodia, Lao PDR and East Timor have over 70 per cent of their populations living on less than USD2 a day.

We should be cautious about patting our own backs too quickly for our human achievements. After the fall of the Berlin Wall and the Soviet Union, capitalism seemed to have won a global victory. But the current Euro zone debt problems, the faltering US financial system and the moribund Japanese economy are wake–up calls and sobering reminders of how fragile the capitalist, laissez-faire model of economic, political and social systems is. As fast as developments have enhanced the betterment of livelihoods in the 20th century, we might all see this slip away rapidly in the 21st century due to our callous behaviour, our inability to manage states and cities, and our indifference to the environmental writing on the wall of major impending problems. Certainly, the impact of climate change is undermining confidence further in whether states can find a sustainable model of development. In short, sustainability is a slippery objective for many states and communities in Asia.

This paper investigates the long-term impacts of climate change and the possibility that environmental issues and their dire ramifications for Asian states in the 21st century. Given the often-talked-about rising Asian zeitgeist in this century, with China and India forming the twin catalyst for global economic growth, such optimistic scenarios must be weighed against the environmental challenges Asian states face intranationally and the global impacts of climate change on Asian states, especially China and India. The specific focus is whether Asian states are able to manage the environmental and climate change challenges at the global, regional, state and urban levels. They undergird the usual twain of 'sustainability and development', a term that many essays in this book focus on and that I have serious reservations about. Two areas need attention in integrating both the adaptive mechanisms and mitigation offerings: by looking at the needs for a new global architecture in handling transnational and global environmental concerns and by critically evaluating the dominant national habitats of human beings, sustainable cities and eco-cities within states in Asia.

II.   Climate Change: The Accumulating Evidence

Climate change across human history has been as much a narrative of both plus points as it has been a negative factor on human development. Over 80,000 years of *Homo sapien* history, the Ice Age made it possible for early human beings to cross over land bridges to explore and experiment as well as populate many areas around the world. Global warming has been taking place over the last 20,000 years and certainly over what scientists call the Holocene (10,000 before present [BP]) period. The global warming Holocene period was one of much human creative activity – agricultural origins, plant and animal domestication, genesis of cities, development of religions, technological developments, pottery and writing. Steven Mithen's (2004: 504) book, *After the Ice*, documents these creative human developments well. He argues that human history between 20,000 and 5,000 BP 'reached a turning point during a period of global warming' and set the stage for the 'origins of human civilization'. The moral of the story is that carbon dioxide (CO_2) is important for two reasons: (a) without it, all vegetative life will be eradicated and (b) CO_2 has helped to enhance global warming for the last 10,000 years to make planet Earth a habitable place for human life, the agricultural revolution, the birth of cities, and the development of civilizations (see Volk, 2008).

Despite the monitoring of the increase of CO_2 since 1958 by scientists from the Scripts Institute, climate change has registered rather slowly in the public imagination. It has taken a series of severe heat waves, prolonged droughts, intense floods, snow storms, heavy rains, typhoons and hurricanes, melting mountain snow caps, sea level changes, the melting of Greenland and arctic ice caps over the last four decades for the mass media and the public to become aware that these are not isolated natural events and phenomena but a product of climate change. Yet, among climate scientists, these erratic and frequent climate disasters define 'weather changes'. It would take a longer historical experience to accept that we are undergoing climate change. It is no surprise that one of the definitive statements by Flannery (2006) avoided the use of 'climate' and entitled his book 'The Weather Makers'.

Yet, there is now a growing body of scientific evidence to demonstrate that our earth is in the throes of CO_2 increases, global warming and climate change. The numerous scientific articles and publications testify to this issue (see Flannery, 2006; DiMento & Doughman, 2007; Intergovenmental Panel on Climate Change [IPCC], 2007; Linden, 2007; Stern, 2007; Walker & King, 2008). Between 1999 and 2010, the world has experienced some of the worst droughts of the century, the worst floods in 500 years, the strongest El Niño in 130,000 years and the worst tropical storms in recorded memory (Linden, 2007: 249). Despite the controversial book, *Skeptical Environmentalist*, by Lomborg (2001), the 935 peer-reviewed scientific papers on climate change between 1993 and 2003 have all accepted that 'humans are changing climate' (Linden, 2007: 228).

What has become alarming for the scientific community and the public is that climate change is not merely a natural phenomenon a product of climate cycles taking place over the centuries. For the first time in 'glacial time' (Castells, 2004: 183), climate change is viewed as a product of human-induced activities. Indeed, human influence on nature has led to scientists coining a new age, the 'Anthropocene' and 'Anthropozoikum' to underscore the need for an Earth science system that positions humankind as an 'integrated part of the Earth System' (Ehlers & Krafft, 2006: 11). At the root of this Anthropocene-based integrated earth system are the rising urban agglomerations around the world. Humankind has changed radically urban environments or what Spirn (1984) called the 'granite garden' to underscore the human-engineered built-up environment. For Martin (2007: 4) the 21st century is the 'critical century', or what he calls the 'make-or-break century', because it is not sustainability but 'survivability' that now matters. The 2007 IPCC Reports and the plethora of books on global warming and climate change portend a dismal scenario for the global climate and its impact on Earth and its varied species this century (Flannery, 2006; Linden, 2007; Stern, 2007; Walker & King, 2008), which will have profound impacts on countries and cities. The IPCC view in the March 2009 meeting in Copenhagen forewarns that climate change might lead to 'abrupt or irreversible climatic shifts' (*The Straits Times*, 14 March 2009: A12). How national and urban populations will have to adapt to the new environmental challenges is an issue that has no precedent in human history.

Clearly, human contributions to an increase in CO_2 and carbon emissions are becoming clearer. It seems evident that developed countries emit more CO_2 and carbon per capita than developing and poorer countries do. In 2002, the average global per capita CO_2 emission was 4 metric tonnes compared to North America's 19.6 metric tonnes and Asia's 2.6 metric tonnes (see Table 1.1).

Table 1.1: World's human–environment indicators data sheet (mid–2008)

Countries	Population mid-2008 (millions)	% urban	Gross national income purchasing power parity per capita (USD) 2007	CO_2 emissions per capita (metric tonnes) 2002	% of natural habitat remaining (2007)
World	6,705	49	9,600	4.0	78
More Developed	1,227	74	31,200	11.7	82
Less Developed	5,479	44	4,760	2.1	76
Less Developed (Excl. China)	4,154	44	4,560	1.9	76
Africa	967	38	2,430	1.1	84
Northern America	338	79	44,790	19.6	85
Latin America/ Caribbean	577	77	9,080	2.5	73
Asia	4,052	42	5,650	2.6	69
Southeast Asia	586	45	4,440	1.7	44
Europe	736	71	24,320	8.4	76
Oceania	35	70	23,910	12.2	89

Source: Population Reference Bureau, 2008 World Population Data Sheet, Washington, DC

Based on the 2006 global emissions of CO_2 of 7.5 billion tonnes of carbon, the global average per capita was 1.25 tonnes, while a US citizen emitted 5 tonnes. In his book, *CO_2 Rising*, Volk (2008: 115–20) noted that there was a distinct correlation between a country's wealth, energy consumption and CO_2 emission. There is a clear correlation between economic growth and CO_2 emission: in Asia, China and Vietnam are countries with vigorous growth in CO_2 emissions (Volk, 2008: 120). How fast can the USA and other developed countries reduce their carbon emissions by economic methods? And how fast will developing countries want to increase their development trajectories that will inevitably increase CO_2 emissions? These are vexing issues with no easy answers in sight. They have been debated vociferously in climate change global forums in Kyoto, Bali, Copenhagen and Cancun, and will continue in Durban in 2012.

III. Asian Century: Climate Change Challenge

Unlike other challenges (population, water, food, urbanization) that have plagued Asian states and communities in the 20th century, the new environmental challenge of the 21st century will certainly be climate change translated from CO_2 increases and global warming – an environmental issue that human societies have little experience in coping with and managing on a global or national scale. Unfortunately, there is no consensus amongst Asian states in finding a unified response to this global challenge.

Prolonged Drought, Heat Waves and Water Scarcity

History has proven that climate change and its unfolding impacts of prolonged drought have been the major explanations for the fall of many civilizations over the centuries (Linden, 2007; Fagan, 2008). In nearly all correlations of environmental causes to the fall of civilizations, protracted drought, fall in precipitation and aridity are the nails in the coffin of civilizations. Linden's (2007) book, *The Winds of Change*, documents several sudden climatic events that have affected the demise of several civilizations. He argues that a sudden hot and dry climatic event could have led to the end of the Akkadian civilization some 4,200 years ago, which was recorded in 'The Curse of Akkad' (Linden, 2007: 50–3). In the Southeast Asian region, the impacts of climate change have also had been known to have caused the expansion and downfall of civilizations (Lieberman, 2003). The current Southeast Asian situation is getting worse, and one sees a rise in droughts plaguing the region (see Table 1.2).

Table 1.2: Climate hazard hotspots in Southeast Asia

Climate hazard hotspots in Southeast Asia	Dominant hazards
Northern Vietnam	Droughts
Eastern coastal areas of Vietnam	Cyclones, droughts
Mekong region of Vietnam	Sea level rise
Bangkok and its surrounding area in Thailand	Sea level rises, floods
Southern regions of Thailand	Droughts, floods
Philippines	Cyclones, landslides, floods, droughts
Sabah, Malaysia	Droughts
Western and eastern area of Java Island, Indonesia	Droughts, floods, landslides, sea level rises

Source: Yusuf and Francisco (2009)

Given that 70 per cent of the world water usage goes into agriculture, any major prolonged drought is going to affect food production severely and lead to starvation for its resident populations. The US intelligence agencies have forewarned that between 120 million and 1.2 billion people in Asia 'will continue to experience some water stress' and that as many as 50 million could face hunger by 2020 (*The Straits Times*, 26 June 2008: 20).

Prolonged drought is a major cause of heat waves around the world. To date, the number of deaths from heat waves is rising: Chicago (700 deaths in 1995), Andhra Pradesh, India (1,000 deaths in 2002), Europe (35,000 in 2003). China is also going through one of the worst droughts in 50 years in eight of its provinces in the northern and central areas of the country – with a reported 4.3 million people and 2.1 million livestock affected as well as 43 per cent of winter wheat supply at risk (*The Straits Times*, 7 February 2009: B5).

In economic terms, however, the 1997–8 El Niño impact was disastrous for Indonesia. Per capita incomes fell by 75 per cent from USD1,200 to USD300; unemployment increased to 40 per cent; in May 1998, food prices escalated by 4 per cent; and poverty levels increased, with 22.5 million people identified as being poor or living at poverty levels in 1997 (before El Niño and the currency crisis) to over 100 million in the spring of 1998 (Linden, 2007: 209–10). The 1997–1998 El Niño spurred on forest fires that burnt some 8 million hectares of land (Schweithelm & Glover, 2006:1) and the total economic damages of haze and fires costs USD4.5 billion (Schweithelm *et al.*, 2006: 133). As more people live in cities, the drought causing El Niño's is likely to have devastating impacts (especially on food security and water supplies) on urban populations in the future. The question is whether countries and cities are fully prepared to handle the future impacts of such a drought.

Sea Level Rises

Evidence shows that sea levels have already risen in the 20th century by 0.1–0.2 metres and are predicted to rise between 0.90 and 1.7 metres by 2100 (Spash, 2002: 66). In Southeast Asia, the Asian Development Bank's (2009: 5) report on climate change notes that the region has been experiencing a rise in sea levels of 1–3 millimetres annually over the last several decades from the 1950s. Over 60 million people in Asia will be affected if the sea level rises by 0.5 metres. Southeast Asian states with 173,251 kilometres of coastlines are likely to be the most affected region in Asia. Already the results of climate change on the Maldives are so visible due to sea level rises that the newly elected president stated that the country needs to save money to buy land elsewhere in order to pursue a mass evacuation programme for its citizens. In the Pacific Ocean, people in the island states are equally worried about their future, as sea level rises are already drowning out their small islands, beaches and coastal areas. Australia and New Zealand are often seen as target countries for the mass evacuation of Pacific Islanders seeking refuge from sea level rises.

Throughout Asia, major coastal cities (Shanghai, Calcutta, Singapore, Ho Chi Minh City, Dhaka) are likely to feel the impact of sea level rises beside the low-lying delta and coastal areas where huge concentrations of population live. Bangladesh is one country where sea level changes will have devastating impacts on its dense coastal populations. It is already well known that the country that is most prone to natural hazards and widespread poverty does not present an encouraging picture that will enable the country to avert a major disaster like the flooding of its low-lying agricultural farmlands.

There seem to be three choices open to states in Asia to avert sea level rises: (a) build dykes or seawalls to prevent seawater from intruding into the land areas; (b) encourage people along coastal areas to migrate to higher areas or inland areas as part of a national development programme; and (c) conduct national mass evacuations of people from islands that are likely to be drowned out by sea level rise. The Maldives is a good national case in point. It seems unlikely that the people in this island country will be able to defend the drowning out of their island world. The question is, who will accept all these climate-change refugees?

Global Warming

The irony of global warming arising from CO_2 increases is that its impacts around the world are likely to have two major outcomes: economic in terms of agricultural activities and tourism venues and health wise in terms of an increase of diseases.

Clearly, a rise in temperatures in higher latitudes is likely to open up vast areas in Russia, Canada, America and northern Europe for agricultural production, provided there are water sources. This would mean that the traditional bread-baskets might be shifting with new beneficiaries while the current food baskets might be losing their agricultural advantage. In tropical Asia, increases in temperature might prove to be disastrous for agricultural productivity, be it rice, palm oil or rubber. Even in hilly areas, rising temperatures could affect the growth of tea, coffee and other highland vegetables and crops. To date, there are a few studies that have looked into the impacts of rising temperatures on agricultural productions.

The consequences of global warming are already visible in the tourism industry. Tourism is a major global industry and is important to many Asian national and local economies: China, India, Thailand, Singapore, the Maldives, Bali and Macau. However, the main academic studies on climate change on the tourist industries seem to be confined to Western countries. Given that Europeans place a lot of emphasis on tourist destinations with good weather (Amelung *et al.*, 2007: 286), global warming can have a detrimental impact on certain current tourist destinations if these sites become hotter or wetter. It is already predicted that, with global warming, northern Europeans will find the Mediterranean areas in summer less attractive. Increasing temperatures mean melting of ice caps in mountainous areas, shorter winters and snow periods, which have affected many ski resorts in Europe, the USA and Canada.

In Asia, global warming impacts on tourism are likely to take place in several decades. If tropical Southeast Asian beach resorts become hotter and coral reefs deteriorate, it seems unlikely that these sites will be able to attract tourists (Amelung *et al.*, 2007: 290). Southeast Asian beach resorts in Bali, Phuket, Krabi, Pattaya, Menado, Palawan, Vietnam, West Malaysia, the Indonesian islands and Sabah are also likely to face sea level rises that may inevitably wipe out their current beaches, coral reefs and diving sites. Chou Loke Ming (in this issue) has spelt out, in no uncertain terms, the massive impact of climate change (heating of the oceans) on the decimation of Southeast Asia's coral reefs and, with it, the rich marine biodiversity.

IV. Global Architecture for Climate Change: How Will Asia Benefit?

Given the above complex political issues, the new global architecture for dealing with climate change and environmental issues needs to have political focus, economic muscle and administrative abilities at two ends of the global political landscape. On the one hand, states in Asia need to do more nationally to develop and enforce controls and regulations to delimit CO_2 emissions. On the other hand, we need an international system that is able to harmonize, integrate and find consensus in tackling the many environmental challenges facing Asian countries. Given that most Asian states are too poor, economically powerless and lacking in expertise to initiate climate change mitigation processes, they need to engage the developed countries to put in motion concrete steps to slow down global warming. Asian states cannot take a back seat position on climate change. They need to be actively involved to ensure that the international community acts responsibly and effectively in halting global warming. If not, developing Asian states will remain the losers in the unrelenting impacts of climate change in the latter half of the 21st century. Specifically, the Asian Development Bank (2009: xxvii) advocates that developed countries need to provide 'adequate transfers of financial resources and technological know-how' to developing countries since all countries must find a common global solution.

Arising from the global economic crisis, there are views that the world might be seeing signs of what Filipino economist Walden Bello calls 'deglobalization' and that would be a worrying sign for all stakeholders. If countries become more economically nationalistic and impose trade protectionism like India, Russia and the European Community, then the global community will be poorer for it. This is an issue that the Association of Southeast Asian Nations has been most worried about because it will certainly hurt many small, developing, export-orientated countries. Given that developing Asian countries cannot overhaul the political ecology of our current 'nation-state' system, one can certainly put in place a more responsive, effective, efficient and comprehensive global system to handle cross-border climate change problems, environmental trade and green investments. Indeed, the reason for fleshing out the impact of the current financial chaos is to demonstrate that the international attempts to forge a common stand in addressing the larger oncoming climate change scenario are clearly more daunting. So long as the European Community, the United States and Japan fall into economic recession, global cooperation and financial underwriting of climate change issues are likely to be postponed and politically marginalized. Asian states are likely to feel the brunt of greater social and economic impacts arising from global warming and climate change.

Finding Global Common Ground for Climate Change

Asian states need to ensure that there is some global consensus on climate change issues in the next meeting in Durban, South Africa, in 2012. Asian states must try to speak with one voice to halt global warming and precipitate mitigation action amongst the global community. However, finding common ground amongst the 196 countries to deal with climate and environmental changes is a daunting task for several reasons.

Firstly, the impact of climate change has winners and losers amongst states, cities and provinces in the short term. Hence, while the losers might want quick redress, the winners might be less interested in quick remedies. Studies show that, with global warming, new breadbaskets in temperate countries will shift further north in favour of Canada, the Scandinavian countries and Russia. Global warming will also affect current tourist meccas like the Mediterranean countries. At the same time, many skiing resorts in North America, Switzerland and Europe will lose their attractiveness as tourist sites because their snow-capped mountains are rapidly melting.

Secondly, many of the losers of climate change are in poor countries in the developing world that have little political and economic clout to change global opinions much less set applied and pragmatic mitigation agendas. In Asia, China, Bangladesh, India and many states in Southeast Asia are likely to be losers as global warming develops. Even in the USA, critics noted that the slow response of the Bush Administration to New Orleans in the aftermath of Katrina was due to the fact that the area was poor, politically marginalized and the Democratic Party's political turf.

Thirdly, the issues of energy changes and reductions are so tied with economic issues and technological systems (transport, air conditioning, heating and electrical generators) that it is difficult for many developed as well as developing countries to wean themselves away from oil and gas systems. Oil- and gas-producing countries (Brunei, Indonesia, Myanmar, Malaysia and many Middle East countries) and companies (Shell, Exxon-Mobil) are enjoying the benefits of the monopoly on energy supplies and the high costs and hence their reluctance to make drastic changes towards a carbon-free environment. It would take decades to change technologies using oil and gas to an alternative environmentally-friendly fuel. Unfortunately, time is running out if change does not take place in the energy sector.

Fourthly, the difficulty in dealing with a coordinated global response to climate and environmental change reflects the way the world has developed its own political ecological system across the natural ecosystem. How is this so? The world as one interconnected and interrelated natural ecosystem or organism – if you like the analogy of Lovelock's (2007) Gaia thesis – has now been subdivided into over 196 political territories, defined by the boundaries of political states. Hence the natural global ecosystem is now a product of subdivided political ecosystems defined by power relationships between the 196 political entities rather than a subdivision brought about by nature. Within these 196 constituent polities lie major political contradictions, economic failures, cultural disjunctions and social upheavals. Hence the national response to climate change challenges is governed more by national leveraging and comparative advantages than by a global perspective for the common good. The challenge for the international community is to accept that climate change needs a global perspective to mitigation and, at the same time, a national environmental adaptive mechanism to buffer climate change impacts at the local level.

Fifthly, as Fareed Zakaria (2007) so eloquently argues in his book, *Freedom Lives,* we are witnessing and experiencing illiberalism, which is undermining the fabric of democracy and liberalism in the USA and around the world. Hence the question of democracy and liberalism is no guarantee that relationships within and between countries would be balanced, rational and respectful in ensuring the common good.

Sixthly, globalization and its implications are seeing not only incredible economic, political and cultural transformations but also a changing global power system where Western political and economic supremacy is likely to have to accommodate the ascendant Asian powers. Over a century ago, the German historian, Georg Wilhelm Friedrich Hegel (1956: 116) in his thought-provoking book, *The Philosophy of History,* underscored his Euro-centric view of world history when he noted: 'China and India lie, as it were, still outside the World's History, as the mere presupposition of elements whose combination must be waited for to constitute their vital progress. The unity of sustainability and subjective freedom so entirely excludes the distinction and contrast of the two elements, that, by this very fact, substance cannot arrive at reflection on itself – at subjectivity'. How very different the Western perception of ascendant China and India currently is. As Huntington (1996: 308) argues, the emerging politics of culture of non-Western civilizations and the 'increasing cultural assertiveness of their societies' are widely recognized: China, India, Japan and the Middle East. Mahbubani (2008), in his book, *The New Asian Hemisphere,* has been less apologetic about this Asian assertiveness which he succinctly summarized as the political and economic rise of Asia, which the Western world needs to accommodate in the currently entrenched Western-dominated global system. Other commentators are less enthused by Mahbubani's analysis of China's rise given her ageing population (334 million over 60 years by 2050), her dependence on non-Chinese-owned factories (60 per cent in 1980) and her dismal environmental record (50 per cent of the population lack clean drinking water and 70 per cent of lakes and rivers are polluted). Furthermore, not only for China but also most of the developing countries, the subtext of globalization and its capitalistic enforcer is its widening disparities of wealth between states and within states and cities. Needless to say, these changes in the developed world (North America, Europe, Japan, Australasia) are likely to have political ramifications and exacerbate realignments of trade flows, migration patterns, geopolitical restructuring, food and natural resource utilization and development patterns.

Seventhly, the costs involved in research and development on climate change are prohibitive and out of the reach of developing Asian countries. While Lomborg (2008) strongly advocates research and development on climate change, the global community will be handicapped in developing the right global and regional responses to climate change because few countries have the expertise and financial abilities. Clive Spash (2002) noted that there were only five major climate change monitoring centres in the whole world, all located in the Western world. Unfortunately, no Asian state has the research expertise and research and development capability to undertake climate change research. But, given that many Asian countries are losers and victims of climate change, it is imperative that Asian states raise their profiles in international forums on climate change. Asian states must voice their concerns and ensure that they are heard when it comes to mitigating climate change or finding adaptive mechanisms to reduce impacts.

Exploring Common Grounds in Asia

Given these major impediments to finding a common ground for climate change, the global community needs to put in place the right infrastructure to deal with the long-term repercussions of climate change. The issues here are too complex, but one can highlight several possible areas:

Firstly, we need to have a global system that actively puts in place cultural adaptive mechanisms and develops long-term mitigation programmes to reduce CO_2 emissions, the source of global warming and other climate changes. For example, the rise in sea levels is likely to make the Maldives and other Pacific Island states uninhabitable and will drown many islands in Southeast Asia. Countries will certainly have to develop mass evacuation programmes, and the global community will need to decide the number of climate change refugees that various countries are willing to accept. If these issues are not resolved in the next couple of decades, many people in the affected tropical islands are likely to be affected. Clearly, many coastal-based Asian cities (Shanghai, Manila, Singapore, Mumbai, Ho Chi Minh City, Penang, Kuching, Jakarta, Surabaya) are likely to face the full wrath of sea level rises.

Secondly, there are several lessons that we can learn from the current global economic chaos. We need to think not only in a global manner but also, on environmental issues, ecologically. The ecological perspective embraces the climate change challenge from a holistic, integrative and global system.

In short, climate change cannot be nationalized or politicized because its impacts observe no political boundaries and national borders. In 2003, 35,000 people in Europe died of a massive heat wave despite being in a regional community that was economically well-off by international standards. If each country keeps defending its national environmental positions, the Asian community will suffer in the long run.

Thirdly, arising from an ecological perspective, the global community needs to think in terms of common-pool resources and the fact that mother earth has finite resources and capabilities. Global warming is about environmental goods and common-pool resources. Even the infinite resources around us (like air, water) are only, relatively speaking, infinite. Indeed, even these resources are being polluted rapidly and hence become less usable. If we do not think in terms of common-pool resources, we risk thinking of selfish, greedy, exploitative and non-conservational ways of exploiting and using the environment around us. All the undergirding issues that have led to the current global economic crisis need to be understood if we want to find a common ground on common-pool resources: trust, reciprocity, social and political capital, holistic vision and sustainability. Without a common-pool vision amongst stakeholders, the global community in the long run will face Hardin's (1980) scenario of the 'tragedy of the global commons'. If 20 million Japanese in Tokyo can observe civic behaviour to keep their city clean and take their city as a common-pool resource, I think there is hope in extending this thinking to all citizens. At the end of the day, Gaia will not be able to support our extravagant consuming and rapacious polluting behaviour. The resulting problem, as Lovelock (2007) notes, is that Gaia is sick with a fever but there is wonder whether she is terminally ill.

Fourthly, given that the biggest culprits of climate change are the developed countries and major corporations, any mitigation issues would have to begin with the right economic incentives and equations. Climate change issues are so enmeshed with global capitalism, world trade and corporate profitability that no major country or corporation wants to take the lead in making changes. For example, in Spain, Tabara (2003) showed that climate change policies lagged behind because they were repeatedly portrayed as being a potential drag on economic growth. This is, in a nutshell, the message of Al Gore's 'inconvenient truth': that, while nations know the impact of climate change, they do not want to do anything because of the economic, political and social inconveniences it is likely to cause. This explains the lack of global leadership and political will to tackle climate change issues.

Economic Equations for Climate Change: Asian Implications

The lessons we are learning from the massive current global economic crisis demonstrate that, while economic activities are increasingly being globalized, the system of dealing with global economic problems is constrained in an outmoded political system of the 20th century. The magnitude of the 2008 subprime crisis is best seen in the fact that USD50 trillion of financial assets was wiped out in 2008, including USD9.6 trillion of losses in Asia alone. The flows of trade, tourism, currencies, industrial relationships and labour are all increasingly moving rapidly across porous borders, but the way governments are handling the economic crisis still remains very much a national affair. Each country has been trying to solve its domestic economic problems and indeed showing signs of being nationalistic in its responses. The global economic system is flawed because one cannot expect the USA to remain the world's buyer of finished products and to remain at the forefront of a never-ending materialistic binge with an insatiable appetite seen in its consumer spending.

The academic discourses from the social sciences on climate change have produced a rather unilateral frame of reference. If one looks at the social science inputs on climate change, it would seem that the major voices have come from only economists. Three books by economists have made interesting contributions to the debate about cultural adaptive and mitigation issues: they are Spash's (2002) *Greenhouse Economics*, Stern's (2007) *The Economics of Climate Change* and Nordhaus' (2008) *A Question of Balance*. While all three books make important contributions to the understanding of the economic costing and equations of climate change, an understanding of the global framework on climate change must certainly be more interdisciplinary in substance and outlook. Clearly, the economic perspective tends to resonate most amongst private-sector entrepreneurs, businessmen, industrialists and governments. Everyone seems to think that climate and environmental change is an economic issue that can be solved by using carrot (economic incentives, tax rebates) and stick (polluter pay, fines, carbon taxes, regulations) economic policies. Certainly, many governments see climate change challenges as either economic opportunities (carbon trading) or liabilities (reductions of CO_2 affecting businesses).

Using his DICE model, Nordhaus (2008: 195) calculates that climate change damages in the baseline (uncontrolled) case will amount to USD22.6 trillion compared with USD17.3 trillion in an optimal case. Stern (2007: 179) has estimated that the economic impact of a baseline-climate scenario would be 2.6 per cent globally but that the mean costs for India and Southeast Asia would be 6 per cent of regional gross domestic product (GDP) by 2100. Based on social costing, Nordhaus (2008: 196–7) using the 2005 baseline of USD28 per metric ton of carbon, calculates that carbon price will rise to USD95 by 2050 and USD202 per tonne carbon by 2100. Depending on the targets of carbon reduction by 2050, economic costs and mitigation measures will need to be calculated accordingly. Stern (2007: 233) suggests a reduction target of 85 per cent to meet its 450 ppm targets by 2050; the 2007 German proposal limits global CO_2 emission by 50 per cent of the 1990 levels by 2050; and Al Gore advocates a 90 per cent reduction of CO_2 emissions from current levels in the USA (Nordhaus, 2008: 201). Hence, based on these targets, Gore's reduction would cost USD17 trillion, while Stern's (2007) would cost USD22 trillion.

V. National Strategies in Adaptation and Mitigation

The importance of the nation-state in managing climate change and environmental problems cannot be ignored because, despite all the talks about globalization, the nation-state in Asia still remains the political controlling body for populations and territorial autonomy. In ecological terms, Asian states are political ecosystems in that their ecosystems are defined by the power of the state governments and competing commercial enterprises. Hence the state is the definer and bearer of national standards, benchmarks and regulations, which are important requisites in containing climate change emissions at the domestic level within national boundaries. Given that the international system is still very much a state-based system, one needs to work with countries to ensure that nations have effective climate change control systems. While Asia is often seen as the global dynamo of the 21st century, the economic catalyst for the rest of the world's economic growth and development, this huge continent also carries the seeds of her own destruction if she does not take into consideration the environmental equation in her modernizing and developmental programmes. What makes the Asian continent so vulnerable? I suggest five factors: population growth, its widening disparities of wealth, its weak governments, urban growth and the growing expectations of its citizens for a better quality of life. All these issues are supported by the environmental equation and climate change complications.

In the Asian landscape, we are already witnessing various categories of states based on their government's ability to manage the state. These include the failed states, the lesser developing and fragile states, the developmental states and the developed states. Several Asian states have been written off as failed states. These include Pakistan and, to some extent, Afghanistan and Iraq. Other Asian countries are in a fragile state that are tottering on the brink of political and economic collapse. These include North Korea, Laos, Bangladesh, East Timor and Nepal. The bulk of the Asian states might be termed developmental states seeking modernization and development. These include China, India, Jordan, Indonesia, Mongolia, Myanmar, Vietnam, Thailand, South Korea, Sri Lanka, Saudi Arabia and the Gulf States. And, finally, we have the more developed countries of Japan, Singapore, Brunei, South Korea and Israel, which have achieved high levels of development, political stability and economic prowess. None of these categories are rigid classifications given the dynamic nature of politics and economic development. Given the wide diversity of states in Asia, it is difficult to apply blanket prescriptions for environmental management and climate change adaptation and mitigation solutions. What every Asian country needs is enlightened political leaders, good governance and non-corrupt government machinery. The tragedy is that state formation in Asia, like elsewhere in the developing world, is of recent vintage, a product of colonialism and hence very much in the throes of political development.

While China and India have become the current poster countries of rapid development and successful capitalism, they have also become the pin-up states of environmental degradation, inequality and poverty. Rapid development has meant that both countries have not been able to consider and reflect on the environmental lapses in their quest for modernization and development. Indeed, what is ironic is that both states have a strong indigenous tradition of ecological principles and practices enshrined in their religions and cultural traditions: the Taoist beliefs and *feng shui* practices in China; the Hindu and Buddhist religions and philosophies; and the *Vastu Sastra* practices in India. Yet, the paths to human-nature relationships have met with different outcomes in both countries. In China, indigenous beliefs and religious worship have not led to a moral responsibility of Earth care and conservation. Tuan (1972) noted that the Chinese landscape was already devoid of forest by the 19th century because of centuries of demand for wood for various purposes. His view is that eco-friendly world views do not necessarily dictate eco-friendly human behaviour in Chinese history. To compound the matter, the communist regime that took over in 1949 changed the eco-centric traditional Taoist culture of the Chinese into a 'possibilistic' ideological credo (environment provides *possibilities* for human usage), where human beings were placed in the centre and above nature. Mao grew the Chinese forests in his ambitious tree-planting programme on the Marxist ideological position that human beings need to conquer nature and take over the power of nature. Such a Marxist credo, if followed blindly, would have undermined China's environment even further. The idea that human technology, intelligence and power are above Mother Nature is clearly what environmentalists feel uneasy about and is in total opposition to ecological thinking.

In India, human-nature relations had a better historical outcome. The Indian culture is embedded with deep reverence for nature in Hinduism and Buddhism, the idea of the transmigration of the soul that unifies all nature with human beings, an adherence to vegetarianism and the respect for cows, which, Harris (2008) has argued, provides important ecological and sustainable ways for living. When the philosopher Passmore (1974) compared India and Australia in the 1970s, he noted that India, despite its 850-million population, conserved and lived in more ecologically correct ways than the 18 million Australians, who, through their mining, pasture systems and agrarian activities, left the landscape heavily degraded. In both the Indian and Chinese cases, we see how environmental relationships are important in understanding the long-term sustainability of productive landscapes.

In the long run, it is difficult not to be worried about the Malthusian equation in assessing the sustainability of Asian states, especially the large states of China, India, Indonesia, Pakistan, Bangladesh and even Japan. Asia accounts for three-fifths of the global population, and a major proportion resides in two countries. Will both China and India, with over one billion population, be able to sustain their population's livelihood, their standards of living, their increasing expectations of a better life and the state's trajectory for national development, modernization and urbanization? I have my doubts.

Besides the broad global initiatives, cities and countries can act collectively to work together on important environmental initiatives arising from climate change challenges in the future through adaptive and mitigation measures. Despite the fact the decision-making on environmental issues lies with governments concerned with their nation-states, major environmental impacts unfortunately will affect cities and urban agglomerations. Hence, unless adaptive measures are put in place for urban communities, the outcomes of climate change and accompanying environmental impacts will have serious consequences for the global urban populations. Adaptation is defined here as 'any adjustment in natural or human systems in response to actual or expected climatic stimuli or their effects, which moderates harm or exploits beneficial opportunities' (Stern, 2007: 458). There is probably an exhaustive list that one could compose, but I will highlight a few important areas that need attention.

Water

Given the onslaught of global warming and climatic change for the rest of the 21st century and the likely outcomes of aridity, water shortages and decreasing precipitation, countries and cities should be mindful that they need to take steps to mitigate the anticipated long drought in the near future. In this area, a tripartite arrangement needs to be forged amongst city governments, companies (multinational companies) and universities/research institutions to research and develop new technologies for increasing drinkable and clean water. Here is one area where environmental concerns can also have positive economic outcomes. The International Water Week (23–25 June 2008) in Singapore, for example, attracted some 390 companies in the water technology business. Singapore's long-term research in water technologies, in conjunction with private companies, is likely to see some interesting results. When our water agreements with Malaysia run out by 2061, the city-state will probably be self-sufficient in water, provided that our annual precipitation rates remain the same. Singapore consumes 1.3 million cubic litres of water, or roughly 550 swimming pools, per day. Through more prudent and efficient usage and higher water charges, water consumption in Singapore is being reduced from 176 litres per day in 1994 to 158 litres in 2006. The target is to bring daily water consumption down to 155 litres a day by 2012. Many cities around the world could embark on a similar venture: using the urban environment as a test bed for many innovative water systems such as desalination, recycling water and reclaiming water. No matter how affluent and economically buoyant a city is, if there is severe and prolonged water scarcity, urban residents will migrate, the urban economy will fizzle out and the city will collapse.

We have already successfully tested methods of making deserts into agricultural oases. The fertile Napa and Santa Clara Valleys were once deserts, but, with a 400-mile irrigation scheme put in place decades ago, the California government has changed these barren areas into one of the most productive agricultural areas in the USA, producing all sorts of vegetables, fruits and so forth.

Food Security

Nothing is more worrying for political leaders than the national threats of food insecurity. If one is without transport, one can walk. If housing is inadequate, one can live as a slum or squatter dweller. But, without food, there is no alternative. Without food, you either die from starvation or become sick and prone to fatal illnesses. But before that, a hungry person is an angry person, willing to revolt and demonstrate violently. The twin problems in Asia and globally of prolonged drought and massive floods are causing havoc to agricultural production and food supplies. With over 45 per cent of the population living in cities in Asia, food scarcity creates inflation and urban dwellers suffer the most. In 2008, the food crisis caused riots and strikes in 30 countries and brought down two governments. In 2010–2011, the food crisis is hitting with greater venom. The Arab Spring is said to be fuelled by inflation in food prices in major cities caused in part by global warming. The result is the collapse of governments in Tunisia, Egypt, Libya and likely also in Syria and Yemen.

Asia, with countries containing states with the world's largest populations (China, India, Indonesia, Pakistan, Bangladesh and Japan), is vulnerable to climate change-induced food insecurity. Rich countries and political entities like Singapore, Brunei, Japan, Hong Kong and Taiwan might view food inflation with less worry. According to former Prime Minister of Singapore, Lee Kuan Yew, price inflation of food was not worrying because, he argued, 'it doesn't matter whether you grow your own food or you buy your food. The question is the price. If there is a food shortage worldwide, the price of food, produce, will go up. Thus, the answer for a country like Singapore is to make sure that our incomes rise, our total GDP rises faster than our food prices'.

Such logic assumes that the world produces enough food to feed every one of the seven billion people and that food-exporting countries do not stop exporting food. Given the climate change problems and food insecurity in 2010–2011, many Asian countries are following strictly national policies with regard to their rice production. India has introduced a rice export ban; Myanmar has curbed rice exports; Cambodia, suffering from its worst floods, has halted rice exports because of domestic price increases; and Vietnam, another rice exporter, has cut rice exports to ease its domestic inflation. In addition, the new Thai Government is adding inflation to food by trying to double its rice price to meet its election promise to increase incomes of farmers. Overseas rice producers like Brazil and Egypt have likewise banned rice exports. These national policies by rice exporters are likely to reverberate amongst Asian rice importers, and the impact is likely to spiral and undermine the already gloomy economic situation.

Agriculture is threatened by two global trends. Firstly, agricultural land use is diminishing. In 1950, per capita arable land was 0.45 hectares per person, but, by 1997, it was 0.25 hectares per person; by 2050, it is estimated to be 0.15 ha per person. The availability of land for meeting human requirements has shrunk from 19.5 acres per person in 1900 to 5 acres per person in 2005 (Laszlo, 2009: 16). The other worrying trend is shrinking water supplies, a major variable in agricultural production. In 1950, fresh water per capita was 17,000 cubic metres but, by 1999, it was reduced to 7,300 cubic metres. The estimate for 2025 is 4,800 cubic metres (Laszlo, 2009: 9). Unless we find new agricultural techniques and technologies to grow crops on less land and with less water, the future global food security situation looks dismal.

Urban Transportation Systems

With the invention of the car, cities of the 20th century have become automobile cities. Cars in cities have grown almost five-fold. In 1970, there were 200 million cars globally. But, by 2006, there were 850 million cars, and it is estimated that, by 2030, there will be 1.7 billion cars (Newman & Kenworthy, 2007: 67). Essentially, cities of the 21st century are going to be shaped more by their transportation systems. This has dire environmental consequences due to carbon emissions from vehicles, and economic consequences, with oil prices hitting over USD140 per barrel in July 2008. The growth of the car population creates other problems in cities. Traffic congestion in Bangkok is the biggest cause of waste of time. The city has 5.7 million vehicles, a figure that is growing at 2,000 vehicles per day. Each day, 700,000 trips are made on Bangkok's metro train system compared to 6.5 million on public buses and 10 million in private vehicles (Beech, 2008: 43). In 1999, the average motorist spent an equivalent of 44 working days sitting in traffic jams (Brown, 2001: 194), which can be translated as USD1 billion lost in productivity annually (Beech, 2008: 44).

Given that cars are one of the most important culprits of CO_2 emissions in cities, it is appalling to read glowing reports about India producing a 'people's car', the Tata Motors Nano car, sold at INR 100,000 (about USD2,000). Right in the heart of climate change debates, the car is applauded for now being affordable to village Indians, riding on the villagers' sentiments that possessing a car is 'a status symbol' (Jyoti, 2009: 40). In China, India, Vietnam, Cambodia, Taiwan and Malaysia, there is a graduation of transportation from bicycles to motorcycles and, finally, cars of all brands and sizes. Hence, in the booming economies of China and India, car populations are likely to increase by 15 times and 13 times respectively over the next 30 years (Schaefer-Preuss, 2008: 26). Cars are marketed not only in India but also marketed around the world. While cars are excellent to the fuel economy (56 miles to the gallon), have Indians and the world community wondered what millions of these cars on the road will do in raising CO_2 levels? Are industrialists, entrepreneurs and government officials oblivious to the fact that public transportation is the most efficient means for transport? Unfortunately, the industrial economy seems to outweigh in many Asian countries a change to mass transportation systems. In Thailand, South Korea, China, India, Indonesia and Malaysia, there are national car industries that need to be protected economically above environmental concerns.

In a study of 15 cities around the world in 1995, the most energy-wasteful cities were Atlanta and New York, cities dependent on cars. The most energy-efficient cities were Ho Chi Minh City and Shanghai. With regard to gasoline usage, each citizen in these two US cities, Atlanta (2,962 litres annually) and New York (1,237 litres), were using 110 times and 45 times more gasoline respectively than a resident of the Vietnamese city of Ho Chi Minh (27 litres annually) (Newman &

Kenworthy, 2007: 68). If urban transport systems are held hostage to oil, then the current steep increases in fuel prices are creating urban inflation across the board for all products, which will make urban living unbearable for the urban poor.

The most important equation in energy-efficient and environment-friendly urban environments boils down to the use of public transportation systems. Tokyo and Hong Kong are highly densely populated cities, but they use 10–25 times less gasoline than Atlanta because of a dense network of public mass transit and rail systems (Newman & Kenworthy, 2007: 68). Based on the various modes of urban transport, a study of 32 cities in 1990 showed clearly that cars have the lowest fuel efficiency and lowest occupancy rates amongst the five main modes of urban transportation: cars, bus, heavy rail on electric, diesel and light rail (see Table 1.3). The economic efficiency of urban transport system demonstrates that, while freeway traffic carries 2,500 people per hour, buses carry 5,000–8,000 persons; light rails support 10,000–20,000 persons and heavy rails carry 50,000 persons per hour, 20 times more than freeways (Newman & Kenworthy, 2007: 83). The economic savings of city residents is enormous in the choice between cars and transit systems: a transit-based urban system spends about 5–8 per cent of GDP on transportation, while car-based cities spend 12–15 per cent or, in the case of Brisbane, 18 per cent of GDP (Newman & Kenworthy, 2007: 82).

Table 1.3: Average Fuel Efficiency and Occupancy by Mode in 32 Cities, 1990

Mode	Average Fuel Efficiency	Measured Average Vehicle Occupancy
	(megajoules per passenger kilometer)	(number of occupants)
Car	2.91	1.52
Bus	1.56	13.83
Heavy Rail (electric)	0.44	30.96
Heavy Rail (diesel)	1.44	27.97
Light Rail/Tram	0.79	29.93

Note: Rail mode occupancies are given on the basis of average loading per wagon, not per train. The average occupancy of cars is a 24-hour figure.
Source: Starke, Linda (ed.) (2007) *State of the World 2007: Our Urban Future,* New York and London: W.W. Norton & Co., p. 73.

Solving Brown Issues: Public Housing

Of the estimated one billion people living in informal settlements worldwide (Chafe, 2007: 115), more than 900 million urban dwellers live in slums in low- and middle-income nations (Satterthwaite & McGranahan, 2007: 28). The rural urban migration that feeds slum and squatter settlement growth in cities explains why cities in poor countries are expanding rapidly. Over and above this, the lack of electricity means that more firewood and charcoal are used for cooking, for example. This creates indoor pollution, which is a major problem in slum and squatter settlements (Lomborg, 2001).

A city that cannot house its population in reasonably good housing can never solve its intra-urban environmental challenges. The brown issues within cities will not be resolved if public housing projects are not addressed for slum and squatter dwellers. Without the ambitious public housing programmes by the Housing Development Board since 1960, which now houses 85 per cent of Singapore's 5.18 million population, Singapore's rapid and outstanding development over the last 40 years would never have been possible. The city-state's transformation from Third World to First World could not have taken place if 35 per cent of its urban population in the 1960s still remains in slum and squatter settlements today. The quality of life and living for the majority of all urban residents is what makes cities the beacons of human civilization and sustainable systems.

The nature of housing policy varies from country to country as it does at different periods of time. In the UK, while sustainable housing has been in place since the early 1970s, there have been changes in the values, direction and technological practices employed over the decades (Lovell, 2004). Clearly, the objectives of sustainable housing are different (smart housing) in the UK in the 1970s from those in the 2000s and onwards. Sustainable housing today is geared towards climate change issues: low-carbon housing and low-carbons discourses. Low-carbon housing is defined as housing that 'has lower greenhouse gas emissions (principally CO_2) compared with an average new house built to the UK 2000 Building Regulations – that is less than one tonne of carbon per year' (Lovell, 2004: 36). In Singapore, on the other hand, public housing has moved from merely adequate shelter at the barest minimum to upgrading programmes that best reflect 'ecological modernism' goals that underscore the use of eco-efficiency and technological solutions in housing while keeping up with the domestic rise of social standards and aspirations of its residents. In Singapore's new housing policies, the issues seem directed less at carbon reductions (energy efficiency, renewal energy) than at saving space and ensuring spatial efficiency in building higher-density housing (40- to 50-storey-high flats). Hence one meets environmental efficiency through economies of scale.

The earlier Singapore model of bare environmental technologies and eco-efficiency is what the United Nations is looking at for the lesser developed countries. So is China. The move to sustainable ways of solving intra-urban housing problems is thus geared to providing public housing flats with clean water, modern sanitation and a well-managed refuse disposal system.

Eco-Cities and Sustainable Cities

The challenge of sustainable living in the light of climate change would lie in how we manage our cities. Given that cities will become the norm of living for the world's population this century, one would need to ask whether Asia's environmentalists, economists, technologists, social scientists and think-tank scholars have paid sufficient attention to how cities are going to become liveable environments and sustainable nodal entities for 50 per cent of Asia's population. Asian states need to pay attention to the pan-European network, referred to as the Building Environmental Quality Evaluation for Sustainability, which foregrounds urban sustainability, provides a road map that links spatial and time frame and adopts assessment methods (Vreeker *et al.*, 2009). Globally, by 2030, four of the five global urban residents will be in what we refer to now as the 'developing' world. This huge urban population will be residing in 377 cities of between one to five million residents spread over Africa (59 cities), Latin America and the Caribbean (65 cities) and Asia (253 cities) (Flavin, 2007: xxiii). The largest expansion of urban population between 2000 and 2030 will be in the less developed countries, which will see a more than 15 per cent increase in urban population as opposed to the 7 per cent increase in the More Developed Countries during the same time frame. In Asia, rapid urbanization means that 44 million people are added to cities annually. This means, on a daily basis, that a further 120,000 people are added to Asian cities, requiring more than 20,000 new dwellings and 250 kilometres of new roads and other infrastructure facilities to supply more than six megalitres of potable water (Roberts & Kanaley, 2006: 1).

Specifically, governments and planners around the world have attempted to green the urban development agenda by building 'eco-cities'. What is an ecological city? According to geographer White (2002: 194), an ecological city is a 'city that provides an acceptable standard of living for its human occupants without depleting the ecosystems and bio-geochemical cycles on which it depends'. China, in particular, seems to be going into partnership with various governments and corporations to build its new green-friendly eco-cities (Tianjin, Guangming, Yangzhou, Changzhou, Gegu and Guiyang). The British Government has scheduled as well to build 10 eco-cities by 2020. But are these eco-cities the panacea for arresting global warming? In my view, they are not. The eco-cities have been maintaining intra-urban environmental standards for some time and have been progressive, but they are less able to deal with reducing extra-urban ecological footprint challenges. Reducing the ecological footprint of cities requires not just technological inventions but also the more difficult changes in our thinking, which should be geared towards an economic system, our capitalistic system of trading, our materialistic value systems, our consumer-based ethos, our endorsement of common-pool resources and our ethical relationships to nature and community. Currently, there are two models of the eco-city: Singapore and British. The Singapore model of the eco-city essentially addresses the intra-urban environmental issues that are often referred to as brown issues (clean water, clean electricity, garbage disposal systems, modern sanitation and sewerage). There is no doubt that the Singapore model provides an apt model for many Asian developing countries in solving their major brown issues, the environmental underbelly for creating sustainable cities. The Singapore model looks at housing through the prism of ecological modernism, in which public housing provides a badge of modern status for its residents. Hence, it is not surprising that given their heavily polluted and unclean cities, the Chinese Government is looking to Singapore as a model for their eco-cities, a model that can solve their brown issues sustainably and cleanly. The reason why Singapore can deal with its intra-urban brown equation is that the country is a city-state. The British model of the eco-city is perhaps more geared towards finding a balance between residents within the city with that of its hinterlands. The British model is thus concerned more about the ecological footprint of its urban residents.

The challenge of Asian urbanization falls within two spheres. Firstly, there is the intra-urban environment and urban space that one needs to address and deal with. Essentially, do we have the ability to develop an expanding and plastic infrastructure that can accommodate an expanding resident and floating population (tourists, mobile labour) within the city? The term 'plastic' is used to emphasize the flexibility at which city officials need to expand essential infrastructures to accommodate increasing urban populations. The academic discourses are on adequate housing, clean water, modern sewerage systems, garbage disposal and efficient public transport networks.

Secondly, with massive populations concentrated in spatial nodes, the city becomes a major importer of many natural resources, food, water and other creature comforts to sustain its population. Based on White's (2002) definition of the eco-city, this is the real critical challenge for many cities in the near future. Are they sustainable in terms of getting supplies to feed their citizens? This becomes a major problem because in many Asian cities, poverty reigns, and a two-tier social system of the haves and have-nots becomes starkly evident. The question of supporting and sustaining urban populations is thus not an issue of the relationships to immediate hinterlands but whether urban populations can afford the imports of food, water, energy supplies and natural resources to sustain their living. Hence, if cities essentially are not economically sustainable entities, it would be difficult for resident populations within cities to lead sustainable livelihoods.

Reducing and Changing Energy Consumption

The whole global economy is lubricated with oil. The consequences of changing to other alternative energy sources are really mind-boggling. Where does one begin unless there is a global mindset change and political will amongst state heads, political leaders, corporate honchos, and entrepreneurs to reduce carbon-emitting sources of energy and switching to other clean-energy sources? Given that the major coal-using energy countries are also the major global powers and countries: China, India, the USA, and Germany, it seems unlikely that changes in energy policies will come rapidly. Despite all the talk of clean coal, that is still yet to be realized. Hence many of these countries have been less than enthusiastic in raising the bar in tackling the problems of CO_2 outputs.

As cities become major population nodes, they are also major consumers of energy due to domestic consumption. Put these figures together and it explains the 140 per cent surge in electricity capacity in China and the 80 per cent rise in India in the next decade (2008–2017) (*The Economist*, 2008: 78). With the estimated

75 per cent rise in electricity demands in the emerging economies over the next decade (2008–2017), all of which are dependent on fossil fuels, the vision of air pollution and global warming in our near future looks grim. Currently, the USA and China are major urban emitters of greenhouse gases because they still use coal for the generation of electricity.

The long dependence on cheap fuels has lulled the global urban community into accepting oil as the main energy provider from power plants to motorcycles. Research on alternative energy sources began only recently. Furthermore, cheap oil means that it is difficult to embark on other alternative fuels, which are more expensive to run than petroleum. Secondly, unlike coal, oil and gas are relatively portable fuels and hence are technologically compatible with mobile vehicles. Oil and gas also are less polluting than coal and hence are seen as environmentally friendly. Thirdly, the non-carbon and renewable energy sources have had limitations. Wind and solar powers are not easily available in many cities and countries because of their less predictable supplies.

The more radical alternative, as seen in the current international political climate, is for urban governments to adopt Lovelock's (2007) prescription of nuclear energy. Given the problems with fossil fuels in a Gaian world, where deteriorating global warming predominates, Lovelock (2007: 196) puts his argument this way: 'Nuclear energy is free of emissions and independent of imports from what will be a disturbed world. We would be right to cut back all emissions to a minimum, and this includes emissions of methane from leaking pipes and landfill sites. But most of all, we need electricity to sustain our technologically based civilization'. After the disastrous Fukushima nuclear plant accident due to the tsunami in 2011, Germany and Italy decided to terminate their nuclear energy programmes in two to three decades. No country in Asia, however, has negated the option of using nuclear energy. Yet, the adoption of nuclear energy is contingent on heavy government financial involvement and management, the question of how cities will be able to treat and store nuclear waste materials safely and effectively, and the vexing issue of security in a terrorist-phobic global environment – hence the developed world's reluctance to approve of Iran and North Korea's nuclear energy programmes.

VI. What is Asia's Response to the American Model of Unsustainable Development?

Huntington's (1996) clash of civilization thesis, like Hardin's (1980) lifeboat ethics, remains an outmoded Western cold-war thinking: that the Western global supremacy or hegemony is being eroded and undermined by new players in the global political landscape. The capitalistic-driven, modern, developed 'civilized' societies now comprise countries in the far-flung corners of the globe. North America, Europe, Japan, Australia and New Zealand that Huntington defined as the 'West' versus the rest of the other civilizations. His grim threats of the impending 'fall' of Western civilization were attributed to Islamic and Confucian states, located in Asia. One can see the ghost of Said's Orientalist ideology being resurrected again.

But the West versus the rest thesis in economic terms lies more in the restricting of the global demand and supply of natural and human resources. The whole economic synergy in the US model of development around the world has led to massive demands of commodities globally. Currently, speculators from the developed countries still control the commodities futures. The commodity index over the last four years (over the period 2003–2008, for example) rose from USD13 billion in 2003 to USD260 billion in March 2008. The price of 25 commodities has been raised to an average of 183 per cent within those four and a quarter years. Such price increases have created major inflation trends worldwide, food shortages and commodity price increases in basic utilities that have become the source of urban riots, strikes and political tensions especially in the developing world. For example, the *China Daily* reported that there were 50,000 environment-related riots, protests and disputes in China in 2005 (Roseland & Soots, 2007: 153). But, if this current scenario plays itself out in the remainder of the 21st century, it can have unhealthy outcomes as alluded to by various academics and policy commentators. Clearly, while Western models of capitalism are criticized for being unsustainable by Speth (2008), Stoll (2008) and Parr (2009), the rest of the developing world seems oblivious to their inherent environmental dangers, economic overkill, unsustainable development and widening disparities of wealth. The poster boy of development, the USA is too shiny a model for any developing, poverty-stricken country to see the downside of the capitalistic model.

On the other hand, the developed Western world is resisting competition and the new threats of the nouveau riche societies especially in Asia – China and India. The main challenge to the Western control of commodities reflects the changing geopolitical and political ecological scenario. The burning question is whether the current developed club of the Western 'world' (North America, Europe, Japan, Australia, New Zealand) will accommodate new entrants from Asia, Africa and Latin America. Specifically, whether there is a global hinterland that can accommodate all the new entrants into consumerism and capitalistic commodification. The complication of globalized hinterlands is that the agricultural produce grown is no more than a product of demand from the immediate cities they used to serve. Today, with demand for various raw materials globally, hinterlands have been turned from food-producing fields (rice, wheat, corn) into biodiesel and biofuel plantation (palm oil, soya bean, corn, sugarcane), livestock foods (maize), raw materials (rubber plantations), beverages (coffee, tea, cocoa), dairy farm (cattle, sheep) lands and timber plantations. Is there any wonder that food prices have escalated, food supplies have dwindled, urban inflation is rampant and food insecurity is becoming a major political and security problem?

The new czars of the global economy are now in Asia. Not surprisingly, *The Economist* (2008: 13) labelled China as the 'new colonialists' on the basis of its voracious appetite for global commodities. China's demand for oil rose from

1.88 billion barrels at the end of 2003 to 2.8 billion barrels in March 2008, an increase of 920 million barrels in just four and a quarter years. China currently consumes 50 per cent of the world's pork, 50 per cent of the world's cement, 33 per cent of the world's steel, 25 per cent of the world's aluminium and it has bought four-fifths of the increase in the world's copper supply since 2000 (*The Economist*, 2008:13). China's development, based on the Western model of development (the fossil-fuel-based, automotive-centred, throw-away economy), will not 'work for China's 1.45 billion in 2031' according to Brown (2001: 11). Given this dependence on fossil fuels, China's oil imports are likely to triple by 2030 (*The Economist*, 2008: 13).

This is not the only worry of the developed countries, but it is a major concern for environmentalists, whether the global hinterland can accommodate the massive consumption and consumerism in the lesser developed world. Will this capitalistic consumer-based economic system work for India or the three billion in the developing world? Unfortunately, the global politics of natural resources and development has less to do with the world's environmental problems and sustainability issues than with the developing countries feeling cheated of not attaining the 'American dream'. If the developed countries in the global community thwart China's or India's purchasing powers, the global competition for commodities is going to be the new fault lines of the international economic and political systems. Can international trade then ever be free, fair and open ended in a fixed pie scenario? As each national stakeholder tries to maximize his share of Gaia's natural resources at whatever costs, the tragedy of the global commons seems an inevitable future scenario. As Cahill (2002: 165) argues, the ecological footprint is meant 'to demonstrate the practicalities of resource conservation, social justice for the world's poor countries today and equity for future generations'. But this might be more of a theoretical statement than an underscoring of practical realities.

VII.   Concluding Thoughts

> What after all today's ecological movement is advocating is a return to isolation, and the abandonment of treasure and knowledge to tribes and nations in foreign lands that pose no threat to us. Consciously or otherwise, this is a death wish. We are not talking here about eschewing food additives and colouring matter, whole food in a whole land, as were the earlier ecologists, but something different – and deathly. For today's ecologists, their hope of regeneration presupposes a return to primitivism, and, thus, whether they wish to enunciate it or not, concomitant anarchy, the burning before the replanting, the cutting down of the dead tree. The father of the movement is an utter rejection of all that is and, for at least three millennia, all that was.
>
> –*Bramwell* (1990: 248)

Bramwell's dismal concluding statement in her history of ecology in the 20th century would come as a surprise to most environmentalists. It seems difficult to imagine that concerned ecologists are actually advocating a return to primitivism or a return to a state of indigenous living by adopting an ecological approach to our way of living. On the other hand, even if the key to sustainable living lies with indigenous peoples, should we write them off as unrealistic examples? What is needed is not to write off the value of indigenous livelihoods, folk culture and folk science as an 'oddity' (Orr, 2004: 10) but to embrace such indigenous knowledge as part of the wider fund of knowledge that can support the building of a sustainable society. And, within the kernel of indigenous folklores and culture lies important knowledge about human-nature relationships and ecological ideas. It seems a myth to believe that education is about giving students the means for upward mobility and success in life framed within a unilineal view of development and state progress. The tragedy, as Orr (2004: 12) notes, is that our planet does not need more 'successful people'. Rather, it needs more 'peacemakers, healers, restorers, storytellers and lovers of every kind'. And Asia still has a storehouse of many of these indigenous healers, folk restorers and storytellers.

While cities have traditionally served as the catalyst for human progress and civilization as well as to sustain high culture, climate change is likely to have major impacts on cities both directly and indirectly. The growing global environmental problems and the unpredictable climate change outcomes point us in another direction of caution, circumspection and reflection of the trajectory of global civilization. Apart from Fernández-Armesto's (2002), Linden's (2007), Fagan's (2008) and Whyte's (2008) books, which discuss the impacts of climate and environmental changes, circumstances and limitations of varied civilizations, few historical studies have considered the environmental dimensions, the abuse of nature in the equation of what makes civilizations sustainable. If one accepts Braudel's (1995: 7) view that civilization is 'the common heritage of humanity', then all cities and civilizations around the world should share the responsibility in sustaining not just their cities but also global Gaian civilization. The software of Gaian civilization should include the way society relates to nature in an ecologically responsible manner as well as the use and conservation of environmental resources. Asia has a lot to share because, on a global scale, it is more than any other region, has produced a plethora of civilizations in the Middle East, South Asia, Southeast Asia and East Asia. Indeed, in his history of civilizational peaks over the centuries the American anthropologist Alfred Kroeber noted that India had more civilizations that peaked than any other region or country in human history.

While the West has given the world modern scientific discoveries and knowledge from the time of Copernicus to the end of the 20th century (see Gribbin, 2003), which enable us to deconstruct our environments and nature, as well as to provide a better understanding of natural processes, it is in Eastern religions and philosophies that the solutions to sustainable living and a harmonious balance in human-nature relationships are likely to be found. The sociologist, Nisbett (2004), in his thought provoking book, *The Geography of Thought*, argues that

Western and Eastern worldviews undergirded different relationships to society and nature. Greek civilization undergirding Western civilization was ethnocentric and concerned with freedom and individuality, while Chinese civilization was concerned with harmony, the collective spirit of family, village and clan (Nisbett, 2004: 4–5). The Greek interest was in defining human and non-human objects, deconstructing nature that finally allowed for identity of atoms and molecules in modern science, while the Chinese mind was trying to maintain harmony in contradictory and dynamic forces of yin-yang relationships and accepting the *resonance* between man, earth and heaven (Nisbett, 2004: 10–17). Maintaining harmony and balance for the Chinese was always predicated in relationships, and, unlike the Greeks, the Chinese could not separate human beings from nature. Hence nature could never be studied independently. In her interesting book, *Being Human,* Peterson (2001: 85) takes this argument further by demonstrating that Asian views of nature and human nature were embedded in the ideas of *relational self* and *relational nature.* The idea is best exemplified in Buddhism where individuals are always relational and hence 'relational nature must be found within man's own nature and not in something external'. The world needs both modern science as developed in the West and Eastern-based philosophies of *relational* nature and self to find the pathways to sustainable life and living. In short, environmental sustainability requires both the 'hard' certainties of science and the soft, plastic and risky uncertainties of the social sciences (economics, sociology, geography, political science).

References

Amelung B, Nicholls S, Viner D (2007) Implications of global climate change for tourism flows and seasonality. *Journal of Travel Research* **45**, 285–96.

Asian Development Bank (2009) The *Economics of Climate Change in Southeast Asia: A Regional Review.* Asian Development Bank, Manila.

Beech H (2008) The Capital of Gridlock. *Time,* **171(12),** 43–4.

Begley S (2009) We can't get there from here. *Newsweek,* **CLIII(12),** 47.

Bowen M (2008) *Censoring Science: Inside the Political Attack on Dr James Hansen and the Truth of Global Warming.* Plume, New York.

Bramwell A (1990) *Ecology in the 20th Century: A History.* Yale University Press, New Haven and London.

Braudel F (1995) *A History of Civilizations.* Penguin Books, New York.

Brown LR (2001) *Eco-Economy: Building an Economy for the Earth.* WW Norton & Company, New York.

Cahill M (2002) *The Environment and Social Policy.* Routledge, London.

Castells M (2004) *The Information Age, Volume II: The Power of Identity.* Blackwell Publishing, Oxford.

Cassidy J (2009) Rational irrationality: The real reason that capitalism is so crash-prone. *Annals of Economics* October 5.

Chafe Z (2007) Reducing natural disaster risk in cities. In: Starke L (ed) *2007 State of the World: Our Urban Future*, pp. 112–33. WW Norton & Company, New York.

DiMento JFC, Doughman P (eds) (2007) *Climate Change: What It Means for Us, Our Children and Our Grandchildren*. MIT Press, Cambridge, Massachusetts.

Ehlers E, Krafft T (2006) Managing global change: Earth system science in the Anthropocene. In: Ehlers E, Krafft T (eds) *Earth System Science in the Anthropocene*, pp. 5–18. Springer, New York.

Escobar A (1995) *Encountering Development: The Making and Unmaking of the Third World*. Princeton University Press, Princeton.

Fagan B (2008) *The Great Warming: Climate Change and the Rise and Fall of Civilizations*. Bloomsbury Press, New York.

Fareed Z (2007) *The Future of Freedom: Illiberal Democracy at Home and Abroad*. WW Norton & Company, New York.

Fernández-Armesto F (2002) *Civilizations: Culture, Ambition, and the Transformation of Nature*. A Touchstone Book, New York.

Flannery T (2006) *The Weather Makers: How Man Is Changing the Climate and What It Means for Life on Earth*. Grove Press, New York.

Flannery T (2009) *Now or Never: Why We Must Act Now to End Climate Change and Create a Sustainable Future*. Atlantic Monthly Press, New York.

Flavin C (2007) Preface. In Starke L (ed) *2007 State of the World: Our Urban Future*, pp. xxiii–xxv. WW Norton & Company, New York.

Geertz C (1973) *The Interpretation of Cultures*. Basic Books, Inc, New York.

Gribbin J (2003) *Science: A History*. Penguin Books, London.

Hardin G (1980) *Promethean Ethics: Living with Death, Competition, and Triage*, University of Washington Press, Seattle.

Harris M (2008) *The Cultural Ecology of India's Sacred Cattle*. Blackwell, Malden, MA.

Hegel GWF (1956) *The Philosophy of History*. Dover Publications, Inc, New York.

Huntington S (1996) *The Clash of Civilizations and the Remaking of the World Order*. Simon & Schuster, New York.

Imanishi K (2002) *A Japanese View of Nature: The World of Living Things*. RoutledgeCurzon, New York.

Intergovernmental Panel in Climate Change Report (2007). University Press, Cambridge.

Jyoti T (2009) Nano Power. *Time* **173(13)**, 39–41.

Laszlo E (2009) *WorldShift 2012: Making Green Business, New Politics & Higher Consciousness Work Together*. Inner Traditions, Toronto.

Lieberman V (2003) *Strange Parallels: Southeast Asia in Global Context, c. 800–1830, Vol. 1: Integration on the Mainland.* Cambridge University Press, Cambridge.

Linden E (2007) *The Winds of Change: Climate, Weather, and the Destruction of Civilizations.* Simon & Schuster Paperbacks, New York.

Lomborg B (2001) *The Skeptical Environmentalist: Measuring the Real State of the World.* Cambridge University Press, Cambridge.

Lomborg B (2008) *Cool It: The Skeptical Environmentalist's Guide to Global Warming.* Vintage Books, New York.

Lovell H (2004) Framing sustainable housing as a solution to climate change. *Journal of Environmental Policy and Planning* **6(1),** 35–55.

Lovelock J (2007) *The Revenge of Gaia.* Penguin Books, London.

Mahbubani K (2008) *The New Asian Hemisphere: The Irresistible Shift of Global Power to the East.* Public Affairs, New York.

Martin J (2007) *The Meaning of the 21st Century: A Vital Blueprint for Ensuring our Future.* Riverhead Books, New York.

Mithen S (2004) *After the Ice: A Global History, 20,000–5000 BC.* Harvard University Press, Cambridge, Massachusetts.

Mulder N (1994) *Inside Indonesian Society: An Interpretation of Cultural Change in Java.* Duang Kamol, Bangkok.

Newman P, Kenworthy J (2007) Greening urban transportation. In: Starke L (ed) *2007 State of the World: Our Urban Future*, pp. 66–89. WW Norton & Company, New York.

Nisbett RE. (2004) *The Geography of Thought.* Free Press, New York.

Nordhaus W (2008) *A Question of Balance: Weighing the Options on Global Warming Policies.* Yale University Press, New Haven and London.

Orr DW (2004) *Earth in Mind: On Education, Environment, and the Human Prospect.* Island Press, Washington.

Parr A (2009) *Hijacking Sustainability.* The MIT Press, Cambridge, Massachusetts.

Passmore J (1974) *Man's Responsibility of Nature: Ecological Problems and Western Traditions.* Charles Scriber's Sons, New York.

Peterson A (2001) *Being Human: Ethics Environment and Our Place in the World.* University of California Press, Berkeley and Los Angeles.

Population Reference Bureau (2008) *2008 World Population Data Sheet.* Population Reference Bureau, Washington, DC.

Rees M (2003) *Our Final Century: Will the Human Race Survive the 21st Century?* Heineman, London.

Roberts B, Kanaley T (eds.) (2006a) *Urbanization and Sustainability in Asia: Good Practice Approaches in Urban Region Development*. Asia Development Bank, Manila, Philippines.

Roberts B, Kanaley T (2006b) Overview. In: Roberts B, Kanaley T (eds) *Urbanization and Sustainability in Asia: Good Practice Approaches in Urban Region Development*, pp. 1–11. Asia Development Bank, Manila, Philippines.

Roseland M, Soots L (2007) Strengthening local economies. In: Starke L (ed) *2007 State of the World: Our Urban Future*, pp. 152–71. WW Norton & Company, New York.

Satterthwaite D, McGranahan G (2007) Providing clean water and sanitation. In: Starke L (ed) *2007 State of the World: Our Urban Future*, pp. 26–47. WW Norton & Company, New York.

Schaefer-Preuss U (2008) Cities can and must cut energy use. *The Straits Times* **25 June,** 26.

Schweithelm J, Glover D (2006) Causes and impacts of the fires. In: Glover D, Jessup T (eds) *Indonesia's Fires and Haze: The Cost of Catastrophe*, pp. 1–13. Institute of Southeast Asian Studies and International Development Research Centre, Singapore and Ottawa.

Schweithelm J, Jessup J, Glover D (2006) Conclusions and policy recommendations. In: Glover D, Jessup T (eds) *Indonesia's Fires and Haze: The Cost of Catastrophe*, pp. 130–43. Institute of Southeast Asian Studies and International Development Research Centre, Singapore and Ottawa.

Simon JL (ed) (1995) *The State of Humanity*. Blackwell, Oxford, UK and Cambridge, USA.

Spash CL (2002) *Greenhouse Economics: Value and Ethics*. Routledge, London.

Speth JG (2008) *The Bridge at the Edge of the World*. Yale University Press, New Haven.

Spirn AW (1984) *The Granite Garden: Human Nature and Human Design*. Basic Books, Inc, New York.

Starke L (ed) (2007) *2007 State of the World: Our Urban Future*. WW Norton & Company, New York and London.

Stern N (2007) *The Economics of Climate Change: The Stern Review*. Cambridge University Press, Cambridge.

Stoll S (2008) *The Great Delusion*. Hill and Wang, New York.

Tabara JD (2003) Spain: Words that succeed and climate policies that fail. *Climate Policy* **3,** 19–30.

The Economist (2008a) Economic focus: Record spending on infrastructure will help sustain rapid growth in emerging economies. **387(8583),** 78.

The Economist (2008b) The new colonialists. *The Economist* **386(8571),** 13.

The Straits Times (2008) Global warming 'will harm national security'. **26 June:** 20.

The Straits Times (2009) Climate change 'near irreversible'. **14 March:** A12.

The Straits Times (2009) Rush to save crops as drought deepens. **7 February:** B5.

Tuan YF (1972) Discrepancies between environmental attitude and behavior: Examples from Europe and China. In: English PW, Mayfield RC (eds) *Man, Space, and Environment*, pp. 68–81. Oxford University Press, New York.

Volk T (2008) *CO$_2$ Rising: The World's Greatest Environmental Challenge.* The MIT Press, Cambridge, Massachusetts and London.

Vreeker R, Deakin R, Curwell S (2009) (eds) *Sustainable Urban Development, Vol. 3: The Toolkit for Assessment.* Routledge, London.

Walker G, King DS (2008) *The Hot Topic: How to Tackle Global Warming and Still Keep the Lights On.* Bloomsbury Publishing, London.

Walsh B (2008) Why green is the new red, white and blue. *Time* **171(16),** 31–42.

White R (2002) *Building the Ecological City.* Woodhead Publishing, Cambridge.

Whyte I (2008) *Worth Without End? Environmental Disaster and the Collapse of Empires.* IB Tauris, London.

Yusuf AA, Francisco H (2009) Climate change and vulnerability mapping for Southeast Asia. Economy and environment program for Southeast Asia [Article on the Internet] January 2009. [Cited 18 Apr 20].

CHAPTER TWO

Understanding Sustainable Development and the Challenges Faced by Developing Countries

Asanga GUNAWANSA
School of Design and Environment,
National University of Singapore

"Thus, true sustainability lies in the ability of governments to pursue sustainable development (SD) in all three areas of environment, society and the economy, by striking an appropriate balance in their sometimes competing interests. For example, a common response for stricter environmental regulation is that it often inhibits growth. Thus, developing countries in particular, may have a tendency to prefer economic growth over environmental regulation. However, the more efficient way of dealing with the competing interests is not to have a trade-off between a healthy environment on the one hand and healthy growth on the other, but to look for forms of development that are environmentally, socially and economically sustainable. Such developments could lead not to a trade-off but to an improved environment, together with economic and social development. Although one might argue that this is easier said than done, the counter argument is that if appropriately planned and implemented, introducing development mechanisms that could cater to the demands of all three limbs of SD is possible."

— Asanga GUNAWANSA

Understanding Sustainable Development and the Challenges Faced by Developing Countries

Asanga GUNAWANSA[1]

School of Design and Environment, National University of Singapore

Abstract

In 1987, the United Nations Commission on Economic Development, in its report, 'Our Common Future', defined 'sustainable development' as one that 'meets the needs of the present without compromising the ability of future generations to meet their own needs'. Since then, hundreds of definitions of sustainable development have been developed by various groups, based more or less on the definition of the United Nations Commission on Economic Development. The United Nations Commission on Economic Development explains that sustainable development 'contains within it two key concepts: the concept of 'needs,' in particular, the essential needs of the world's poor, to which overriding priority should be given; and the idea of limitations imposed by the state of technology and social organization on the environment's ability to meet present and future needs'. However, this full definition is seldom quoted and is often ignored. This chapter explores the development of the concept and the international legal and policy architectures that have provided the foundation for its development, with special emphasis on the challenges faced by developing nations in connection with their sustainable development goals. Two key issues explored are whether the current global interest in sustainability meets the expectations of all targeted beneficiaries, especially the world's poor, and whether all stakeholders are effectively and efficiently engaged in the process of development. Another issue explored is the competing nature of traditional sustainable development goals such as eradication of poverty with the new sustainable development goals such as mitigation of climate change and adaptation to climate change.

Key words: climate change, developing countries, Kyoto Protocol, poverty eradication, sustainable development, UNCED

---

[1] School of Design and Environment, National University of Singapore. 4 Architecture Drive, Singapore 117566. Email: bdgasan@nus.edu.sg.

I. Introduction

We live in a world in which almost every person who can read and hear has encountered the term 'sustainable development' (SD). A simple Google search of the term would show that SD is featured on approximately 27.8 million web pages. When a group of university freshmen who had signed up for a freshmen course on SD were tested on the knowledge of the term, the majority response was short: 'It means development that lasts'. When asked to explain 'development that lasts', two different responses were received. One explanation was that it meant the sustainable economic development of countries. The other referred to it as achieving economic development without exhausting natural resources. A key element that was missing from both explanations was the concept of social sustainability. However, this lack of comprehensive understanding of the concept of SD is not a weakness confined only to this group of students.

If one starts to delve into public rhetoric, a number of serious questions about our general understanding and acceptance of the concept of SD will emerge. For example, bureaucrats might argue that SD is an empty concept, too vague or ill-defined to be of any use in practical decision-making and real-life policy implementation (Jacobs, 1999). The environmentalists might argue that the notion of 'sustainable use' of resources is a dangerous influence that is a threat to natural resources worldwide (Patterson, 1998). According to Hattingh and Attfield (2002), instead of contributing towards the protection of nature and ensuring a continued availability of resources, some claim that 'sustainable use' is nothing but a green mask used by industries and governments to justify and continue the ruthless exploitation of natural resources as it has always been done. Some philosophers also point out that, in some of the interpretations, the notions of sustainability and SD rest on highly dubious assumptions that do not help us to curb our unrestrained exploitation of nature, but rather stimulate and accelerate it (Norton, 1999).

As Dobson (1999) points out, like many other political concepts such as democracy, liberty and social justice, the meaning of SD is up for grabs. This is because, while general agreements could be reached on a rhetorical level on the core notions of the concept, political and ideological battles exist between different concepts of sustainability. For example, for one of the least developed countries in the world, SD may mean economic development that would create jobs, eradicate poverty, increase literacy and improve health care. Thus, environmental sustainability, which might slow down economic progress, may be a less important concern.

The sustainability requirements of industrialized countries that have achieved economic development are likely to be different from those of developing countries. For such countries, sustainability would mean efficiently managing scarce resources whilst ensuring the continuity of current economic growth. Thus, in their situation, environmental sustainability would be superior than, if not ranked equal to, economic sustainability. Likewise, if individual perceptions are considered, the sustainability requirements of a refugee in Gaza or a slum dweller in India would be very different from those of an average person in a developed country, such as the USA or Australia (Gunawansa, 2009). In the circumstances, looking beyond the definition of the United Nations Commission on Economic Development (UNCED), it could be said that sustainability has different meanings to different people, depending on their economic, social and environmental priorities.

The remaining parts of this chapter are organized as follows: Part II briefly examines the development of the concept of SD and presents the argument that SD, as a concept, encompasses environmental, social and economic sustainability. Part III deals with SD and eradication of poverty. Part IV discusses SD and climate change mitigation. Part V concludes *inter alia* that true sustainability lies in the capacity of countries to strike a balance between sometimes competing needs of the aforesaid three elements of SD.

II. Development of the Concept of SD

Historical Development

According to environmental historian Guha (2003), the intellectual concern for the protection or conservation of nature that is characteristic of recent environmental movements goes back to at least the last decades of the 18th century. He points out that during that time, the Industrial Revolution had caused countries and their citizens to realize that human behaviour and modernization could destroy the environment. Further, according to Guha, these sentiments, which emerged in the late 19th and early 20th century, caused the 'first wave of environmentalism', in which intellectuals, such as William Wordsworth, John Ruskin and Also Leopold, criticized the Industrial Revolution, searched for scientific techniques to manage nature and aimed to preserve 'untouched' areas.

The concept of the preservation of the environment has found favour in several early 19th-century international conventions. For example, when colonial leaders of France, Germany, Great Britain, Italy, Portugal and Spain met in London in 1900 for the Convention for the Preservation of Wild Animals, Birds and Fish in Africa, one of the key objectives was to prevent... uncontrolled massacre [of wildlife] and to ensure the conservation of diverse wild animal species in their African possessions, which are useful... or inoffensive... to man' (Madonna, 1995). This treaty was replaced in 1933 by the London Convention Relative to the Preservation of Flora and Fauna in their Natural State.

These early ideas of sustainability were subsequently adopted by 19th- and 20th-century thinkers, such as Leopold, and led to a movement from species protection to habitat protection and on to the protection of all forms of biological diversity (Guha, 2000). A key weakness of the early ideas on sustainability is that they focused almost entirely on the need for environmental sustainability and ignored the need to promote economic and social sustainability for all. Coming from

40

an era in which colonialism was big and the basic rights of humans for equal treatment, political independence and economic stability were yet to be formally recognized, this is not a surprise.

Bringing SD to the Mainstream Global Agenda

It could be argued that the modern environmental sustainability movement did not emerge until the early 1960s. There is general agreement that the modern movement was propelled in 1962 by the publication of Rachel Carson's (1962) *Silent Spring* in which she wrote about the harmful effects of pesticides, which sparked controversy about the way in which humans were altering the environment.[2] The growing concerns regarding the environment that was triggered by this seminal text, in turn, caused like-minded works to be published, and prompted the establishment of environmental groups like the Sierra Club and Greenpeace (Sale, 1993).

It could be said that the above developments led to the increased number of studies of ecological systems and the establishment of international conferences and agreements on environmental management and conservation. The United Nations Conference on the Human Environment held in Stockholm in 1972 and the Convention on International Trade in Endangered Species of Wild Fauna and Flora (CITES)[3] held in 1973 are two good examples. The Stockholm Declaration that was made after the 1972 conference aimed to put forward 'common principles to inspire and guide the peoples of the world in the preservation and enhancement of the human environment'.[4] The key aim of CITES was to combat the commercial exploitation of endangered species through international trade restrictions. The relative failure of CITES led to the negotiations of the 1992 United Nation Convention on Biological Diversity (CBD).[5] In 1982, the United Nations World Charter for Nature adopted the principle that every form of life is unique and should be respected regardless of its value to humankind. It called for an understanding of human dependence on natural resources and the need to control the exploitation of natural resources.[6]

In addition to the above, it could be said that the interest on sustainability that grew during this era also led to several multilateral agreements and consensus that focused on SD as a concept that went beyond environmental sustainability. For example, the World Conservation Strategy released by the International Union for Conservation of Nature and Natural Resources in 1980 identified *inter alia* the main agents of habitat destruction as poverty, population pressure, social inequity and trading regimes. It called for a new international development strategy to

---

[2] This book was first published by Houghton Mifflin in 1962. The 4th edition of the book was published by the same publisher in 2002.

[3] For more information on CITES and the text of the convention, see http://www.cites.org.

[4] Declaration of the United Nations Conference on the Human Environment (Stockholm, Sweden: 1972), see http://www.intfish.net/treaties/unche.htm.

[5] For more information on CBD and the text of the convention see http://www.cbd.int/.

[6] For more information see www.un.org/documents/ga/res/37/a37r007.htm.

redress inequities. In 1984, the International Conference on Environment and Economics concluded that continued environmental improvement and sustained economic growth are essential, compatible and interrelated, and that the environment and the economy, if properly managed, are mutually reinforcing (Organisation for Economic Co-operation and Development, 2009) and are supportive of and supported by technological innovation. Further, the Conference concluded that environmental considerations should, as a matter of priority, be brought effectively into the centre of national decision-making on overall economic policy. It provided further that they also needed to be fully integrated with other policies, such as agriculture, industry, energy, transportation and land use management.[7]

Over the years many different definitions have been given to SD, but the most frequently quoted definition is from Our Common Future, also known as the Brundtland Report, the Report of the UNCED in 1987, which attempted to weave together social, economic, cultural and environmental issues and the related global solutions in defining SD in the following words:

> 'Sustainable development is development that meets the needs of the present without compromising the ability of future generations to meet their own needs. It contains within it two key concepts:

> - The concept of needs, in particular, the essential needs of the world's poor, to which overriding priority should be given; and
> - The idea of limitations imposed by the state of technology and social organization on the environment's ability to meet present and future needs'.

Thus, the idea of SD grew from numerous environmental movements in earlier decades, which focused almost exclusively on environmental sustainability to something that encompasses a much wider concept. This was clearly visible from the consensus reached in 1992 during the UNCED held in Rio de Janeiro (Earth Summit).[8] At this Summit, the heads of states from 108 countries adopted the following three major agreements aimed at changing the traditional approach to development (UN, 1997):

- Agenda 21 – a comprehensive programme of global action in all areas of SD;
- The Rio Declaration on Environment and Development – a series of principles defining the rights and responsibilities of States; and
- The Statement of Forest Principles – a set of principles to underlie the sustainable management of forests worldwide.

---

[7] For more information see http://sedac.ciesin.columbia.edu/entri/texts/oecd/OECD-4.02.html#fn0.
[8] The Earth Summit was unprecedented for a UN conference, in terms of both its size and the scope of its concerns. According to the UN it drew "hundreds of thousands of people from all walks of life". It is recorded that Some 2,400 representatives of non-governmental organizations (NGOs) attended the Summit. Further, 17,000 people attended the parallel NGO Forum. For more information on the Earth Summit see http://www.un.org/geninfo/bp/enviro.html.

In addition to the above, two legally binding Conventions aimed at the prevention of global climate change and the eradication of the diversity of biological species were opened for signature at the Summit, giving high profile to these efforts:

- The United Nations Framework Convention on Climate Change (UNFCC);[9]
- CBD.[10]

The Earth Summit called on the UN General Assembly to establish a means of supporting and encouraging actions by governments, businesses, industries and NGOs to bring about the social and economic changes needed for SD. This was fulfilled in December 1992 with the establishment of the Commission on Sustainable Development (CSD)[10] under the Economic and Social Council.[11] CSD was charged with reviewing the implementation of the Earth Summit agreements. It provides policy guidance to governments and major groups involved in SD, and strengthens Agenda 21 by devising additional strategies where necessary (UN, 1997). Reports submitted annually by governments are the main basis for monitoring progress and identifying problems faced by countries in relation to their SD approaches. According to UN, by mid-1996, some 100 governments had established national SD councils or other co-ordinating bodies. More than 2,000 municipal and town governments had each formulated a local Agenda 21 of its own.[12] Thus, Earth Summit could be credited for bringing SD into the mainstream.

In 1995, at the World Summit for Social Development held in Copenhagen, the largest gathering of heads of state at that time with 117 members, the participating countries reached a new consensus on the need to put people at the centre of development. It pledged to make the conquest of poverty the goal of full employment, and for the fostering of social integration to override the objectives of development. At the conclusion of the Summit, a Declaration on Social Development[13] and a Programme of Action,[14] which represented a new consensus on the need to put people at the centre of development, were adopted.

---

[9] UNFCC calls for the avoidance of "dangerous anthropogenic interference with the climate system". Its key objective is to consider what can be done to reduce global warming. In 1997 a number of nations approved an addition to the treaty: the Kyoto Protocol, which introduced legally binding measures to deal with climate change. For more information on the UNFCC and the Kyoto Protocol see http://unfccc.int/2860.php.

[10] The key objective of CBD is to objective is to develop national strategies for the conservation and sustainable use of biological diversity. It recognized for the first time in international law that the conservation of biological diversity is "a common concern of humankind" and is an integral part of the development process. For more information see http://www.cbd.int/.

[11] For more information on CSD see http://www.un.org/documents/ga/res/47/ares47-191.htm.

[12] For more information on CSD activities see http://www.un.org/esa/dsd/csd/csd_index.shtml.

[13] UN (2008), World Summit for Social Development Copenhagen 1995, Copenhagen Declaration on Social Development; see http://www.un.org/esa/socdev/wssd/copenhagen_declaration.html.

14 UN (2008), Programme of Action of the World Summit for Social Development; see http://www.un.org/esa/socdev/wssd/pgme_action.html.

The principles and goals set in the Declaration on Social Development that was made at the Summit included the following:

'Fulfil our responsibility for present and future generations by ensuring equity among generations and protecting the integrity and sustainable use of our environment'[15]

'Recognize that, while social development is a national responsibility, it cannot be successfully achieved without the collective commitment and efforts of the international community'[16]

'Recognize that the achievement of sustained social development requires sound, broadly based economic policies'[17]

'Promote democracy, human dignity, social justice and solidarity at the national, regional and international levels; ensure tolerance, non-violence, pluralism and non-discrimination, with full respect for diversity within and among societies'[18]

'Ensure that disadvantaged and vulnerable persons and groups are included in social development...'[19]

'Promote universal respect for, and observance and protection of, all human rights and fundamental freedoms for all, including the right to development; promote the effective exercise of rights and the discharge of responsibilities at all levels of society; promote equality and equity between women and men; protect the rights of children and youth'[20]

'Reaffirm the right of self-determination of all peoples...'[21]

'Recognize and support indigenous people in their pursuit of economic and social development, with full respect for their identity, traditions, forms of social organization and cultural values'[22]

The declaration made during the 1995 Summit and the Programme of Action that was agreed on by the participating countries clearly established that the concept of SD is far reaching than mere environmental sustainability and that it encompasses economic and social development, including human rights and the rights for self-determination as well as of women, children and indigenous groups.

---

[15] Art 26 (b), Declaration on Social Development.
[16] Ibid, Art 26(c).
[17] Ibid, Art 26(e).
[18] Ibid, Art 26(f).
[19] Ibid, Art 26(i)
[20] Ibid, Art 26(j).
[21] Ibid, Art 26(k).
[22] Ibid, Art 26(m)·

Further, it established the importance of bilateral, regional and international co-operation for achieving SD, thus making it clear that SD is a global concern.

In the year 1997, when progress made after the Earth Summit in 1992 was assessed at the Summit held in New York (Rio + 5), a number of gaps were identified, particularly with regard to social equity and poverty. This was largely reflected by falling levels of official development assistance (ODA) to developing countries and growing international debt, along with failures to improve technology transfer, capacity building for participation and development, institutional co-ordination and reduction of excessive levels of production and consumption. The review meeting called for the ratification, reinforcement and stronger implementation of the growing number of international agreements and conventions that referred to environment and development (Gardiner, 2002).

In the year 2000, during the UN Millennium Summit that was held to discuss the role of the UN at the turn of the 21st century, the heads of states of the UN member countries, 189 at that time met and ratified the UN Millennium Declaration.[23] One of the key agreements reached was to help citizens in the world's poorest countries to achieve a better life by the year 2015. Further, these agreements covered several other global issues, such as AIDS, and the fair distribution of benefits of globalization among the member countries (UN, 2000a). In order to achieve the agreements reached, the framework for the progress of the goals established was outlined in the following Millennium Development Goals:[24]

- Goal 1 – Eradicate extreme poverty and hunger;
- Goal 2 – Achieve universal primary education;
- Goal 3 – Promote gender equality and empower women;
- Goal 4 – Reduce child mortality;
- Goal 5 – Improve maternal health;
- Goal 6 – Combat HIV/AIDS, malaria and other diseases;
- Goal 7 – Ensure environmental sustainability; and
- Goal 8 – Develop a global partnership for development.

Under each of the above goals, various targets were set. For example, under the first goal, which dealt with eradication of poverty and hunger, the targets set included the following:

- Halving, between 1990 and 2015, the proportion of people whose income was less than $1 a day;
- Achieving full and productive employment and decent work for all, including women and young people; and
- Halving, between 1990 and 2015, the proportion of people suffering from hunger.

---

[23] UN, Millennium Declaration, A/res/55/2; see http://www.un.org/millennium/declaration/ares552e.htm.

[24] For more information on background, reports, and statistics on the Millennium Development Goals see http://www.un.org/millenniumgoals/.

In the year 2002, when the World Summit on SD was held in Johannesburg, marking 10 years since UNCED (Rio + 10), in a climate of frustration at the lack of progress made by governments in achieving the sustainability goals that were set in the previous summits, the Summit identified several key aims (UN, 2003). These include:

- To cut by half, by 2015, the proportion of people living on less than USD1 a day;
- To cut by half, by 2015, the number of people suffering from hunger;
- To cut by half, by 2015, the 1.1 billion people without access to safe drinking water;
- To cut by a significant amount the 2.4 billion people living without inadequate sanitation, to improve sanitation in institutions, such as schools, and to promote safe hygiene; and
- To launch an action programme to reduce the number of people lacking access to modern energy.

Apart from the plan of action above, a significant outcome of the Summit was the recognition of the importance of the integration of the three components of SD, namely economic development, social development and environmental protection as interdependent and mutually reinforcing pillars (UN, 2003). Further, poverty eradication, changing unsustainable patterns of production and consumption as well as protecting and managing the natural resource base of economic and social development were recognized as overarching objectives of, and essential requirements for, SD.

Although the 2002 Summit covered everything from measures to cut poverty, improve sanitation and ecosystems, reduce pollution, and improve energy supply for the poor, the absence of the USA rendered the Summit partially impotent. Further, although the Summit took note of the lack of achievements since various SD goals were set during the 1992 Earth Summit, the failure to agree on an effective mechanism to implement the plans and provide a foundation for countries in need of assistance to implement the SD goals, could be identified as a major shortcoming of the 2002 Summit.

In February 2005, the Kyoto Protocol entered into force and became legally binding on its 128 parties at that time.[24] It gave effect to the mandatory emission reduction targets for 30 Annex 1 countries during the operational period on 2008–2012,[25] and amongst other things, giving effect to the Clean Development Mechanism (CDM), which encourages investments in developing-country projects that limit emissions while promoting SD.

---

[24] The Kyoto Protocol is a protocol to the United Nations Framework Convention on Climate Change. The UN Framework Convention on Climate Change (UNFCCC) of 1992 calls for the avoidance of "dangerous anthropogenic interference with the climate system". To date, approximately 180 countries have ratified it.

[25] Under the Kyoto Protocol, industrialized countries are to reduce their combined emissions of six major greenhouse gases during the five-year period 2008-2012 to below 1990 levels.

In May 2005, prior to the Summit (Millennium + 5), to evaluate the progress made in connection with the Millennium Development Goals, UN (2005a) published a report detailing the progress, or rather the lack of progress, made in connection with the eight goals, and how large an effort was needed to achieve them.[26] For example, concerning the first goal, eradication of extreme power and hunger, the report stated that, although the global poverty rates were falling, led by Asia, millions more people had sunk deep into poverty in sub-Saharan Africa, where the poor were getting poorer. It stated further that, although progress had been made against hunger, slow growth of agricultural output and expanding populations had led to setbacks in some regions. Since 1990, millions more people have been chronically hungry in sub-Saharan Africa and in Southern Asia, where half the children under age five are malnourished.

To give another example, in connection with the seventh goal, ensuring environmental sustainability, the report stated that, although most countries had committed to the principles of SD, this did not result in sufficient progress to reverse the loss of the world's environmental resources. It added further that, although access to safe drinking water had increased, half the developing world still lacked toilets or other forms of basic sanitation. Furthermore, nearly one billion people were living in urban slums because the growth of the urban population was outpacing improvements in housing and the availability of productive jobs. Thus, the report concluded that achieving the goal would require greater attention to the plight of the poor, whose day-to-day subsistence was often directly linked to the natural resources around them, and an unprecedented level of global co-operation (UN, 2005a).

In September 2005, when Millennium + 5 was held, world leaders, by adopting the original declaration in 2000, affirmed their faith in UN and its Charter 'as indispensable foundations of a more peaceful, prosperous and just world'.[27] The world leaders who attended the Summit resolved to meet a number of Millennium Development Goals (MDGs), which included halving the proportion of people living in poverty and hunger by 2015, ensuring primary schooling for all children and reversing the spread of HIV/AIDS, malaria and other major diseases. However, noting the slow progress made during the period 2000–2005, many states, rich and poor alike, were not keen to raise any further hopes or make any further promises.

According to UN, although the 2005 Summit passed a 40-page outcome document that reflected the minimum consensus within reach at the time between the 191 UN member-states in the areas of development, peace and security, human rights and UN reform,[28] it fell far short of overcoming the global co-operation deficit

---

[26] Millennium Development Goals Report.
[27] According to the UN, 154 heads of state and government and over 900 ministers came together at the summit, to take stock of progress so far on the implementation of the MDGs, and to decide on concrete steps towards the realization of the MDGs and the reform of the UN.
[28] The full text of the document is available on the Summit website; see www.un.org/summit2005.

documented in numerous reports in the run-up to the Summit (e.g. in the HDR and the UN pre-Summit report of 2005 above). Thus, not surprisingly, the immediate reactions to the Summit outcomes were of disappointment. In rare unanimity, not only NGOs and the media but also many heads of governments and ministers from the North and South criticized the weak outcomes of the negotiations (Martens, 2005). Even the Secretary-General of UN at that time, Kofi Annan, expressed his disappointment to the assembled heads of states and governments, saying (UN, 2005b):

> '[...] let us be frank with each other, and with the peoples of the United Nations. We have not yet achieved the sweeping and fundamental reform that I and many others believe is required. Sharp differences, some of them substantive and legitimate, have played their part in preventing that'.

In the year 2008, during the emergency UN Summit of MDGs, the British Prime Minister at the time, Gordon Brown, commented in his address that the world had to face the shameful truth that, despite all the promises, MDG on infant mortality would not be met by 2050, let alone 2015.[29] Urging the world leaders not to use the credit crunch as an excuse to abandon pledges made by UN back in 2000, with the world only halfway to the 2015 deadline, he stated: 'I say to the richest countries of the world, "The poorest countries have been patient, but 100 years is too long to wait for justice. So to make poverty history we need to make new history today and make it happen now"'.

The MDG report in 2011 communicated that, although some achievements, especially in the area of poverty reduction, have been made mainly as a result of rapid economic growth in eastern China, the child mortality rate has been improved, and more children get universal education than those a decade ago, many of the eight MDGs and linked targets are in danger of becoming unmet by the deadline year of 2015 without redoubled efforts in developing countries, a sustained favourable international environment for development and increased donor support. It identifies the following as the key remaining challenges:

* Nearly a quarter of children in the developing world are underweight, with South Asia being the region most affected due to shortage of quality food and poor feeding practices, combined with inadequate sanitation.
* Wide gaps remain in women's access to paid work in at least half of all regions.
* The proportion of people going hungry has plateaued at 16 per cent, despite general reductions in poverty.
* Advances in sanitation often bypass the poor and those living in rural areas.

---

[29] The text of the speech by Prime Minister Gordon Brown, at the UN summit on the Millennium Development Goals, September 25, 2008, is available online on the website of the Internationale Politik, the Journal of the German Council on Foreign Relations; see http://www.ip-global.org/archiv/2008/winter2008/.

- Over 2.6 billion people still lack access to flush toilets or other forms of improved sanitation. And where progress has occurred, it has largely bypassed the poor.
- The development of facilities to provide access to clean drinking water has been unsatisfactory. In all regions, coverage in rural areas lags behind that of cities and towns.
- Global carbon dioxide emissions have continued to increase despite the international timetable for addressing the problem. The largest increase is seen in eastern Asia, where the emissions have increased from 3 billion metric tonnes in 1990 to 7.7 billion metric tonnes in 2008.

In the circumstances, it could be said that SD as a concept has grown from its initial recognition as a need for environmental protection into something much wider, encompassing the need to integrate social, economic and environmental protection as interdependent and mutually reinforcing pillars in achieving SD. The economic aspect of SD is geared mainly towards improving human welfare, primarily by increasing the consumption of goods and services. The environmental domain focuses on the protection of the integrity and resilience of ecological systems. The social domain emphasizes the enrichment of human relationships, the achievement of individual and group aspirations as well as strengthening of values and institutions. However, as observed above, it is clear that, despite the various initiatives taken to implement SD goals, they have not been adequately achieved, especially in the developing countries.

III. Sustainable Developments and Poverty Eradication

Nature of the Problem

The UN (2008a) report on the status of MDG implementation in 2008 noted that, although some gains against poverty had been achieved since 1990, with the number of people living in extreme poverty falling by 130 million, this progress had taken place against the backdrop of overall population growth of more than 800 million people in the developing regions. As a result, 1.2 billion people were still living on less than a dollar a day. Further, it was reported that half the developing world lacked access to sanitation and that, every week, 200,000 children under five died of diseases and 10,000 women died giving birth. In addition, it stated that 'we need to adjust ourselves to the new geography of poverty. Some regions score highly on most of the goals, whereas sub-Saharan Africa is lagging behind. In a few years' time, for the first time in history, there will be more people, in absolute figures, living in extreme poverty in Africa than in Asia'.

The latest UN (2011) report on MDG states that, despite significant setbacks after the 2008–9 economic downturn, exacerbated by the food and energy crises, the world is on track to reach the poverty-reduction target. As a result, it is expected that the global poverty rate will have fallen below 15 per cent by 2015, well under the 23 per cent target. This global trend, however, mainly reflects rapid growth in Eastern Asia, especially China. Further, the said report also states that, despite real progress that has been made, the development has not effectively reached the most vulnerable. As the following table adapted from the UN Millennium Development report of 2011 shows, the poverty reduction achieved in Southern Asia, Caribbean, sub-Saharan Africa and Latin America is marginal. Further, the percentage of poor in Caucasus, Central Asia and Western Asia has increased during the period concerned.

Table 2.1: Proportion of people living on less than $1.25 a day, 1990 and 2005 (percentage)

Region	1990	2005
Sub-Saharan Africa	58	51
Southern Asia	49	39
Southern Asia (excluding India)	45	31
Caribbean	29	26
Caucasus and Central Asia	6	19
Southeastern Asia	39	19
Eastern Asia	60	16
Latin America	11	7
Western Asia	2	6
Northern Africa	5	3

The global economic crisis that is still ongoing is seen as one of the major reasons behind the slow progress that has been made in some regions. According to the World Bank (2009a), the rise in food prices during the period from 2005 to 2008 is estimated to have pushed between 130 and 155 million more people into extreme poverty. According to the President of the World Bank Group, Mr Robert B Zoellick, 'a one per cent decline in developing country growth rates will trap 20 million ... people in poverty' (UN, 2008).

There is much hype now about the rapid development in China and India. However, if one considers the wealth distribution in these two countries and in other developing nations, the great divide between the rich and the poor is clearly visible. For example, in India, according to the national budget in 2006, the economic growth was 8.1 per cent. According to the Minister of Finance, India had completed its 'Golden Quadrilateral', a multilane highway that links New Delhi in the north, Calcutta in the east, Chennai in the south and Mumbai in the west (Hallinan, 2006). Further, the collective wealth of India's 311 billionaires had jumped by 71 per cent in 2005. However, for India's rural and urban poor, the chasm between them and the wealthy became wider and deeper during the same period. In 2005, India slipped from 124 out of 177 countries to 127, according to the United Nations Human Development Index (UN, 2006). Life expectancy in India was recorded as being 7 years less than that in China and 11 years less than in Sri Lanka, another developing country in the same South Asian region as India. Further, according to UN, mortality of children under five years is almost three times China's rate and almost six times Sri Lanka's, and even greater than in Bangladesh and Nepal (Hallinan, 2006). This goes to show that the true SD of a country cannot be measured merely by looking at the economic growth alone, as that may not reflect the true distribution of wealth.

Need to Look Beyond ODA for Eradication of Poverty

International efforts in poverty eradication have long been defined in primarily material terms, with transfer of finances from developed countries and international development agencies, such as the World Bank and the Asian development Bank, to developing countries being used as the main mechanism. Easterly (2006), a professor of economics at New York University and a former senior research economist at the World Bank for more than 16 years, argues that, although the developed countries spent approximately USD2.3 trillion on foreign aid over the last five decades to help eradicate poverty, very little has changed. Easterly credits the real help in fighting against poverty to people that look for a specific problem in a country and then attempt to help that country in a very practical way, instead of offering services for free. He argues that the open economic system could deal with poverty if efficiently managed to address specific issues. Examples he gives include selling bug nets to a population to prevent malaria or providing medical services to the poor for a modest fee. He argues that funding corrupt governments will not help eradicate poverty.

It is not difficult to agree with Easterly (2006) on his point that external development assistance could be of little help when dealing with poverty if the funds received are not efficiently managed and put to good use by the recipient governments. In addition, sometimes foreign aid, far from ushering in greater self-sufficiency, could have a detrimental effect on recipient countries by encouraging them to depend on foreign assistance without focusing adequately on necessary domestic reforms to eradicate poverty. Further, too much reliance on foreign aid could also lead to externally dictated priorities and misappropriation of funds.

Need for Transparency and Policy Reforms

It could be argued that the fact that the international financial agencies and the donor countries have historically negotiated projects and loans with the executive branch of developing country governments, excluding the elected parliaments, the legislative branch, from the development process, has contributed to misappropriation of funds and corruption in allocating funds for development activities. Transparency International claims that, according to its estimates, at one time, over USD30 billion in aid for Africa, an amount twice the annual gross domestic product of Ghana, Kenya and Uganda combined, ended up in foreign bank accounts (UN, 2000b).

According to Bello (2006), Professor of Sociology at the University of the Philippines and Executive Director of the Bangkok-based Focus on the Global South:

> 'About one out of every $3 that the World Bank gave to Indonesia during Suharto's government over a 30-year period from the mid-60s to the mid-90s went to the pockets of Suharto's people. This came to about USD10 billion of the USD30-billion World Bank lending program'.

As the United Nations Development Programme (UNDP, 2002) observed in its seminar on Human Development Report on democracy in 2002, 'the deeper is their intervention in sensitive governance reforms in developing countries, the greater is the need for international organizations to be open and accountable'. Thus, in order to ensure that room for corruption and misappropriation of funds is minimized, reforms at the level of international organizations and donor countries alone will not be good enough. Initiatives will have to be taken to ensure that recipient countries put in place mechanisms that are transparent in accepting funds, in negotiating projects, in procuring for project implementation and in distributing the funds.

The World Bank and International Monetary Fund (2008), which were created after the Second World War with a mandate 'to help prevent future conflicts by lending for reconstruction and development and by smoothing out temporary balance of payments problems' (George, 1999). They clearly have an incredible influence over the domestic policy agendas of developing countries (Stiglitz, 2003). The same could be said of the regional development banks, such as the Asian Development Bank and the African Development Bank. In addition, bilateral and regional donors also have considerable influence over the policies of countries that rely on them for financial assistance. Furthermore, regional integration movements such as the European Union, the Association of Southeast Asian Nations and the South Asian Association for Regional Cooperation also have considerable influence over the domestic policies of their member countries. Thus, these banks and organizations could play an effective role in influencing the necessary reforms in the policy and administrative architectures in developing countries, which could help in the efficient and transparent management of funds for the eradication of poverty.

A key mechanism to promote transparency and greater accountability in ODA is to provide a right to information within the national frameworks, whilst implementing information disclosure policies within international financial agencies and donors. Providing a right to information locally would ensure that the intended beneficiaries of ODA have the right to find out about ODA received and the manner in which ODA is utilized for development. In turn, this would lead to better accountability by the executive branch of developing countries to the public and would facilitate public participation in policymaking. This in turn would contribute to good governance by fostering greater transparency in policymaking; more accountability through direct public scrutiny and oversight; enhanced legitimacy of decision-making processes; better-quality policy decisions based on a wider range of information sources; and, finally, higher levels of implementation and compliance, giving greater public awareness of policies and participation in their design (Caddy, 2005). However, unfortunately, although the right to information has been recognized under international law[30] and has been recognized as a basic human right by many countries, giving it constitutional protection, it has still yet to be fully harnessed as a tool for promoting transparency, accountability and public participation at the national and international levels (Rodrigues, 2006).

In the circumstances, eradication of poverty is an issue that should be dealt with not only by providing ODA to countries in need, but also by ensuring that effective and efficient policy frameworks and administrative structures are put in place to ensure that externally provided finances and other forms of assistance are, in fact, put in good use to improving the lives of the poor. Such mechanisms should be sustainable. Thus, necessary developments should take the form of providing education, health care, vocational training and financial assistance for starting industries at individual and community levels. The Grameen Bank system introduced by the Nobel Peace Prize winner Professor Muhammad Yunus (1999), also known as the 'banker to the poor', for making small loans in impoverished countries, is a good example to follow.

The Grameen Bank system has reversed the conventional banking practice by removing the need for collateral for providing financial assistance, and has created a banking system based on mutual trust, accountability, participation and creativity. It provides credit to the poorest of the poor in rural Bangladesh without any collateral.[31] It looks at credit as a cost-effective weapon to fight poverty, which serves as a catalyst in the overall development of socioeconomic conditions of the poor, who have been kept outside the banking orbit on the ground that they are poor and hence not bankable (Yunus, 1999). According to Professor Yunus, if financial resources can be made available to the poor people on terms and conditions that are appropriate and reasonable, 'these millions of small people with their millions of small pursuits can add up to create the biggest development wonder'.[32]

---

[30] In 1946, the United Nations General Assembly recognised that "Freedom of Information is a fundamental human right and the touchstone for all freedoms to which the United Nations is consecrated" (UN General Assembly, (1946) Resolution 59(1), 65th Plenary Meeting, December 14). Later, the right to information was enshrined in Article 19 of the International Covenant on Civil and Political Rights which provides inter alia "Everyone shall have the right to hold opinions without interference. Everyone shall have the right to freedom of expression; this right shall include freedom to seek, receive and impart information and ideas of all kinds, regardless of frontiers, either orally, in writing or in print, in the form of art, or through any other media of his choice".
[31] As of March, 2009, Grameen Bank has 7.80 million borrowers, 97 percent of whom are women. With 2,548 branches, GB provides services in 84,096 villages, covering more than 100 percent of the total villages in Bangladesh.
[32] The Grameen Bank website; see http://www.grameen-info.org/.

The developing countries with a good agricultural base could also focus more on developing their agricultural sector. This will not only increase the food production but also provide a source of income for the poor. In fact, food production and agriculture is the world's single largest source of employment, with nearly 70 per cent of the poor in developing countries who live in rural areas depending on agriculture for their livelihoods (Dixon *et al.*, 2001). Thus, farmers must be accorded a rightful place in the processes of development, and their capacity for self-improvement and for contributing to food production as well as national development should be enhanced.

The developing countries should also re-evaluate their current spending so that more attention and dedication could be given to eradication of poverty. Today, some of the economic activities and their institutional contexts are clearly at odds with SD. For example, military expenditure in the world is exceeding USD1 trillion (UN, 2005c) which far exceeds the estimated costs of meeting the MDGs. According to the UN's Millennium Project estimates the costs of meeting the MDGs in all countries will be approximately USD189 billion in 2015 (UNDP, 2005), a fraction of the sum spent on military activities.

However, it should be noted that, if eradication of poverty is to effectively contribute to SD, then it cannot be done as a standalone measure, separated from other SD goals. For example, it has been widely acknowledged that economic prosperity has come at a tremendous cost to our natural environment (Intergovernmental Panel on Climate Change [IPCC], 2007). If we consider developed countries, it is difficult to debate that any of them emerged as a major industrial power without a legacy of significant environmental damage. In fact, this has been the central argument between the developed and developing nations in negotiating responses for climate change under UNFCC and the Kyoto Protocol (Gunawansa & Kua, 2009). Thus, when initiatives are taken to put mechanisms in place for eradicating poverty, it is important to ensure that the related economic and industrial activities meet the environmental sustainability requirements. In addition, it is also important to ensure that necessary social structure reforms are undertaken to safeguard equality for all, irrespective of their economic, social or ethnic status.

From the SD viewpoint, eradication of poverty has not only economic but also social and environmental dimensions. Therefore, it needs to be assessed using a comprehensive set of indicators that go beyond income distribution alone. For example, economic policies seek to emphasize means of expanding employment and gainful opportunities for poor people through growth, improving access to markets and increasing both assets and education. Social policies would focus on empowerment and inclusion by making institutions more responsive to the poor and by removing barriers that exclude disadvantaged groups. Environment-related measures to help poor people should seek to reduce their vulnerability to resource depletion and natural disasters, crop failures, loss of employment, sickness, economic shocks and others. Thus, an important objective of poverty alleviation is to provide poor people with enhanced physical, human and financial resources that will reduce their vulnerability.

IV. Climate Change and Sustainable Development Challenges for Developing Countries

Vulnerability to Climate Change

There is an overwhelming scientific consensus that the climate is changing and that the earth is warming up due primarily to human-induced activities. According to IPCC, a scientific intergovernmental body set up by the World Meteorological Organization and by the United Nations Environment Programme (UNEP) to provide decision-makers and others interested in climate change with an objective source of information about climate change, since pre-industrial times, increasing emissions of greenhouse gases (GHGs) have led to a marked increase in atmospheric GHG concentrations, causing global warming (IPCC, 2007). In its Third Assessment Report in 2001, IPCC (2001) warned that global warming, even at existing levels, had already impacted a number of important physical and biological systems. It predicted significant further impacts, including:

- increased risk of flooding for tens of millions of coastal dwellers worldwide;
- increased incidence of extreme weather events;
- reduced yields of the world's food crops; and
- decreased water availability in many water-scarce regions.

Further, according to IPCC (2001), climate change is likely to threaten all life forms on earth, with the extent of vulnerability varying across regions and populations within regions. The impacts, however, are likely to fall disproportionately upon developing countries, in particular, the poor living within them. In 2007, IPCC (2007) projected that, at the current level of emissions, summer sea ice could vanish completely anytime from 2040 to beyond 2100. In recent times, the world has witnessed some of the predictions made by IPCC come true. The tsunami in Asia, hurricanes in the USA and flooding and extreme heat in Europe are good examples.

Today, the majority agree with the evidence presented by ICC in its reports, and accept as well as act based on the predictions made concerning the threats posed by climate change. Thus, it could be said that climate change is one of the biggest environmental problems the world will face in the coming decades. The changing climate will disrupt complex environmental, social and economic systems that have been built up over the centuries, and that cannot withstand rapid fundamental change. The poorest countries in the world are likely to be the worst affected due to their slow reactive capacity to adapt to changing climate conditions (Gunawansa, 2009).

Need to Harmonize Climate Change Mitigation Efforts with SD Goals

Since the Earth Summit in 1992, several nations of the world have been working towards reducing GHG emissions. The Kyoto Protocol, which was agreed on 11 December 1997 at the 3rd Conference of the Parties to the UN Framework Convention on Climate Change (UNFCCC), was envisaged to be a first step on the road to reducing GHG emissions amongst developed nations.[33] The Kyoto Protocol was created as an effort to force action by the international community as, despite the establishment of the UNFCCC in March 1994, very little action had been taken as of December 1997 at the international level to reduce GHGs as a global response to climate change. According to UNEP, currently, over 200 international environmental agreements and an uncountable number of bilateral agreements on the subject of environment exist. However, the Kyoto Protocol is the only multilateral framework we have to address climate change. Under this, the industrialized nations have agreed to cut GHG emissions to an average of 5.2 per cent below the 1990 levels by 2012.

When UNFCCC was ratified, the principle of common but differentiated responsibilities to reduce 'dangerous anthropogenic interference with the climate system' was generously acknowledged. This principle recognized that:

- the largest share of historical and current global emissions of GHGs has originated in developed countries;
- per capita emissions in developing countries are still relatively low; and
- the share of global emissions originating in developing countries will grow to meet their social and development needs.

Thus, developing countries were not included in any numerical limitation of the Kyoto Protocol because they were not considered to be amongst the main contributors to the GHG emissions during the pre-treaty industrialization period. Further, their per capita GHG emissions were considered to be much lower than those of developed nations. This decision also took into account the fact that poorer economies of developing countries would be unable to absorb the costs of switching from a fossil fuel-based system to cleaner fuels. The aim was to bring poorer countries into future climate change agreements as cleaner technologies develop and become less expensive.

However, although the Kyoto Protocol does not require developing countries to reduce the emissions of GHGs, it recognizes that developing countries do share the common responsibility that all countries have in reducing GHG emissions. Therefore, they are encouraged, under CDM, to benefit from transfer of technology and foreign investments into sectors, such as renewable energy, energy generation and afforestation projects, which will contribute to mitigate GHG emissions and to achieve the ultimate objective of UNFCCC, whilst at the same time contributing to SD in developing countries. In brief, CDM project activities

---

[33] The Kyoto Protocol entered into force on 16 February 2005. To date, approximately 180 countries have ratified it.

are expected to generate Certified Emissions Reduction Units created through CDM investments of Annex 1 countries (industrialized countries with GHG reduction targets) in Non-Annex 1 countries (developing countries without any commitments to reducing GHG emissions). Such investments are expected to lead to the reduction of emissions, which otherwise would occur in the absence of such projects.[34] However, in practical terms, this aim may be difficult to achieve because of the competing interests of maintaining economic development by the industrialized countries and of achieving economic development by the developing countries, on the one side, against the cutbacks in economic development in order to achieve environmental sustainability on the other.

Moreover, it should also be pointed out that the CDM mechanism is burdened by potentially unrealistic expectations of what it can deliver. For example, one of the key aims of CDM is to facilitate technology transfer from Annex 1 countries to developing countries and to contribute to SD of the latter. As pointed out in this paper, SD, as a concept, goes beyond environmental sustainability and consists of broader issues, such as eradication of poverty and promotion of education, sanitation and health care. Thus, concerns here are whether the developing countries would be receptive to the type of CDM projects Annex 1 countries might be interested in; further, whether the developing countries have the capacity to receive and benefit from the technology that will be transferred to them under CDM projects and whether they could engage such technology towards meeting their SD goals.

Technology transfer between countries is an area that has generated special interest in international as well as local policymaking and legislation. As a result, in addition to most countries having national policy frameworks and legislation dealing with technology transfer, there are many agreements at the multilateral and regional as well as bilateral levels that operate, and apply to technology transfer across national borders. At the very core of these legal and policy frameworks as well as the multilateral, regional and bilateral agreements is the interest of protection of the rights of technology owners (intellectual property rights or IPR) against unfair and unauthorized use. Other interests include enabling the developing countries to benefit from technology transfer and the protection of the interests of the recipient countries.

One key impediment in cross-border technology transfer is that the bulk of technology transfers occur within the countries that generate them. According to Pachauri and Bhandari (1994), the provision of a favourable environment for technology transfer must be based on equity concerns and on participatory decision-making to improve the chances that such enabling environment will be sustainable. Thus, technology transfer from the countries and companies that develop them to other countries would require effective and efficient legislative and regulatory frameworks that recognize the ownership of technology and prevent the abuse of ownership rights. In the absence of such environments

---

[34] See Article 12 of the Kyoto Protocol.

conducive to facilitating technology transfer, some developing countries may miss out on the opportunity to benefit from the CDM mechanism.

One of the main difficulties concerning international technology transfer is that multilateral, regional and bilateral treaties may have imposed various conditions relating to such transfers and the related IPRs. For example, Annex 1C to the multilateral agreement establishing World Trade Organisation, the Agreement on Intellectual Property Rights (TRIPS)[35] sets down minimum standards that should be met by national laws for many forms of IPR regulations, including copyright, geographical indications, industrial designs, integrated circuit layout designs, patents and trademarks, and confidential and undisclosed information. Further, it also specifies the enforcement procedures, remedies and dispute resolution procedures.[36]

In the circumstances, it could be said that, although CDM aims to contribute to SD of developing countries by enabling them to inter alia benefit from technology transfer, the current IPR regimes might act as a hindrance to some developing countries attracting CDM investments. This may be a factor that has contributed to only a few developing countries benefitting from CDM project investments. The following figure, which shows the distribution of CDM projects across the developing countries as of 31 October 2011, clearly establishes that not all developing countries in the world have benefitted from CDM projects.

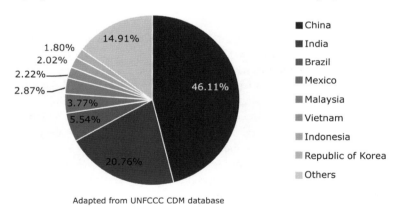

Adapted from UNFCCC CDM database

Figure 2.1: Registered project activities by host party (total: 3,559)

Of the 3,559 CDM projects that have been registered as of 31 October 2011, China is the biggest beneficiary, with more than 46 per cent of the projects. India, Brazil and Mexico are the next biggest beneficiaries, with 20.76 per cent, 5.54 per cent and 3.77 per cent of projects respectively. Of the 72 developing countries that have so far participated in CDM projects, 556 projects are shared

---

[35] See http://www.wto.org/english/res_e/booksp_e/analytic_index_e/trips_e.htm.
[36] TRIPS remains the most comprehensive international agreement on intellectual property to date.

by 62 countries, whereas 3,003 projects are shared by eight developing countries.[37] As the following table shows, the developing countries with rapidly growing economies have benefitted the most from the CDM mechanism, whereas the poorer countries have not caught the attention of the CDM investors.

Table 2.2: Countries with highest and lowest number of CDM projects

China – 1641	Albania – 1
India – 739	Azerbaijan – 1
Brazil – 197	Ethiopia – 1
Mexico – 134	Guyana – 1
Malaysia - 102	Jamaica – 1
Vietnam – 79	Lao People's Democratic Republic – 1
Republic of Korea – 60	Liberia – 1
Thailand – 60	Madagascar – 1
Philippines – 57	Mali – 1
Chile – 51	United Republic of Tanzania – 1
Colombia – 35	Zambia – 1

Adapted from UNFCC (2011): Registered Project Activities by Host Party.

This supports the argument that not all countries have benefitted from the current global efforts to deal with climate change by recognizing the common but differential responsibility of nations and implementing mitigation measures, ensuring the participation of both developed and developing countries in a way that is mutually beneficial.

Another point that should be made is that current investment promotion and economic development policies in most developing countries are more focused on one aspect of SD, that is economic sustainability. Thus, environmental sustainability has taken a back seat. In the circumstances, if developing countries are to choose between traditional investment projects and CDM projects, chances are high that traditional projects would be picked even though CDM projects would have the added benefits, such as environmental sustainability and technology transfer, if the traditional projects are likely to bring higher economic benefits to developing countries.

One key reason for this problem, especially from the point of view of democratic systems of government, is that a government's capacity to make effective policy

---

[37] See UNFCC CDM statistics at http://cdm.unfccc.int/Statistics/RegistrationNumOfRegisteredProj ByHostPartiesPieChart.html.

and implement them depends on being elected to do so by the people. When it comes to democratic elections in developing countries, promises of economic development as against the creation of environmentally sustainable societies might guarantee them being elected to power. Thus, in the absence of consensus between political parties on national development strategies and balancing the needs of economic development and the need to achieve environmental sustainability, developing countries would face the challenge of putting in place efficient long-term development policy frameworks that would help them achieve SD.

As far as the on-going activities of various countries in dealing with climate change are concerned, a point that should be made is that some of these activities may overlook the needs of the most vulnerable people in society. In particular, some of the current efforts in countries to address climate change, such as energy efficiency in industries, introduction of green construction standards and development of eco-cities, to name a few, focus more on climate change mitigation by the industries and on promoting environmentally sustainable lifestyles to the more affluent sections of society. Whilst the countries that engage in such initiatives should be commended for their action, they should also be encouraged to give adequate attention to the adaptability of the less affluent to changing climate conditions.

According to the information available on the project, the planned eco-city in Tianjin, 150 kilometres southeast of Beijing, which will tackle the growing problems of pollution by providing a 'green lung' and eco-corridors with extensive greenery for 110,000 energy-efficient homes is expected to cost approximately CNY30 billion (USD4 billion) (*The Straits Times*, 17 April 2008). At such a cost, it is likely that paying for such housing is likely to be beyond the capacity of China's poor. For example, eco-friendly houses planned in China's eco-towns for Huangbaiyu and Dongtan provinces were planned to cost approximately USD3,600 per unit, thus reasonably affordable. However, they ended up costing approximately USD20,000 per unit, making them unaffordable for most of the villagers who were expected to move into the new eco-friendly housing (Ethical Corporation, 2009). Thus, it is important that, whilst countries undertake commendable projects such as developing eco-cities, they should also focus on developing the living conditions of the poor who might not be able to afford the luxuries of an eco-city dwelling.

According to the United Nations Human Settlement Programme (UN-Habitat, 2003), close to one billion people live in slums across the world, and this figure is expected to double by the year 2030. According to the same source, this means that, currently, almost half of the world's urban population live in slums, or one in every six persons is a slum dweller. The table on the next page shows the slum population in different regions of the world as of 2003, according to U-Habitat.

Table 2.3: Slum population in different regions of the world

Region	Slum Population %
Sub-Saharan Africa	71.9
North Africa	28.2
Oceania	24.1
South-central Asia	58
East Asia	36.4
Western Asia	33.1
Southeast Asia	28
Caribbean	31.9

As noted above in this chapter, in UN's MDG, the international community recognizes the need to halve the proportion of people living in extreme poverty by 2015. If this is to be achieved, it is fundamentally necessary to directly address the needs of the burgeoning population of poor people living in cities. The current lifestyles of slum dwellers are far from environmentally friendly. For example, although they may not consume as much energy as the more affluent sections of society do by using too much electricity for lighting, heating and air conditioning, the poor sanitation, health issues and poverty amongst slum dwellers are SD issues that cannot be ignored. Further, in the event of climate change-related disasters, such as flooding and extreme weather conditions, these people will be the most vulnerable in society.

In 1994, the Programme of Action of the International Conference on Population and Development called on governments to 'respond to the need of all citizens, including urban squatters, for personal safety, basic infrastructure and services, to eliminate health and social problems ...' (UN, 1995). More recently, the United Nations Millennium Declaration drew attention to the growing significance of urban poverty, specifying in Target 11 the modest ambition of achieving by 2020 'a significant improvement in the lives of at least 100 million slum dwellers' (United Nations Millennium Project, 2005). If these goals are to be achieved, it is important that countries give adequate attention to the issues affecting the less affluent groups and strike a balance between their efforts towards climate change mitigation by focusing on all sectors of the society.

According to United Nations Population Fund (2005), between 2000 and 2030, Asia's urban population will increase from 1.36 billion to 2.64 billion, Africa's from 294 million to 742 million, and those of Latin America and the Caribbean from 394 million to 609 million. As a result of these shifts, developing countries will have 80 per cent of the world's urban population in 2030. By then, Africa and Asia will include almost seven out of every ten urban inhabitants in the world. It is likely that this growth will be mostly caused by migration of people into urban areas in search of better living conditions and economic benefits. Thus, the developing countries will have to find ways to address the economic needs of growing urban populations whilst dealing with other equally important issues, such as improving the living conditions of the poor, generating employment, reducing the ecological footprint to respond to climate change and improving governance to administer increasingly complex urban systems.

V. Conclusion

The key arguments made in this chapter are that sustainable development should benefit all and that governments should not focus their development initiatives only on economic development, neglecting environmental and social sustainability and vice versa. Thus, true sustainability lies in the ability of governments to pursue sustainable development in all three areas of environment, society and the economy by striking an appropriate balance in their sometimes competing interests. For example, a common response for stricter environmental regulation is that it often inhibits growth. Thus, developing countries in particular may have a tendency to prefer economic growth over environmental regulation. However, the more efficient way of dealing with the competing interests is not to have a trade-off between a healthy environment on the one hand and healthy growth on the other, but to look for forms of development that are environmentally, socially and economically sustainable. Such developments could lead not to a trade-off but to an improved environment, together with economic and social development. Although one might argue that this is easier said than done, the counter-argument is that, if appropriately planned and implemented, introducing development mechanisms that could cater to the demands of all three limbs of SD is possible.

Education of SD, the resource crunch, the need to meet the challenges posed by climate change and other global challenges could cultivate social responsibility. This could be complemented by the development of environmentally safer materials and processes for consumers. The result would be the development of a 'green consumer base' and a green industrial sector that will contribute to environmental sustainability and social as well as economic growth. Educating the public would pave the way to understand the long-term benefits of SD as against short-term economic gains that sometimes prevent engagement in SD activities. For example, if we consider the construction sector, one of the biggest GHG-emitting industries in the world, when energy-efficient technologies are used in developments, due to their low energy consumption, savings from low energy usage would pay off for any initial additional cost involved in development or purchase of energy-efficient buildings. As the Building and Construction Authority (BCA, 2007) of Singapore, which has taken measures to introduce minimum green building standards for construction in Singapore, points out, although environment-friendly buildings will cost about 5–10 per cent more upfront, they will bring future saving to about 10–15 per cent on energy expenses. Further, green buildings could yield up to

30 per cent savings in energy consumption through green features such as building envelope designs which reduce heat absorption, provide more day light, maximize natural ventilation, and use more energy efficient air-conditioning systems and light fittings (BCA, 2007).

When the demand for environmentally sustainable products and services grows, this will induce technological advancement and research, which in turn will contribute to the development of sustainable products and services. Such development would also create new jobs and develop social responsibility and corporate responses, thus contributing to environmental, social and economic sustainability.

Another point that is made in this paper is that SD needs collective responsibility. What this means is that, in a national setting, if SD goals are to be achieved, there should be co-operation amongst the government, corporate sector and the general public. The government should be active in identifying national SG goals and putting in place administrative, legislative and regulatory mechanisms that will facilitate the implementation of such goals. As noted in this paper, achievement of all the MDGs set by UN by the target year of 2015 seems unlikely. The capacity of each country to meet such globally set goals is unique. Thus, whilst not ignoring such goals, it is important that countries set their own goals and support the implementation of such. The corporate sector should contribute to SD by ensuring that they co-operate and co-ordinate with the relevant public-sector entities in discussing appropriate policy options for SD and by ensuring that they comply with and support the administrative, legislative and regulatory mechanisms that will be put in place to implement the SD goals.

The corporate sector should move out of purely profit-oriented corporate culture and engage in research and development of sustainable products and services, even if these may not be as profitable as traditional products and services. The efforts of the corporate sector should be supported, not only by corporate management but also by the employees, trade unions and labour organizations. Further, the initiatives taken by the public sector and the corporate sector towards SD need to be supported by appropriate social responsibility. As pointed out in this paper, in the absence of social support, democratic governments would find it difficult to implement SD goals balancing the competing demands of the three limbs of SD.

A final point that should be made is that SD is not something that can be achieved by countries individually in the globalized world in which we live today. Not a single country in the world can claim that it is self-sufficient and does not rely on international transfer of goods and services for development. Given the resource imbalance between the countries and the disparity in economic development and technological advancement in countries, in order to achieve SD, there needs to be co-operation at the multilateral, regional and bilateral levels to support the initiatives that will be taken at the national levels.

In the circumstances, the various initiatives taken at the national level should be supported by bilateral, regional and multilateral co-operation between nations. ODA for developing countries should keep flowing, but with better transparency to ensure that such assistance goes into SD activities and is not diverted for other purposes. The developed countries should share the technologies they develop, and management and practices they successfully employ to meet social, environmental and economic development in their countries with developing countries. Some of the current multilateral legal frameworks that may act as hindrances when it comes to such sharing of knowledge will have to be reconsidered. As the current CDM project distribution in the world shows, clearly, not all countries have benefitted from the Kyoto Protocol's aim to help developing countries participate in climate change mitigation activities whilst benefitting from technology transfer that would contribute to SD. Thus, developed and developing countries should take measures to facilitate transfer of technology and knowledge across borders in a mutually beneficial way and without discrimination. Further, measures should be taken to ensure that, whilst corporate entities in developed countries comply with laws and regulations in their home countries that would introduce environmentally sustainable industrial and service standards, they do not dump their outdated and environmentally harmful products and services into developing countries.

The way to go forward is to realize that we cannot revert to living in caves, and it will be difficult to convince masses in developing countries to give up their economic development dreams. We cannot condemn the poor to aspire to less prosperity. We need to become more innovative, find new and sustainable resources, and engage in research and advancement of technology so that SD can benefit all present and future communities, whilst preserving our resources and minimizing the harm we cause to nature.

References

Building Construction Authority (BCA) (2007) Green Buildings. An advertorial. *The Straits Times*, 21 March 2007.

Bello W. Critics plan offensive as IMF-World Bank crisis deepens [Article on the Internet] 2 May 2006. [Cited 14 November 2010]. Available from: http://news.inq7.net/viewpoints/index.php?index=2&story_id=73871.

Caddy J. Public sector modernization: Open government. OECD Policy Brief [Article on the Internet] Feb 2005. [Cited 14 November 2010]. Available from: http://www.oecd.org/dataoecd/1/35/34455306.pdf.

Carson R (1962) *Silent Spring*. Houghton Mifflin, Boston, USA.

Dixon J, Gulliver A, Gibbon D (2001) *Farming Systems and Poverty: Improving Farmers' Livelihoods in a Changing World*. A joint study by the Food and Agriculture Organization and the World Bank, FAO, Rome and Washington, DC. Available from: ftp://ftp.fao.org/docrep/fao/003/y1860e/y1860e00.pdf.

Dobson A (1999) *Introduction, Fairness and Futurity, Essays on Environmental Sustainability and Social Justice.* Oxford University Press, Oxford.

Easterly W (2006) *The White Man's Burden: Why the West's Efforts to Aid the Rest Have Done So Much Ill and So Little Good.* The Penguin Press, New York.

Ethical Corporation. Special report: China's eco-towns [Article on the Internet] Feb 2009. [Cited 12 November 2011]. Available from: http://www. ethicalcorp.com/environment/china%E2%80%99s-eco-towns-green-communities-%E2%80%93-go-eco-think-small.

Gardiner R (2002) *Earth Summit 2002 Briefing Paper.* UN, New York. Available from: http://www.earthsummit2002.org/Es2002.pdf.

George S (1999) *A Short History of Neo-Liberalism: Twenty Years of Elite Economics and Emerging Opportunities for Structural Change.* Paper presented at the Conference on Economic Sovereignty in a Globalising World, 24–26 March. Bangkok. Available from: http://globalexchange.org/resources/econ101/neoliberalismhist.

Guha R (2000) *Environmentalism: A Global History.* Longman, New York.

Gunawansa A (2009) *Sustainable Development and the World's Poor.* Paper presented at the National Sustainable Development Conference, March 2009, Singapore.

Gunawansa A, Kua HW (2009) *Lessons on Climate Change Mitigation and Adaptation Strategies for the Construction Industries in Three Coastal Cities.* Working Paper. Department of Building, National University of Singapore, Singapore.

Hallinan C (2006) *India: A Tale of Two Worlds.* Foreign Policy in Focus, Silver City, NM, Washington, DC.

Hattingh J, Attfield R (2002) Ecological sustainability in a developing country such as South Africa? A philosophical and ethical inquiry. *The International Journal of Human Rights* **6(2),** pp. 65–92.

Intergovernmental Panel on Climate Change (IPCC) (2001) *Climate Change 2001: The Scientific Basis.* Cambridge University Press, Cambridge, United Kingdom and New York, NY, USA.

Intergovernmental Panel on Climate Change (IPCC) (2007) *The Physical Science Basis; Impacts, Adaptation and Vulnerability; and Mitigation of Climate Change.* Cambridge University Press, Cambridge, United Kingdom and New York, NY, USA.

International Monetary Fund. World economic outlook update [Article on the Internet] 6 November 2008. [Cited 14 November 2010]. Available from: http://www.imf.org/external/pubs/ft/weo/2008/update/03/pdf/1108.pdf.

International Union for Conservation of Nature (1980) *World Conservation Strategy.* IUCN, Geneva. Available from: http://data.iucn.org/dbtw-wpd/edocs/WCS-004.pdf.

Jacobs M (1999) Sustainable development as a contested concept. In: Dobson A (ed) *Fairness and Futurity: Essays on Environmental Sustainability and Social Justice*, xx–xx. Oxford University Press, Oxford.

Martens J (2005) The Development Agenda after the 2005 Millennium+5 Summit. A briefing paper. Friedrich-Ebert-Stiftung, Bonn. Available from: http://www.globalpolicy.org/socecon/un/reform/2005/05summitgpf.pdf.

Madonna KJ (1995) The wolf in North America: Defining international ecosystems: Defining international boundaries. *Journal of Land Use & Environmental Law* **10(2)**, 26.

Norton B (1999) Ecology and opportunity: Intergenerational equity and sustainable options. In: Dobson A (ed) *Fairness and Futurity: Essays on Environmental Sustainability and Social Justice*. Oxford University Press, Oxford.

Organisation for Economic Co-operation and Development. Development aid at its highest level ever in 2008 (Press Release) [Article on the Internet] 30 March 2008. [Cited 14 November 2010]. Available from: http://www.oecd.org/document/35/0,3343,en_2649_34487_42458595_1_1_1_1,00.html.

Pachauri RK, Bhandari P (1994) *Climate Change in Asia and Brazil: The Role of Technology Transfer.* Tata Energy Research Institute, New Delhi.

Patterson G (1998) *Dying to be Free: The Canned Lion Scandal and the Case for Ending Trophy Hunting in Africa.* Viking, London.

Rodrigues C (2006) Promoting Public Accountability in Overseas Development Assistance: Harnessing the Right to Information. Commonwealth Human Rights Initiative. Commonwealth Human Rights Initiative, New Delhi, India. Available from: http://www.humanrightsinitiative.org/programs/ai/rti/articles/cald_conf_paper_rti_oda_may06.pdf.

Sale K (1993) *The Green Revolution: The American Environmental Movement, 1962–1992.* Hill and Wang, New York.

Stiglitz J (2003) *Globalization and Its Discontents.* Penguin Publishing, New Delhi.

United Nations (1995) *Population and Development, Vol. 1: Programme of Action Adopted at the International Conference on Population and Development: Cairo: 5–13 September 1994.* UN, New York.

United Nations (1997) *The World Conferences: Developing Priorities for the 21 Century.* United Nations Publications, UN, New York.

United Nations (2000a) *We the Peoples: The Role of the United Nations in the 21st Century.* United Nations Department of Public Information, UN, New York.

United Nations (2000b) Tenth United Nations Congress on the Prevention of Crime and Treatment of Offenders, Press Kit Backgrounder No3. UN, New York. Available from: http://www.un.org/events/10thcongress/2088b.htm.

United Nations (2003) *Plan of Implementation of the World Summit on Sustainable Development*. Department of Economic and Social Affairs Division for Sustainable Development, UN, New York.

United Nations (2005a) *Millennium Development Goals Report*. UN. New York.

United Nations (2005b) Secretary-General: Address to the 2005 World Summit. New York, 14 September 2005. UN, New York. Available from: http://www.un.org/apps/sg/sgstats.asp?nid=1669.

United Nations (2005c) *United Nations Peacekeeping Operations Background Note*. United Nations Department of Public Information, New York.

United Nations (2006) *Human Development Report 2006, United Nations Development Programme*. UN, New York.

United Nations (2008a) *Millennium Development Goals Report 2008*. UN, New York.

United Nations (2011) *Millennium Development Goals Report 2011*. UN, New York.

United Nations (2008b), New World Bank facility to fast-track $2 billion to help world's poorest countries News Centre [Article on the Internet] 10 December 2008. [Cited 12 November 2011. Available from: http://www.un.org/apps/news/story.asp?NewsID=29250.

United Nations Commission on Environment and Development (UNCED) (1987) Our Common Future. Oxford University Press, Oxford.

United Nations Development Programme (UNDP) Human development report 2002: Deepening democracy in a fragmented world [Article on the Internet]. [Cited 14 November 2010]. Available from: http://hdr.undp.org/reports/global/2002/en/pdf/chapterfive.pdf.

United Nations Development Programme (UNDP) (2005a) *UN Millennium Project 2005, Investing in Development: A Practical Plan to Achieve the Millennium Development Goals. Overview*. UN, New York.

UN-Habitat (2003) *The Challenge of Slums*. Global Report on Human Settlement 2003, UN Human Settlement Programme. Earthscan Publications Ltd, London and Sterling, Virginia.

UN Millennium Project (2005) *Investing in Development: A Practical Plan to Achieve the Millennium Development Goals*. Report to the UN Secretary-General. Earthscan Publications Ltd, London and Sterling, Virginia.

United Nations Population Fund (2005) *World Population Prospects: The 2005 Revision*. UNDESA Population Division, UN, New York.

World Bank (2009a) *2009 Global Economic Prospects*. World Bank, Washington, DC.

Yunus M (1999) *Banker to the Poor: Micro Lending and the Battle Against World Poverty*. Public Affairs, Perseus Book Group, New York.

Environmental Outcomes, Human Impacts

Chapter 3: An Environmental Health Investigation of
Skin Disorders: Southeast Asian Perspectives
David KOH and *Judy SNG*

CHAPTER THREE

An Environmental Health Investigation of Skin Disorders: Southeast Asian Perspectives

David KOH and Judy SNG
Centre for Environmental and Occupational Health Research,
Saw Swee Hock School of Public Health, National University of Singapore

"High rates of diarrhoea among village children along the Kampar River in Sumatra are probably linked to the faecal coliform content in the river water, which is due to activities related to human habitation along the river, rather than effluent from the pulp and paper mill. This can be addressed by having an improved sanitation system, provision of clean water sources, health education and enhancement of personal hygiene.

Other public health issues of concern among the villagers can be met by increasing vaccination coverage among children, providing smoking cessation programs for smokers and preventing the start of smoking among non-smokers."

– *David KOH* and *Judy SNG*

An Environmental Health Investigation of Skin Disorders: Southeast Asian Perspectives

David KOH[1] and Judy SNG[2]
Centre for Environmental and Occupational Health Research,
Saw Swee Hock School of Public Health, National University of Singapore

Abstract

Villagers living downstream from a pulp and paper mill complained of skin rashes. They believed that this occurred because of water pollution caused by the mill. A community-based approach was adopted to investigate the complaint.

The mill used an Elemental Chlorine Free (ECF) bleaching process and an environmental assessment showed that it had adequate environmental pollution controls. River-water quality was measured upstream and downstream from the mill and water quality at these points was found to be similar in physical, chemical and biological characteristics. River ecology findings indicated adequate biotic diversity to support continued productivity of the river for freshwater fish and maintenance of other aquatic life.

The prevalence of dermatitis in the villages was used as a marker for environmental skin disease due to exposure to waterborne pollutants. A total of 917 adults and 432 children in one village upstream and two villages downstream from the mill were examined. The prevalence of dermatitis amongst residents was low (<5 per cent) and differences between the three villages were minimal and non-statistically significant. Lifestyle habits with regard to exposure to river water amongst those with and without dermatitis were similar.

Examination of adults and children revealed that >60 per cent did not have significant skin conditions. The most common skin disorder was fungal infection. No cases of skin cancer were diagnosed. Other health issues, such as low immunization coverage (<50 per cent), prevalence of diarrhoea (7–20 per cent), fever amongst children and smoking amongst adults, were identified as important health priorities. The findings were communicated to the villagers and health education and activities to increase immunization coverage in the villages were subsequently carried out.

[1] Professor and Director
[2] Assistant Professor
Centre for Environmental and Occupational Health Research, Saw Swee Hock School of Public Health, National University of Singapore. MD3, 16 Medical Drive, Singapore 117597. Correspondence to: Professor David Koh. Email: david_koh@nuhs.edu.sg.

This investigation illustrates the importance of a multidisciplinary community-based approach when population clusters of environmental health disorders are managed.

Key words: environmental health, public health, pulp and paper mill, skin disorders, water pollution

I. Introduction

The industrialization of developing countries has raised issues of environmental damage. There are many examples of this in both developed and developing nations (World Health Organization, 1992a). While efforts from international and local government bodies have led to greater control over industrialization in an effort to reduce environmental damage (World Health Organization, 1992a, 1992b), changes do occur regardless of the care taken.

An example is the building of a pulp and paper mill adjacent to a site of forestation used to produce paper. The pulp and paper industry is a major industry globally, with pulp and paper mills being located in more than 100 countries across every region of the world (Keefe & Teschke, 1998).

In the 1990s, a large pulp and paper mill was sited along the Kampar River in the province of Riau in Sumatra (Figure 3.1). Construction of the mill began in 1993. Commercial production of pulp started in January 1995, while paper production commenced in April 1998.

Figure 3.1: Pulp and paper mill situated along the Kampar River in Riau, Sumatra

When the mill began its operations, villagers living downstream complained that effluent from the mill that was discharged into the river caused skin rashes. This was brought to the attention of the foreign media, which reported that the villagers had become ill because of alleged water pollution from the mill and that cases of skin cancer had been found in the villages.

As the concerns of the villagers needed to be investigated, an environmental health survey was conducted. The aim was to assess the villagers primarily for any skin conditions and whether these were associated with river-water pollution from effluent produced by the mill. A secondary aim was to assess the general health of the villagers.

The investigation involved representatives from Riau Mandiri, a local non-governmental organization (NGO), environment and health staff of the mill (PT Riau Andalan Pulp and Paper), academics from the fisheries department of the local Universitas Riau (UNRI) and external consultants from the Finnish Environmental Research Group (FERG) and the National University of Singapore (NUS). Funding for the investigation was provided by the mill.

Each group had an important and complementary role. The NGO, a community self-help and development group, oversaw water sample collection and coordinated field work in the river and villages. They were involved in all stages of the studies in order to ensure transparency of the process. An aquatic environmental impact assessment was made by FERG, and river ecology studies were initiated by UNRI. At the same time, a community health survey of villages upstream and downstream from the mill was conducted. This was a joint effort by NUS consultants, the health staff from the mill, and NGO representatives.

This chapter will focus on the findings of the health survey of villagers, the results of the environmental assessment of the treated effluent discharged by the mill, river-water quality and river ecology findings.

II. Materials and Methods

Survey Area

The locations of the villages in relation to the pulp and paper mill are shown in Figure 3.2. The flow of the Kampar River is in a generally west-to-east direction. Rantau Baru is 45 kilometres upstream from the mill. Sering and Pelalawan villages are 7.7 and 24.8 kilometres downstream from the mill respectively.

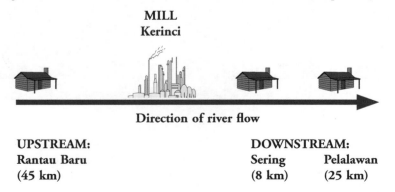

MILL
Kerinci

Direction of river flow

UPSTREAM:	DOWNSTREAM:	
Rantau Baru	**Sering**	**Pelalawan**
(45 km)	**(8 km)**	**(25 km)**

Figure 3.2: Locations of Rantau Baru, Sering and Pelalawan in relation to the mill

Study Design

A cross-sectional health survey was conducted in three villages (Rantau Baru, Sering, Pelalawan) along the Kampar River. There were 149 houses in Rantau Baru, 203 houses in Sering and 223 houses in Pelalawan, each with approximately five inhabitants. Sample size estimation was computed using the following estimates for alpha, power, matching and proportions of dermatitis in the villages ($\alpha = 0.05$; Power = 0.8; Match = 1; $P_0 = 0.05$; $P_1 = 0.1$). The calculations yielded a sample size requirement for 474 persons in each village.

A one-stage cluster (of people in a house) sampling per village was performed. Each sampling frame consisted of all the houses in each of the villages. Houses were mapped and numbered. 100 houses were then randomly chosen from a list of random numbers generated for each village. The sampling fractions were 1/1.49, 1/2.03 and 1/2.23 for Rantau Baru, Sering and Pelalawan respectively (Saw *et al.*, 2001).

Fieldwork

Field trips made by trained public health personnel included discussions with villagers and the local health team about possible health issues and practices. Information was also obtained about pre-existing health facilities and services as well as the general socioeconomic status and living conditions of the villagers. Two questionnaires were designed: one for the household and another for individuals.

Household Questionnaire

A questionnaire for each household was designed to determine the number of occupants, duration of the occupants' residence in the village, literacy, household income and household possessions (e.g. radio, television and motorized boat). Information on access and use of basic amenities, such as electricity, rubbish bin, and water from sanitary sources, such as a well, was gathered. Residents were also asked about their use of toilet facilities, including unsanitary (using the river) and sanitary (public and individual) use, and sources of drinking water (river, well, pond or rainwater). This included whether water was either always or seldom boiled before drinking.

Individual Questionnaires

Separate questionnaires were developed for adults and children under the age of 12 years. To reduce bias, these were designed as for a general health survey, without specific reference to the river or illness associated with exposure to the river water.

Questions derived from a previous health survey conducted on an indigenous population in Malaysia (Wong & Chen, 1991) were adapted and modified, with the knowledge acquired from the field trips.

Questions were translated into the Indonesian language. They were again checked with the local health team to ensure correct translation and clarity, and were translated back. They were also checked to ensure that they were socio-culturally correct and appropriate. The survey was administered by personnel from the local health-care team. For children under the age of 12 years, questions were answered by the primary person caring for the child, after verbal consent was obtained.

Physical Examination

Medical doctors examined the adults and children for any skin conditions (Figure 3.3). Before the survey, the doctors were trained specifically in the diagnosis of skin conditions. All skin findings were photographed and reviewed by the researchers in Singapore.

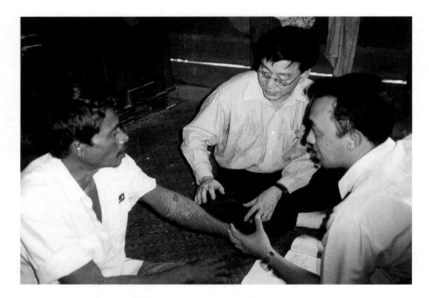

Figure 3.3: Examination of villagers for skin disorders

Training of Fieldworkers

All personnel were intensively trained to ensure consistency in the survey. A pilot study was carried out before the survey proper to estimate the duration of the interview and examination (about 15 minutes) as well as to determine the clarity and face validity of the questionnaire. The local doctors and fieldworkers were aware that a health survey was being conducted. However, they were blinded to the hypothesis that exposure to effluent from the mill was causing skin problems and ill health.

Survey

The survey was conducted within the period of November to December 1999. Three attempts on three different occasions were made before a household was deemed to be a non-response. Attempts were also made to conduct the survey on days such as market day, when it was known that most people in the household were present. However, if a house was found to be vacant, it was replaced by another house with a higher map number but not on the list.

Survey quality was assessed by regularly checking questionnaires for incomplete responses or coding errors as well as direct observation of interviewers by the external NUS observers, who accompanied the teams on the field. Medical treatment for villagers was also initiated on an as-needed basis.

Ethical Approval

Approval for the study was obtained from a committee comprising community representatives and members from local non-government organizations. Completion of the examination and survey was done on a voluntary basis, with participants having the right to decline.

Water Sampling and Environmental Impact Evaluation

At around the time of the village survey (8 November 1999 and 14 December 1999), water samples were collected simultaneously at several locations including 1 kilometre upstream from the water intake point, 1 kilometre downstream from the waste discharge point of the mill and at Sering village (8 kilometres downstream from the mill). Physical, chemical and biological variables of the river water were analyzed. Scientists who conducted the water analysis were blinded to the source of river water collection points and to the results of the health survey.

An aquatic environmental impact evaluation by an international environmental research group was performed in August and September 1999 (Lehtinen, 1999). This included a survey of effluent discharge from the mill, river water and fish biota. All samples were frozen and subsequently analyzed at the Institute of Environmental Chemistry at the University of Abo Akademi, Finland.

Data Analysis

Statistical Package for the Social Science 10.0 was used to analyze the data.

III.   Results

Community Health Survey

The household response rates for all three villages were 100 per cent (Figure 3.4) while the individual response rates for the villages were above 95 per cent. In Rantau Baru, the individual response rate was 423/440 (96.1 per cent); in Sering, it was 454/469 (96.9 per cent); and in Pelalawan, it was 465/488 (95.3 per cent).

The average monthly household income of the three villages was less than IDR1,000,000 (USD75) for most households (Rantau Baru – 95.8 per cent, Sering – 94 per cent, and Pelalawan – 98.7 per cent). Many households had an average monthly income of less than IDR200,000 (USD15) (Rantau Baru – 76.4 per cent; Sering – 42.8 per cent; Pelalawan – 35.2 per cent). In each village, more than half of the household members were able to read in >50 per cent of households. A total of 21 households (14.6 per cent) in Rantau Baru, 73 (44 per cent) in Sering and 70 (56.9 per cent) in Pelalawan were supplied with electricity. Nearly all households used unsanitary toilets located over the river (Sering and Pelalawan – 100 per cent; Rantau Baru – 95.1 per cent).

Figure 3.4: The response rate amongst the households in the villages was 100 per cent and, amongst individuals, it was >95 per cent.

Socio-Demographic Characteristics and Two-week Health Recall of Health Events

Socio-demographic information of the adult villagers is presented in Table 3.1, on the next page. The age and gender distributions in all villages are similar. Smoking prevalence is approximately 40 per cent.

Table 3.1: Socio-demographic characteristics and smoking history of adults in the three villages

	Rantau Baru (n = 293)	Sering (n = 287)	Pelalawan (n = 337)	Two-sided p (chi-square test)
Age in years (mean, SD)	32.9 (16.3)	31.1 (14.9)	33.2 (17.2)	0.22*
Gender (No. [%]) Male Female	142 (47.8) 151 (50.8)	141 (49.1) 146 (50.9)	162 (48.1) 175 (51.9)	0.97
Occupation (No. [%]) Homemakers Fisherman Farmer Wood collector Factory worker Others	67 (22.9) 107 (36.5) 33 (11.3) 10 (3.4) 1 (0.3) 75 (25.6)	70 (24.4) 36 (12.5) 38 (13.2) 53 (18.8) 14 (4.8) 75 (26.1)	75 (22.3) 19 (5.6) 27 (8.0) 57 (16.9) 8 (2.4) 151 (44.8)	<0.001
Cigarette smoking (No. [%]) Current Non-smoker Ex-smoker	117 (39.4) 162 (54.5) 14 (4.7)	127 (44.3) 150 (52.3) 10 (3.5)	128 (38.0) 207 (61.4) 2 (0.6)	0.006

* Two-sided p from one-way analysis of variance

A two-week recall of acute health conditions such as diarrhoea, cough/cold or fever is presented in Table 3.2. Sering residents had a higher reported rate of diarrhoea than residents of other villages did. The two-week-period prevalence of cough/cold and fever in all villages was similar.

There were 145, 167 and 126 children under 12 years of age in Rantau Baru, Sering and Pelalawan respectively. There were more reported cases of diarrhoea (20.4 per cent vs 17 per cent and 8 per cent) and the passing out of worms (7.2 per cent vs 3.5 per cent and 2.4 per cent) amongst the Sering children than the Rantau Baru and Pelalawan children. The children in Rantau Baru had a higher reported rate for fever (35 per cent versus 20 per cent in the other two villages) (Table 3.2).

Table 3.2: Two-week recall of selected health conditions amongst adults and children

In the past two weeks:	Rantau Baru No. (%)	Sering No. (%)	Pelalawan No. (%)	Two-sided p (chi-square test)
Adults	(n = 293)	(n = 287)	(n = 337)	
Diarrhoea	20 (6.8)	50 (17.4)	26 (7.7)	<0.001
Cough/Cold	70 (23.9)	77 (26.8)	61 (18.1)	0.049
Fever	41 (14.0)	39 (13.6)	56 (16.6)	0.44
Children	(n = 145)	(n = 167)	(n = 126)	
Diarrhoea	24 (16.6)	34 (20.4)	10 (7.9)	0.013
Cough/Cold	46 (31.7)	52 (31.1)	28 (22.2)	0.32
Fever	51 (35.2)	33 (19.8)	26 (20.6)	0.013
Pass out worms	5 (3.5)	12 (7.2)	3 (2.4)	<0.001

Information on the immunization status of the children is presented in Table 3.3. The highest immunization rates were in Pelalawan. Immunization rates in Sering and Rantau Baru were similar.

Table 3.3: Immunization history of children above one year but less than 12 years in three villages

	Rantau Baru (n = 129) No. (%)	Sering (n = 152) No. (%)	Pelalawan (n = 119) No. (%)	Two-sided p (chi-square test)
BCG*				
Yes	44 (34.1)	47 (30.9)	66 (55.5)	<0.001
No	78 (60.5)	77 (50.7)	49 (41.2)	
Unknown	7 (5.4)	28 (18.4)	4 (3.4)	
First DPT# dose				
Yes	37 (28.7)	39 (25.7)	62 (52.1)	<0.001
No	86 (66.7)	82 (53.9)	52 (43.7)	
Unknown	6 (4.6)	31 (20.4)	5 (4.2)	
First polio dose				
Yes	39 (30.5)	50 (32.9)	65 (54.6)	<0.001
No	83 (66.7)	83 (54.6)	50 (42.0)	
Unknown	6 (4.7)	19 (12.5)	4 (3.4)	

* BCG – Bacillus Calmette-Guérin (or Bacille Calmette-Guérin), a vaccine against tuberculosis
# DPT – a combination of vaccines against diphtheria, pertussis (whooping cough) and tetanus

Self-Reported Itchy Skin Rashes and Precipitating Factors

The self-reported rates for itchy skin rashes in the last year and the precipitating factors were analyzed separately for adults and children under the age of 12 years (Table 3.4).

The rates for itchy skin rash in the last year were 48 per cent, 56 per cent and 61 per cent for adults in Rantau Baru, Sering and Pelalawan. These differences were not statistically significant. For children, the rates in the villages were highest in Rantau Baru (78 per cent), followed by Sering (55 per cent) and Pelalawan (50 per cent). The most frequently reported precipitating factors that were listed by the residents were plants and food.

Table 3.4: Self-reported itchy skin rashes in the last year and precipitating factors

Adults	Rantau Baru (n = 293) No. (%)	Sering (n = 287) No. (%)	Pelalawan (n = 337) No. (%)	Two-sided p (chi square test)
Itchy skin rashes occurred in the last 12 months				
Yes	57 (47.9)	38 (55.9)	49 (60.5)	0.65
No	62 (52.1)	30 (44.1)	32 (39.5)	
Precipitating factors				
Food	25 (8.5)	16 (5.6)	31 (9.2)	0.19
Plant	35 (11.9)	25 (8.7)	23 (6.8)	<0.001
Animal	1 (0.3)	3 (1.0)	4 (1.2)	0.36
Soap	16 (5.4)	17 (5.9)	10 (3.0)	0.80
Medicine	6 (2.0)	4 (1.4)	4 (1.2)	0.52
Children	**(n = 145)**	**(n = 167)**	**(n = 126)**	
Itchy skin rash in 12 months				
Yes	25 (78.1)	22 (55.0)	9 (50.0)	0.07
No	7 (21.9)	18 (45.0)	9 (50.0)	
Precipitating factors				
Food	4 (2.8)	6 (3.6)	3 (2.4)	0.82
Plant	7 (4.8)	6 (3.6)	4 (3.2)	0.76
Animal	0 (0.0)	1 (0.6)	1 (0.8)	0.59
Soap	0 (0.0)	2 (1.8)	0 (0.0)	0.09
Medicine	3 (2.1)	0 (0.0)	0 (0.0)	0.05

Physical Examination Findings

(i) Skin conditions amongst adults

The majority of adults in the three villages did not have any noteworthy skin conditions (Table 3.5). The most common skin condition diagnosed was fungal infection of the skin. No cases of skin cancer were detected.

Table 3.5: Physical examination for skin conditions amongst adults in the three villages

	Rantau Baru* (n = 293) No. (%)	Sering (n = 287) No. (%)	Pelalawan (n = 337) No. (%)	Two-sided p
Types of skin lesions (First provisional diagnosis)				
None	219 (77.7)	179 (62.6)	253 (75.1)	<0.001
Fungal	22 (7.8)	(13.3)	46 (13.7)	
Dermatitis	10 (3.6)	25 (8.7)	11 (3.3)	
Others	31 (11.0)	44 (15.4)	27 (8.0)	
Types of skin lesions				
None	219 (77.7)	179 (62.6)	253 (75.1)	<0.001
Dermatitis	(3.6)	27 (9.4)	12 (3.6)	
Others	53 (18.7)	80 (28.0)	72 (21.4)	
Types of skin lesions				
No dermatitis	272 (96.4)	259 (90.6)	325 (96.5)	0.002
Dermatitis	10 (3.5)	27 (9.4)	12 (3.6)	
Dermatitis unspecified	1	-	-	
Atopi dermatitis	-	4	3	
Neuro dermatitis	-	2	5	
Contact dermatitis	7	19	4	
Other endogenous dermatitis	2	1	-	
Dermatitis exfoliatica	-	1	-	

*Eleven cases in Rantau Baru were not examined by the doctor but data was collected via the questionnaire.

As for dermatitis (skin inflammation) amongst adults, the highest point prevalence was found in Sering (9.4 per cent). The other two villages had a rate of 3.6 per cent. This difference was statistically significant (p = 0.002) (Figure 3.5) on the next page.

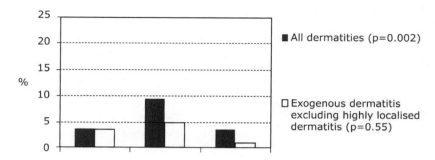

Figure 3.5: Distribution of all dermatitis (endogenous and exogenous) and exogenous dermatitis (excluding highly localized dermatitis) in adults in the three villages

Cases of constitutional or endogenous dermatitis (e.g. atopic or seborrhoeic dermatitis, lichen simplex chronicus) and cases of highly localized dermatitis of the fingers and ears) are unlikely to be due to any agents in the river water. Exclusion of such cases resulted in the contact dermatitis prevalence rate of 10/282 (3.5 per cent) in Rantau Baru, 14/287 (4.9 per cent) in Sering and 4/337 (1.2 per cent) in Pelalawan. The prevalence of such dermatitis in Sering and Rantau Baru was similar (p = 0.55) (Figure 3.5).

(ii) Skin conditions amongst children
The majority of children in the three villages did not have any significant skin conditions (Table 3.6) on page 84. The most common skin condition that was observed was fungal skin infection. No cases of skin cancer were detected.

Table 3.6: Physical examination of children aged less than 12 years in the three villages

	Rantau Baru n = 142* No. (%)	Sering n = 164† No. (%)	Pelalawan n = 126 No. (%)	2-sided p (chi-square test)
Types of skin lesions (First provisional diagnosis)				
None	113 79.5)	104 (63.4)	107 (84.9)	
Fungal	3 (2.1)	(8.5)	8 (6.3)	<0.001
Dermatitis	3 (2.1)	8 (4.9)	2 (1.6)	
Miliaria (heat rash)	11 (7.8)	18 (11.0)	2 (1.6)	
Others	12 (8.5)	20 (12.2)	7 (5.6)	
Types of skin lesions				
Dermatitis	3 (2.1)	(4.9)	2 (1.6)	0.13
No dermatitis	139 (97.9)	156 (95.1)	124 (98.4)	
Dermatitis				
Atopic dermatitis	1	2	2	
Neuro dermatitis	-	1	-	
Contact dermatitis	2	4	-	
Diaper dermatitis	-	1	-	

*Three children in Rantau Baru were not examined by the doctor but data about them was collected via questionnaire.
†Three children in Sering were not examined by the doctor but data about them was collected via questionnaire.

As for dermatitis (skin inflammation) amongst children, the highest prevalence rate was found in Sering (5.5 per cent). The other two villages had prevalence rates of 2.1 per cent (Rantau Baru) and 1.6 per cent (Pelalawan). This difference was not statistically significant (p = 0.11)

Figure 3.6 on the next page shows some of the skin disorders found amongst the children.

Lifestyle Habits of Residents

Lifestyle habits with regard to exposure to river water, such as bathing and swimming in the river, washing of clothes, fishing or use of the boat, were recorded (see Figure 3.7 on page 86). The habits of persons with and without dermatitis in the three villages were similar.

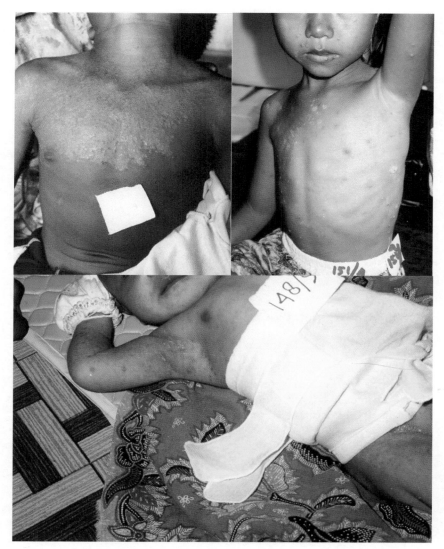

Figure 3.6: Skin disorders seen in the children included fungal infections (top left), vesicles from chicken pox (top right) and heat rash (bottom)

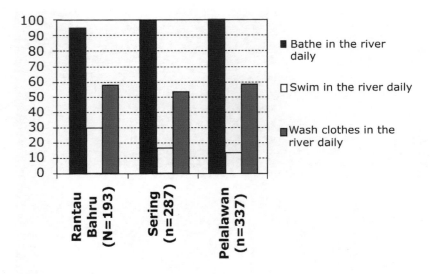

Figure 3.7: Lifestyle factors of residents in the three villages

IV. Environmental Assessment

The Mill Processes and Treatment of Effluent

The mill uses an ECF bleaching process with chlorine dioxide and oxygen in the production of pulp, thus producing only small amounts of polychlorinated phenolic compounds. Pulp production is integrated with a paper machine producing wood-free, uncoated paper.

Combined effluent from pulp and paper production is first treated in a primary clarification stage. It is next led through an equalization pond. After the primary clarification, the effluent is biologically treated in an activated sludge treatment plant, where it is retained for approximately 24 hours (Figure 3.8). The effluent is then collected in a second clarifier and, after treatment, the combined effluent is transported along a 5-kilometre-long open channel and discharged into the Kampar River.

Figure 3.8: Treatment of mill effluent prior to discharge

Assessment of Treated Effluent

Findings showed that the absorbable organic halogen in the final treated effluent was 0.11 kilogram/tonne pulp, low from an international perspective (Axegárd *et al.*, 1993). The international toxic equivalent expressing the overall toxicity potential of the analysis of chlorinated dioxins and furans was 0.72 picogram/ litre. According to the analyzing laboratory, the Finnish National Health Institute, the level of analytical accuracy is ±100 per cent at international toxic equivalent level of <1 picogram/litre, indicating that the concentrations of toxic chlorinated dioxins were at the limit of detection.

Chlorophenolic substances were found in the untreated effluent but at very low concentrations. This included 3.1 microgram/litre 4,5 dichloroguaiachol and traces of 2,4,6 tripentachlorophenol in the sample collected after the primary clarifier to only 0.07 microgram/litre of 2,4,6 tri-pentachlorophenol in the final treated effluent. Of these, 2,4,6 tri-pentachlorophenol is formed naturally in humid environments (Grimvall *et al.*, 1994), and 2,4,6 tripentachlorophenol occurs ubiquitously as a contaminant where wood is burnt. Total concentrations of potentially toxic resin acids, fatty acids and phytosterols were all very low.

The data on quality of river water did not show any differences between upstream and the downstream points closest to the effluent discharge. Chlorinated phenols were not occurring above the level of detection in river water. Consistent with these findings, no chlorinated phenols exceeding the detection limit were found in fish bile samples. FERG researchers thus concluded that, overall, the risk for humans being in contact with water from the Kampar River may be considered practically negligible or non-existent.

Quality of River Water

Table 3.7 shows the parameters of physical, biological and chemical variables of water samples collected before the waste point of the mill and at Sering village (Lee *et al.*, 2002). The parameters were essentially similar and within acceptable limits for river water.

Table 3.7: Physical, biological and chemical parameters of river water samples at the same period of time as the survey

Water Parameters	Waste Point		Sering
	−1 km	+1 km	
Physical and biological			
Temperature (°C)	29	29	29
Total suspended solid (mg/l)	15	14	11
Turbidity (NTU)*	15.9	18.5	14.2
Coliform MPN/100 ml†	8	9	10
Faecal coliform MPN/100 ml†	Nil	Nil	6
Chemical			
pH	5.8	5.8	6.0
Total N as NH_3	0.28	0.31	0.25
Nitrite	0.01	0.01	0.01
Nitrate	0.3	0.3	0.2
Phosphate	0.01	0.01	0.01
Alkalinity	10.0	8.5	8.2
Ca hardness	2.5	2.1	2.5
Chloride	4.52	4.96	4.62
Residual Cl_2	Nil	Nil	Nil
COD‡	19	18	14
BOD§	4	3	4

*NTU, nephelometric turbidity unit
†MPN/100ml, most probable number/100 millilitre
‡COD, chemical oxygen demand
§BOD, biological oxygen demand

River Ecology Findings

Findings from the UNRI Fisheries department showed that there was adequate biotic diversity to support the continued productivity of the river for freshwater fish and the maintenance of other aquatic life (Figure 3.9).

Figure 3.9: The river had adequate biotic diversity.

IV. Discussion

Health Survey Findings

The appropriate sampling strategy, excellent response rates (100 per cent of houses, >95 per cent individuals) and standardized data collection by trained personnel support the validity of the findings.

In terms of the socioeconomic status of the households, as measured by household income, and the ownership of radios and televisions, Pelalawan was ranked the highest, followed by Sering and Rantau Baru. The age and sex distributions in the villages were similar.

The two-week recall of acute health complaints was highest in Sering for diarrhoea in both adults (17 per cent compared to 7 per cent) and children (20 per cent vs 17 per cent in Rantau Baru, and 8 per cent in Pelalawan), as well as for the passage of worms in children (7 per cent). Fever in children was highest in Rantau Baru (35 per cent vs 20 per cent in the other villages), while the period prevalence of cough and cold was similar. The immunization rates amongst the children in all villages were low (Pelalawan = >50 per cent; Sering and Rantau Baru = around 30 per cent). Ideally, coverage should be 100 per cent.

Self-reported rates for itchy skin rashes in the last year were 48 per cent, 56 per cent and 61 per cent for adults in Rantau Baru, Sering and Pelalawan respectively. These rates were not significantly different. For children, the self-reported rates were highest (78 per cent) in Rantau Baru, followed by Sering (55 per cent) and Pelalawan (50 per cent). Physical examination of adults and children revealed that the majority (>60 per cent) were free from significant skin conditions. The commonest skin disorder was fungal infection. No cases of skin cancer were found amongst the residents.

As for dermatitis (skin inflammation), the highest rate amongst adults was in Sering (9.4 per cent). Exclusion of endogenous dermatitis and highly localized dermatitis of the ears and fingers resulted in comparable dermatitis rates of 3.5 per cent and 4.9 per cent in Rantau Baru and Sering respectively. The prevalence rates of dermatitis amongst children were lower than those of adults, and not significantly different (p = 0.13) in the three villages.

Lifestyle habits with regard to exposure to river water were similar amongst persons with and without dermatitis. It is possible that residents avoided exposure to river water because they had previously experienced ill health or skin disorders. This would lead to the underestimation of the possible effects of the river water.

Environmental Assessment

The pulp and paper mill used an ECF bleaching process and had adequate environmental pollution controls. Physical, chemical and biological parameters for upstream water (from the plant) and at Sering village, measured on three different occasions, were essentially similar and acceptable for river-water quality.

However, the exception was the presence of faecal coliforms (6 MPN/100 millilitre) found in the water sample collected at the village of Sering. This is likely to be due to activities arising from human habitation at this point of the river along the riverbanks. As information on coliform counts is not available from either Rantau Baru or Pelalawan, it would not be possible to relate this finding to the higher prevalence of diarrhoea amongst residents in Sering during the survey period to coliform counts.

In combination, the findings from the health survey and the environmental assessment suggest that it is unlikely that the river water could be a cause of inflammatory skin disorders in the villagers living downstream of the pulp and paper mill.

In addition, the health survey found that skin rashes and skin cancer from polluted river water do not seem to be a major health issue for the local community to be excessively concerned about. However, other health issues amongst the villagers have been identified and have to be addressed. Some of these issues are the high smoking rates amongst adults, incomplete immunization coverage of children and the relatively high reports of diarrhoea amongst the residents.

Study Limitations

Despite the strengths of a good response rate and the fact that all participants were examined for skin conditions by a medical doctor, there were several limitations to this study.

Recall bias might have occurred in assessing episodes of ill health. An attempt was made to reduce this by asking for only a two-week recall period. An attempt to reduce bias in claiming an association of ill health with river water was also made by asking general-health questions rather than direct questions pertaining to health and river water.

Furthermore, residents who had previously experienced ill health or skin disorders might have intentionally avoided exposure to river water because this would aggravate their condition. Additional studies using a cohort design are needed to confirm the findings.

Three local doctors interviewed and examined only the participants within the village who were known to them. This was done to maintain rapport and familiarity with the villagers and achieve a high response rate. To minimize observer bias, all the doctors participated in two training sessions in dermatology with public and family medicine health specialists before the fieldwork was conducted. Field trips were also made to each of the villages by NUS staff to assess the manner of examination and confirm the diagnosis of the local doctor (Figure 3.3).

Finally, diagnoses of skin conditions were made clinically without further investigations, such as patch testing, skin biopsy and microscopic examination of skin scrapings. However, most of the skin conditions were familiar to the doctors and were confidently diagnosed.

Follow-up Activities

Following the health survey, the mill staff continued their work in community health care with the local authorities and NGOs, but with an emphasis and focus on the health issues of incomplete immunization coverage and health education for the prevention and management of diarrhoea. A list of children identified as not having been immunized was prepared, and efforts were made to immunize these children. The ultimate aim was to increase immunization coverage for all children.

Health education was given on personal measures that can be taken to prevent diarrhoea, and for the management of diarrhoea (e.g. oral rehydration solution). The village of Sering was provided with a clean water source with a deep well and a sand/carbon filter. Other sanitary measures, such as toilets with proper sewerage disposal, were explored.

The findings of the health survey were disseminated to the local community to reassure all of their concerns of river water contamination (Figure 3.10). This was carried out through the community health outreach programmes established by the mill in conjunction with the local government.

The monitoring of the treated effluent and river water on a periodic basis continues (Figure 3.11). The persons doing the monitoring were trained to collect the samples properly. and samples were sent to a reliable laboratory (with good-quality control) for analysis.

Figure 3.10: Findings of the survey were disseminated to the local community

Figure 3.11: Periodic monitoring of the treated effluent and river water

V. Conclusion

This study has shown that river-water pollution due to the effluent from the mill is unlikely to be the cause of skin conditions amongst the villagers living downstream from the mill. However, as the production of the mill continues to increase, periodic monitoring of the effluents by independent and trained observers should continue.

Other health concerns, such as a high smoking prevalence and incomplete immunization amongst children, were found in the villages. High rates of diarrhoea amongst village children along the Kampar River in Sumatra are probably linked to the faecal coliform content in the river water, which is due to activities related to human habitation along the river, rather than the effluent from the pulp and paper mill. This can be addressed by having an improved sanitation system, providing clean water sources, health education and enhancing personal hygiene.

Other public health issues of concern amongst the villagers can be met by increasing vaccination coverage amongst children, providing smoking cessation programmes for smokers and preventing the start of smoking amongst non-smokers.

Acknowledgements

We thank the local health teams who assisted in the survey, including staff from the Universitas Riau, members of Riau Mandiri, a local non-government organization and the pulp and paper mill (Asia Pacific Resources International Holdings Limited [APRIL], RAPP) for their hard work and support.

This study was supported by financial assistance from PT Riau Andalan Pulp and Paper, a member of APRIL.

References

Axegárd P, Dahlmann O, Haglind I, Jacobson B, Morck R, Stromberg L (1993) Pulp bleaching and the environment – The situation 1993. *Nordic Pulp and Paper Research Journal* **4**, 365–78.

Grimvall A, Laniewski K, Borén H, Jonsson S, Kaugare S (1994) Organohalogens of natural or unknown origin in surface water and precipitation. *Toxicology Environment and Chemistry* **46**, 183–96.

Keefe A, Teschke K (1998) Paper and pulp industry; environmental and public health issues. In: Stellman JM (ed.) *Encyclopedia of Occupational Health and Safety*, 4th ed, p. 7. International Labor Organization 72, Geneva.

Lee J, Koh D, Andijani M, *et al.* (2002) Effluents from a pulp and paper mill: A skin and health survey of children living in upstream and downstream villages. *Occupational and Environmental Medicine* **59,** 373–9.

Lehtinen KJ (1999) *Aquatic Environmental Impact Evaluation of the APRIL Riau Industrial Complex. APRIL Riau.* Finnish Environmental Research Group, Indonesia.

Saw SM, Koh D, Adjani MR, *et al.* (2001) A population-based prevalence survey of skin diseases in adolescents and adults in rural Sumatra, Indonesia, 1999. *Transactions of the Royal Society of Tropical Medicine Hygience* **95,** 384–8.

Wong ML, Chen PCY (1991) Self-reliance in health among village women. *World Health Forum* **12,** 43–8.

World Health Organization (1992a) *WHO Commission on Health and Environment Report of the Our Planet, Our Health.* WHO, Geneva.

World Health Organization (1992b) *WHO Commission on Health and Environment Report of the Panel on Industry.* WHO, Geneva.

Environmental Assessments and Approaches

CHAPTER FOUR

The Economic Valuation of Urban Ecosystem Services

Dodo J THAMPAPILLAI
Lee Kuan Yew School of Public Policy, National University of Singapore

"'Would the benefits of development exceed the benefits of preserving the wooded area'? The indiscriminate adoption of the cost–benefit criterion in such instances is inevitably flawed. This is because decision-making of this vein overlooks important linkages between the ecosystem and the (urban) economy. The most vital linkage to be acknowledged is the premise that any economy – urban or otherwise – could not exist without a minimal threshold level of ecosystem support. Very often the value of urban ecosystem services is elicited by recourse to the valuation of visible uses such as recreation and nature appreciation. The dominant valuation approaches in such contexts are methods based on contingent valuation, travel costs and hedonic prices."

– Dodo J THAMPAPILLAI

The Economic Valuation of Urban Ecosystem Services

Dodo J THAMPAPILLAI[1]

Lee Kuan Yew School of Public Policy, National University of Singapore

Abstract

The economic valuation of ecosystem services in cities often relies on the concept of willingness to pay. Such reliance invariably leads to the adoption of methods that have several inherent limitations. The objective of this paper is to present both a conceptual and methodological framework within which valuation could proceed. The conceptual premise rests on the principle that several ecosystem services cannot be compromised if the urban economy is to be sustainable. Hence methodologically, the valuation of such services could be approached by recourse to the opportunity cost concept, namely, the levels of income that need to be sacrificed in order to retain the sustainability of the urban economy. In the long run, the enhanced productivity and sustainability of the urban systems would offset the short-run losses that are incorrectly magnified.

Key words: ecosystem services, degradation, opportunity costs, sustainability, threshold values, valuation

I. Introduction

The valuation of ecosystems that support urban areas is primarily for policy formulation and decision-making. For example, consider a decision to clear a wooded area within an urban precinct for some form of development that yields monetary returns. A question that policymakers often ask in such an instance is, "Would the benefits of development exceed the benefits of preserving the wooded area?" The indiscriminate adoption of the cost-benefit criterion in such instances is inevitably flawed. This is because decision-making of this vein overlooks important linkages between the ecosystem and the (urban) economy. The most vital linkage to be acknowledged is the premise that any economy – urban or otherwise – could not exist without a minimal threshold level of ecosystem support. Very often, the value of urban ecosystem services is elicited by recourse to the valuation of visible uses such as recreation and nature appreciation. The dominant valuation approaches in such contexts are methods based on contingent valuation, travel costs and hedonic prices. For example, see Pearce (2006), Thampapillai (2006)

---

[1] Alternate Affiliation: Graduate School of the Environment, Macquarie University, NSW 2109 Australia.

and Sinden and Worrell (1979). Apart from the myriad issues associated with such methods[2] (Knetsch, 1994), the monetary estimates elicited can never depict the true value of ecosystem services. For example, consider the case of the aforementioned wooded area. The value of the ecosystem service provided by this area can extend far beyond visible uses. It could, for instance, include watershed services, biodiversity benefits and microclimate regulation. Further, an acknowledgement of the premise that some threshold level of ecosystem services needs to be retained for sustaining the (urban) economy implies that the value of ecosystem services would exponentially increase with their utilization.

The aim of this paper is to present a valuation framework based on the opportunity cost method. The main argument herein is that, at any time, one could only assess the minimum value of ecosystem services. Such minimum value has to be equated to the monetary benefits that need to be foregone in order to preserve ecosystems. The same way as subjectivity is central to methods such as contingent valuation, it is central to opportunity cost methods as well. The subjective assessment is guided by the notion of an 'acceptable sacrifice' and added scientific enquiry. Returning to the case of the wooded area, it is possible, in the first instance, to ascertain the value of income that needs to be foregone in order to preserve this area. If the size of income sacrifice is (subjectively) deemed small, then the preservation option may be preferable in the light of the broad spectrum of ecosystem services that are not readily visible. Alternatively, even when the income sacrifice is substantially large, scientific enquiry may dictate that preservation is in order.

The following section of this paper deals with a conceptual framework that provides the basis for eliciting the opportunity cost value within the urban context. This is followed by the consideration of a quasi-hypothetical case study.

II. The Conceptual Framework

The framework for valuation rests on the concepts of entropy, assimilative capacity and assimilative ability from environmental science. For example, see Daly (1992) and Leandri (2009). Urban ecosystems are generally highly entropic with limited assimilative capacity and ability. Hence urban development will inevitably entail an erosion of assimilative capacity (and ability) alongside the raising of entropy. This is illustrated in Figure 4.1.

---

[2] Contingent valuation (CV) rests primarily on the elicitation of willingness to pay by recourse to survey methods. The major difficulties encountered with CV are issues pertaining to bias and embedding. The latter issue is the inability of the respondents to separate specific issues that they are questioned about from an overall set of general issues. The travel cost (TC) method attempts to value an environmental asset/amenity in terms of the cost of travel to the location of the asset/amenity. A major difficulty with the TC method is that it is applicable to only those locations that are accessible for visitation. The hedonic price (HP) method attempts valuation of an environmental asset/amenity by estimating differentials in property values in the vicinity of the asset/amenity. The major difficulty with the HP method is the influence of other (non-environmental) factors on property value differentials.

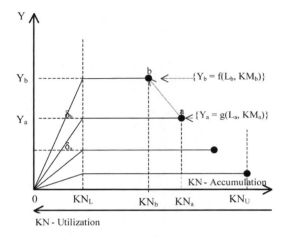

Figure 4.1: Conceptual framework for relating ecosystem end income

Consider Figure 4.1. Suppose that some metric is available for the quantification of the ecosystem that supports the urban environment; that is, all components and attributes of the ecosystem can be aggregated into a single numerical scale such as an index and are represented by KN along the horizontal axis. The accumulation of KN runs from left to right, whilst its utilization is represented from right to left. It is also assumed that, in the primitive state, where no urban development has occurred, the maximum capacity of the ecosystem is KN_U and the domain of assimilative ability is represented by $\{0 \leftrightarrow KN_U\}$. This state is consistent with the lowest level of entropy for this system and the absence of any income (Y), which is represented on the vertical scale. Urban development would entail the generation of Y. Higher levels of Y are enabled by utilizing more labour (L) and capital (KM). For example, in Figure 4.1, the level of income Y_b is due to the utilization intensity (L_b, KM_b), which exceeds (L_a, KM_a) that is associated with income level Y_a. But, as illustrated, the higher level of Y would also entail a contraction in the domain of assimilative ability, including a reduction in the maximum capacity. Further, higher levels of Y also prompt an increase in the gradient of degradation (δ), namely the rate at which Y falls per unit loss of KN. In Figure 4.1, it is supposed that degradation commences when the level of utilization exceeds some minimum threshold level of KN, namely KN_L; that is, once the size of the ecosystem falls below KN_L, assimilative ability is lost and degradation sets in. The essence of the conceptualization presented in Figure 4.1 is that the ecosystem becomes more fragile with higher entropy as resource utilization and income increase.

For example, suppose that the present position of an urban economy is point **a** in Figure 4.1 and that a commercial development project will take this economy to point **b**. In such an instance, an increase in income $(Y_b - Y_a)$ is associated with:

- contraction in the domain of assimilative ability by $\{KN_a - KN_b\}$; and
- an increase in the gradient of degradation by $\{\delta_b - \delta_a\}$.

Note that $\{KN_a - KN_b\}$ is the amount of KN that gets utilized or lost when income increases from Y_a to Y_b. However, some uncertainty exists as to whether the increase in income $(Y_b - Y_a)$ can be maintained. This is because higher levels of income are associated with higher levels of ecosystem fragility and entropy. In such a context, the increase in the degradation gradient $\{\delta_b - \delta_a\}$ can be regarded as an indicator of the higher level of risk of ecosystem failure. Hence, the opportunity cost (OC) of preventing the loss and enduring a higher level of risk is the increase in income less the potential to lose income because of higher entropy and fragility. This can be expressed as:

$$OC = \{Y_b - Y_a\} - \lambda\{[\delta_b - \delta_a]*[KN_a - KN_b]\} \tag{1}$$

In equation (1), the increase in potential income loss per unit of KN lost, namely $\{\delta_b - \delta_a\}$, is taken as a proxy for the cost of risk induced by the enhanced fragility of the ecosystem, and λ is a risk aversion coefficient that ranges between 0 and 1; that is, if the urban planner is a pure risk-taker, then $\lambda = 0$ and OC amounts to $\{Y_b - Y_a\}$. If the policymaker is fully risk averse, then $\lambda = 1$, OC is reduced in full by the cost of risk. Also note that, in terms of the premises presented in Figure 4.1:

$$\{\delta_b - \delta_a\} = \left(\frac{(Y_b - Y_a)}{KN_L}\right) \tag{2}$$

Hence the expression given in equation (1) is also equivalent to:

$$OC = \{Y_b - Y_a\} - \left(\lambda * \left[\frac{[KN_a - KN_b]}{KN_L}\right] * [Y_b - Y_a]\right) \tag{3}$$

Note that besides λ, the ratio $\left[\frac{[KN_a - KN_b]}{KN_L}\right]$ also scales the cost of risk, that is, for example:

$$\left(\left[\frac{[KN_a - KN_b]}{KN_L}\right] > 1\right) \text{ for } ([KN_a - KN_b] > KN_L) \tag{4}$$

OC is maximum only when $(\lambda = 0)$, that is, when risk aversion is totally absent. At least two implications emerge from the conceptual analysis presented thus far. The first is that when many of the invisible contributions of KN are known and yet cannot be readily quantified, decision-makers would prefer to err on the side of caution and forego potential income gains. The legitimate question in such a context is whether the OC of the preserving KN is an acceptable sacrifice. When the risk of system failure is also included, as in equations (1) and (2), the size of OC becomes smaller. As a result, the task of subjectively assessing whether the sacrifice is acceptable or not becomes easier; that is, the smaller the sacrifice, the easier it is to err on the side of caution. The second implication is the need to consider the prospect of stabilizing income levels at a certain magnitude rather than expanding them. This would be particularly pertinent to highly urbanized centres where both entropy and ecosystem fragility are high. In such a context, the portfolio of urban activities would have to include the rehabilitation of KN and various measures and innovations to maintain the prevailing levels of KN.

III.   A Quasi-Hypothetical Case Study

This case study is partially hypothetical because some of the data are assumed. Hence the case is more illustrative than real. The Central Catchment Nature Reserve of Singapore spans some 3,000 hectares and contains a set of five reservoirs that form part of the urban area's water supply. At least 1,000 hectares of this space represents a catchment area for the reservoirs. Some hydro-geologists claim that whilst the catchment terrain assists with surface run-off to feed the reservoirs, the parkland is also a source for recharging the ground-water system. The recharge attribute contributes to the maintenance of water levels in reservoirs not only in the vicinity of the parkland but also elsewhere. Further, suppose that some hydro-geologists have estimated that the parkland contributes to the supply of some 60 million gallons of water per day. Owing to the pressure of population growth and increasing demand for housing, there are some proposals for converting the parkland into a housing estate. There are some suggestions that around 450,000 dwellings in the context of high-rise complexes with some open space could be erected. Further, it is suggested that the loss of water supply through the catchment characteristics could be offset by the construction of a desalination plant. Some data for the analysis are as follows:

- The construction cost of a single dwelling is assumed to be $140,000, and construction is spread over three years.

- Average returns from housing are approximated to be $2,000 per dwelling per month over a 25-year time period.

- The desalination plant with a capacity of 60 million gallons per day could cost $890 million to construct and thereon $3 million per year to maintain.

- The metric for KN in the conceptual framework proposed above is reduced to water-supply capacity. KN_a and KN_b are assumed to be respectively 60 and 40 million gallons of water per day. This is because, despite the housing construction, the residual land area would retain some catchment characteristics and the housing infrastructure itself would have drainage designs to feed the reservoirs. KN_L is assumed to be 30 million gallons per day.

With reference to this illustration, the basic opportunity cost of preserving the 1,000 hectares of parkland, that is $(Y_b - Y_a)$, would amount to:

(Rental returns from Housing) – (Housing construction costs)
   – (Costs of desalination)

As illustrated in Table 4.1 below, this opportunity cost is approximately $154.6 billion in present value terms.

Table 4.1: Summary of housing benefits and costs

Present value of housing benefits	$217,307,186,190.91
Present value of costs (housing + desalination)	$62,703,461,912.67
Net present value of housing development	$154,603,724,278.24

The risk-adjusted value of opportunity cost as per equations (2) and (3) earlier will depend on the magnitude of the risk aversion coefficient, as illustrated in Table 4.2 and Figure 4.2.

Table 4.2: Opportunity cost of preservation including risk

λ	Risk cost (R)	OC-R
1.0	$103,069,149,519	$51,534,574,759
0.9	$92,762,234,567	$61,841,489,711
0.8	$82,455,319,615	$72,148,404,663
0.7	$72,148,404,663	$82,455,319,615
0.6	$61,841,489,711	$92,762,234,567
0.5	$51,534,574,759	$103,069,149,519
0.4	$41,227,659,808	$113,376,064,471
0.3	$30,920,744,856	$123,682,979,423
0.2	$20,613,829,904	$133,989,894,374
0.1	$10,306,914,952	$144,296,809,326
0.0	$0	$154,603,724,278

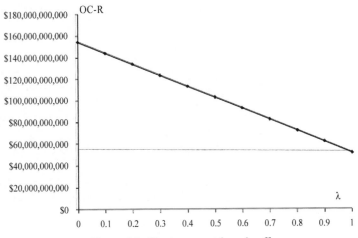

Figure 4.2: The income–risk trade-off

As illustrated, the opportunity cost of preservation is reduced from $154.6 billion in the context of pure risk-taking to about $51.53 billion in the context of pure risk aversion. In the event that the decision-maker is risk neutral (λ = 0.5), then he will opt for the choice of $103.07 billion as preferred income and this could be tantamount to preserving 50 per cent of the land area.

The question of acceptable sacrifice for preserving the parkland is a difficult issue. This is because even with complete risk aversion (λ = 1), the value of income benefits to be given up is $51.53 billion. In this context, the assessment of acceptability could be aided by applying the concept of TV (Krutilla & Fisher, 1975). A definition of TV to the context of the case study considered here is as follows: *The minimum value of the benefits of preserving the parklands in the initial year and that would grow at a specific rate such that the present value of preservation is at least equal to the net present value of housing development.* That is, TV is the initial year's minimum value for preservation benefits that would render preservation just as desirable as housing development. Following Krutilla and Fisher, TV can be estimated as:

$$TV = \left[\frac{\text{(Present value of housing benefits)}}{\text{(Present value of \$1 growing at the rate of growth of the preservation benefits)}} \right] \quad (5)$$

The nomination of an appropriate rate of growth for preservation benefits is a difficult task and has to be guided by scientific enquiry. For illustrative purposes, a growth rate of 0.5 per cent per year is assumed below, and TV for different levels of risk aversion is illustrated in Table 4.3 on the next page.

Consider the extreme scenario of pure risk-taking (λ = 0). If those who seek preservation could clearly demonstrate that the value of preservation benefits as of now is in excess of $6.51 billion, then there is a case for preservation taking precedence over housing development. The question is: Is the Central Catchment Nature Reserve worth $6.51 billion today?

Table 4.3:   Threshold values and risk aversion

λ	TV
1.0	$2,169,876,832
0.9	$2,603,852,198
0.8	$3,037,827,565
0.7	$3,471,802,931
0.6	$3,905,778,298
0.5	$4,339,753,664
0.4	$4,773,729,030
0.3	$5,207,704,397
0.2	$5,641,679,763
0.1	$6,075,655,130
0.0	$6,509,630,496

An alternative line of reasoning could proceed as follows. The preservation benefit per unit of housing in the context of ($\lambda = 0$), namely ($6.51 billion/450,000), is $14,500. Property analysts have claimed that the land value that is built into residential house prices is in excess of $100,000 per dwelling. That is, home buyers would pay at least $100,000 for the purchase of a single dwelling. The TV analysis reveals that an initial outlay of $14,500 per home buyer on the nature reserve with nature benefits growing at 0.5 per cent per annum would yield the same net present value as an outlay of $100,000 per dwelling. Hence there exists a case for arguing in favour of preservation. Such a line of reasoning is, no doubt, tenuous in the context of scarce housing, high house prices and rentals.

IV.   Conclusion

As indicated, the opportunity cost criterion dictates that the value of the ecosystem must *at least equal* the value of development if preservation is to be given precedence. The example considered above dealt with the issue of water conservation versus housing development. It is true that, in most urban contexts, housing is scarce and the provision of shelter is a noble objective. Nevertheless, water is also a basic need. Uncertain regimes of climate change could render vulnerabilities of water shortages that could be difficult to contend with. Also note that the adoption of desalination as a substitute for traditional water supply is not without difficulties, especially with reference to brine disposal. For an example, see Lattemann and Höpner (2007).

Acknowledgements

With the usual disclaimers, I remain grateful to Cheng Kim and Charles Adams for their valuable comments.

References

Daly HE (1992) Is the entropy law relevant to the economics of natural resource scarcity? Yes, of course it is! Comment. *Journal of Environmental Economics and Management* **23(1),** 91–5.

Knetsch J (1994) Environmental valuation: Some problems of wrong questions and misleading answers. *Environmental Values* **3,** 351–68.

Krutilla JV, Fisher AC (1975) *The Economics of Natural Environments: Studies in the Valuation of Commodity and Amenity Resources.* Johns Hopkins University Press, Baltimore.

Lattemann S, Höpner T (2007) Environmental impact and impact assessment of seawater desalination. *Desalination* **220(2008),** 1–15.

Leandri M (2009) The shadow price of assimilative capacity in optimal flow pollution control. *Ecological Economics* **684(4),** 1020–31.

Pearce D (ed) (2006) *Environmental Valuation in Developed Countries: Case Studies.* Edward Elgar, Cheltenham.

Sinden JA, Worrell AC (1979) *Unpriced Values: Markets Without Prices.* Wiley, New York.

Thampapillai DJ (2006). *Environmental Economics: Concepts, Methods and Policies,* Oxford University Press, Melbourne.

CHAPTER FIVE

The Material Consumption of Singapore's Economy: An Industrial Ecology Approach

Marian CHERTOW, Esther S CHOI and Keith LEE

"Singapore, as a highly urbanized island city-state, provides a particularly interesting study site for both urban metabolism and island sustainability. One of the more distinctive findings of the study speaks to that point directly: as an island with limited land area, Singapore needs to import land itself – in the form of sand – in order to support its economic growth. That one figure alone caused great volatility across the study years. Using an MFA (materials flow analysis) approach also uncovered a trend of increased impact from 'virtual water' imports as, increasingly, Singaporeans appear to favor meat over vegetable products in their diets measured on a per capita basis. Finally, fluctuations in Singapore's import and export activity were shown to affect its Domestic Material Consumption inversely as compared to global economic activity."

– Marian CHERTOW, Esther S CHOI and Keith LEE

The Material Consumption of Singapore's Economy: An Industrial Ecology Approach

Marian CHERTOW[1], Esther S CHOI[1] and Keith LEE[1]
Industrial Environmental Management Program,
Yale School of Forestry & Environmental Studies

Abstract

In a world deeply concerned about future accessibility of physical resources, especially materials, water and energy, the young field of industrial ecology brings resolute attention to tracking the flows of these resources through systems at different scales. Material flow analysis is an industrial ecology tool used to examine system metabolism by tracking the input, output, conversion and accumulation of materials, water, energy or selected substances, helping to inform decisions about resource availability, waste management and pollution reduction at local, regional or global levels.

With the exception of several nature reserves, the island city-state of Singapore is highly urbanized, and has been dependent upon industrial and manufacturing activities for its national standing and economic growth. As such, mapping Singapore's 'urban metabolism' through a material flow analysis (MFA) approach, with a particular focus on materials catalogued by international trade databases, is useful for understanding the island's level of sustainable resources use within the context of the global economy and ecosystem.

This study compares relevant data describing Singapore's material flows for the years 2000, 2004 and 2008. Domestic material consumption on the island is found to be highly variable across the three study years. This primarily reflects Singapore's levels of construction activity, including considerable additions of actual land area to the island. The flux in the quantity of sand imports to Singapore illustrates the volatility that one category of goods can introduce to the material record of an otherwise stable economy. For countries that are becoming increasingly dependent on imported products, the implication is that a growing share of the impact is taking place in other countries. Global economic trends are also determined to have

[1] Center for Industrial Ecology, Yale School of Forestry & Environmental Studies. 195 Prospect St, New Haven, CT06511, USA. Emails: marian.chertow@yale.edu, keith.cl.lee@gmail.com, esther.sk.choi@gmail.com; Tel: 203-436-4421; Fax: 203-432-5556.

an effect on Singapore's additions to stock vs. its exports, as its gross domestic product is highly reliant on trade activity.

Key words: industrial ecology, material flow analysis, Singapore, urban metabolism, urban systems

I. Introduction

Industrial Ecology and Metabolism in the Island City-state

In a world deeply concerned about future accessibility of physical resources, especially materials, water and energy, the young field of industrial ecology brings resolute attention to tracking the flows of these resources through systems at different scales, making an analogy with the metabolic flows into and out of the bodily system (Ayers & Ayres, 2002; Graedel & Allenby, 2009). Indeed, an early notion of 'metabolism' is shown in Figure 5.1 on the next page, dating back to 1614, when an Italian physician came to realize, through his own experiments, sitting in a balance and weighing himself after eating, the small amount of weight his body would lose in an hour or more after each meal (Brunner & Rechberger 2004). Industrial ecology, as an interdisciplinary systems science, is engaged with the changes in metabolism in the social–industrial system brought about by human production and consumption interacting with those of nature. The former president of the US National Academy of Engineering defined industrial ecology as:

> 'the study of the flows of materials and energy in industrial and consumer activities, of the effects of these flows on the environment, and of the influences of economic, political, regulatory and social factors on the flow, use and transformation of resources' (White, 1994).

Singapore has been the subject of several research projects within industrial ecology that use various analytic tools. Kannan *et al.* have conducted Life Cycle Assessments on various power-generation technologies (2005), solar photovoltaic systems (2006) and an oil-fired steam turbine power plant (2004) in Singapore. Yang and Ong (2004) focused their study on the integration of landscape ecology and industrial ecology for the industrial development of Singapore's Jurong Island. Schulz (2007) conducted a material flow analysis (MFA) on Singapore that provided the baseline for this study.

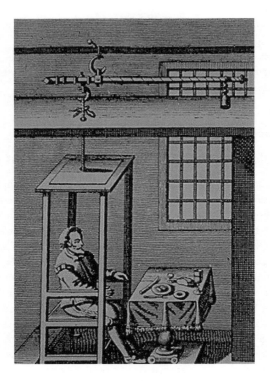

Figure 5.1: Frontispiece of Medicina Statica by Sanctorius in 1614, the Italian physician who experimented with his own bodily metabolism by weighing himself in this balance machine after eating (Source: Science Photo Library, http://www.sciencephoto.com/media/228232/enlarge)

Singapore is both an urban and an island system. Urban areas are characterized by a high concentration of economic activities, a large population and large material stock densities with high levels of energy and material flows (Graedel, 1999). These areas harbour about half of the global population on less than 3 per cent of the terrestrial surface, and the global trend of urbanisation is anticipated to add two billion people to Earth's cities in the coming 25 years (United Nations Department of Economic and Social Affairs, 2002; Schulz, 2007). The increasing rate of urbanisation suggests that examining the external impacts of urban systems, and the materials and energy flows within them is critical for understanding how cities function, what kinds of resources are used as well as how much and in what ways the wastes and emissions are produced and managed. This concept is described by the field of industrial ecology as urban metabolism.

The urban metabolism concept is built upon close analogies with biological metabolism. Like a living organism, urban systems need material and energy inputs, consume these inputs to produce goods and services, and generate wastes, which include both physical wastes and emissions. Kennedy *et al.* (2007)

succinctly described urban metabolism as 'the sum total of the technical and socio-economic processes that occur in cities, resulting in growth, production of energy and elimination of waste'. It is a model that can be used to describe and analyze the flows of materials and energy within the system.

In addition to urban areas, islands, too, have emerged as valuable for industrial ecology studies. On the one hand, islands are defined by obvious borders and have easily identifiable points of entry/exit for tracking material flows, making them practical study subjects (Deschenes & Chertow, 2004). On the other hand, given the tenuousness of resource availability on islands and the recognition that the isolation of islands constrains resource choices, a thorough understanding of material stocks, flows and use rises in importance so that island sustainability can be evaluated.

(i) Material Flow Analysis

Material flow analysis (MFA) is an industrial ecology tool used to examine system metabolism by tracking the input, output, conversion and accumulation of materials or selected substances within specific systems, such as a factory, an industry, a city, a country or globally. According to a recent publication from the Organisation for Economic Co-operation and Development (2008), 'MFA is the only tool that can:

• provide an integrated view of resource flows through the economy;
• capture flows that do not enter the economy as transactions but are relevant from an environmental point of view; and
• reveal how flows of materials shift within countries and amongst countries and regions, and how this affects the economy and the environment within and beyond national borders'.

MFA is based on the concept of mass balance that leads to a simple equation showing that whatever enters a particular system is either accumulated as stock or flows through the system and becomes an output, as seen below:

$$\textbf{total inputs} = \textbf{total outputs} + \textbf{net accumulation} \qquad (1)$$

At the country level appropriate for Singapore, economy-wide MFA is the European name for the type of material accounting that focuses on providing a picture of a society's overall material metabolism based on statistical data. Figure 5.2 gives a sense of the methodology of economy-wide MFA by showing the flow of inputs into an economy. In this illustration, there are imports and 'materials domestically extracted". Within the part of Figure 5.2 labelled 'Economy' are the materials that are accumulating as stocks and the materials that flow from inputs through to outputs. Anything recycled and used within the economy would also be captured here. The outputs are the exports and the total amount of wastes disposed and emitted to the air, land and water. This shows that whatever leaves one economy is going either to either another economy or back into natural systems.

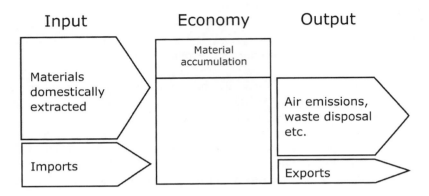

Input   Economy   Output

Materials
domestically
extracted

Material
accumulation

Air emissions,
waste disposal
etc.

Imports

Exports

Figure 5.2: Economy-wide MFA system diagram (Eurostat, 2001)

The ability to map the stocks and flows of a given system is crucial for evaluating society's progress towards sustainability. MFA provides an overview of the urban system by linking regional processes and activities. It also directs attention towards upstream industrial activities and the causes of resource depletion and emissions (Schulz, 2007), contributing to the formation of effective policy measures. MFA not only determines the material consumption of the economic system for a particular year by assessing the static analyses of material flows, but also evaluates the trends of material use by producing 'multiple snapshots' of development when analyzed as a time series (Niza *et al.*, 2009). In this way, one can understand how the composition and extent of material use within the system have changed over time. Dynamic modelling of systems can be applied to further explain the relationships between resource stocks and flows at the city and sectoral scales, as noted by Abou-Adbo *et al.* (2011) in their study of the flows of construction materials and electricity in Singapore's public housing sector.

In terms of policy, MFA can be used for recognition and priority setting to analyze and improve the effectiveness of ongoing policy measures (Hendriks *et al.*, 2000; Wernick & Irwin, 2005). This, in turn, can facilitate the design of efficient and sustainable management strategies. Using MFA, policymakers can both foresee and understand potential environmental problems and the source of future resources because it provides insights into the current and potential material stocks. By highlighting future environmental or resource problems, MFA can assist precautionary policy making (Hendriks *et al.*, 2000). Therefore, MFA is a critical tool that provides an explicit map of material and energy flows, offering a better understanding of urban metabolism and a solid platform to build upon when attempting to achieve a sustainable system.

Since Wolman's (1965) pioneering work on the metabolism of cities, many MFAs have been conducted around the globe on varying scales. Singapore represents an interesting case for MFA, as it is one of the 'Four Asian Tigers' that experienced exceptionally high economic growth and rapid industrialization between the early 1960s and 1990s. Schulz (2007) conducted a similar MFA for Singapore on a national scale to examine the change in direct material inputs over the period of 41 years, from 1962 to 2002.

In this MFA we conducted for Singapore, we used a modified and more comprehensive methodology to build upon Schulz's (2007) work to include both direct and indirect flows. While Schulz, for example, focused on exports as the main output, we have included emissions and wastes in our system. While Schulz found Singapore's domestic extraction of biomass and minerals to be significant historically, we assume in this study that, compared to trade flows, domestic extraction would be residual and, therefore, less significant in the present. Since the scope of Schulz's study does not extend beyond 2000, we hope to continue conducting MFAs for Singapore and extend these to comprehend a greater scope and depth of Singapore's metabolism.

(ii) Singapore

The defining boundary of our study is the spatial scale provided by Singapore's national borders. As a small city-state with 274 square miles (710 square kilometres) of land area and over five million residents as of June 2011 (http://www.singstat. gov.sg, accessed 15 October 2011), Singapore's characterization by some as a 'little red dot' denotes its economic achievement and global standing despite its small size and lack of natural resources. Singapore is a unique example of an urban system. The entire island is highly urbanized, with the exception of several forests and natural reserves maintained by the government. In fact, according to the United Nations (UN), the percentage of Singapore's urban population reached 100 per cent even before its independence in 1963 from the United Kingdom (Globalis, 2010). Given Singapore's predominantly urban environment and its reliance on industrial and manufacturing activities for its national standing and economic growth, mapping its urban metabolism through MFA is especially useful for understanding the present and future of the country.

Abou-Abdo *et al.* (2011) list several favourable characteristics of Singapore that make it very useful and appropriate for urban metabolism research: data availability, a recent history of urbanisation from a small port to a global metropolis and a planning horizon that, from the outset, has been shaped by resource scarcity. Indeed, as an island city-state with physically defined national boundaries, everything crossing the Singaporean boundary is recorded as international trade, providing detailed inventories of material flows into and out of the system. With a strong focus on the manufacturing and service sectors, Singapore has used the powers and capacities of the nation-state – in material and discursive senses – to transform society with the aim of embedding Singapore within the network of relations that propel the world economy (Olds & Yeung, 2004). In other words, Singapore has positioned itself as a major trader to compensate for its lack of natural resources. These attributes of Singapore shaped our methodology to conduct MFA on a national scale.

II. Methodology

Framework

A recently completed MFA of Lisbon, Portugal, by Niza *et al.* (2009) helped to frame ideas for the Singapore MFA. There are many similarities between these two regions. First, both are major urban areas attractive for economic activities and employment, playing a central role in the international activities and trade for each respective country. Second, as with other major cities, both consume mostly final products and have minimal domestic extraction, with the exception of some construction raw materials such as sand and gravel (Niza *et al.*, 2009). Third, while many MFAs tend to focus on only the major material flows or a specific individual substance like metals, Niza *et al.* focus on explaining the complete system of material flows of a particular region, corresponding to our objective.

Naturally, some differences exist between the two cities. The most apparent feature is Singapore's clear boundary in contrast with Lisbon's absence of. Singapore's unique attributes facilitates our approach to identifying and quantifying the amounts of products passing through the system and those that remain for endogenous consumption. In addition, Niza *et al.* (2009) assumed that low levels of local production in Lisbon and exports of materials were residual or non-existent in terms of material flows. This is not the case for Singapore, which actively imports raw materials that are transformed into intermediate and final manufactured goods for export to the world. These differences are addressed in our methodology.

UN Comtrade

The import and export data were obtained from the United Nations Commodity Trade Statistics Database (UN Comtrade; http://comtrade.un.org/db/), a website that provides access to the world's largest depository of international trade data and is the work of the International Merchandise Trade Statistics Section of the United Nations Statistics Division. More than 170 reporter countries provide their annual international trade statistics data detailed by commodities and partner countries. Users can find information on any commodity from any country, including large overviews of imports and exports by country and individual commodity categories. All commodity values are converted from national currency into US dollars using exchange rates supplied by the reporter countries or derived from monthly market rates and volume of trade. The material weight of imports and exports is also provided in metric units. Coverage varies by country but dates as far back as 1962 for some nations.

Commodities are classified according to the Standard International Trade Classification (SITC) and the Harmonized System (HS) code classifications systems. The main difference between these two classifications is that SITC is focused more on the economic functions of products at various stages of development, whereas HS deals with a precise breakdown of the products' individual categories (International Trade Centre, 2010). Since SITC is focused on the stage of processing and economic functions to facilitate economic analysis, the HS code was more pertinent to our study. The HS classification is 'harmonized' in relation to the classification of the traded goods. These are classified according to objective criteria and applications.

III. Other Data Sources

For the emissions and waste data, we referred to the Singapore government websites, mainly the National Environment Agency (NEA: http://app2.nea.gov.sg/index.aspx) and the Ministry of the Environment and Water Resources (MEWR: http://app.mewr.gov.sg/web/Common/homepage.aspx) websites The World Resources Institute Earthtrends (2007) database supplemented the emissions data. Invaluable insights and feedback from colleagues at the Center for Industrial Ecology at Yale University and at the School of Design & Environment at the National University of Singapore facilitated the shaping of our methodology and analysis.

IV. Approach

Singapore's import and export data for 2000, 2004 and 2008 were downloaded from the UN Comtrade website by sectors according to the HS code, 'as reported' and tabulated in spreadsheets. Because of a substantial amount of missing data, those before 2000 were considered unreliable for our study. UN Comtrade provides the data such that the aggregated category is presented first. Then the breakdown of that category's items, values and weight is presented accordingly. Figure 5.3 differs from Figure 5.2 in that it is missing the section for 'materials domestically extracted', since we assumed that this is negligible in Singapore. Therefore, domestic material consumption (DMC) of Singapore is equal to imports minus the sum of exports, emissions and waste for any given year.

V. Results

The full results of the MFA are shown in Figure 5.3, which schematically represents the inflows and outflows for Singapore in 2008. Total material input was approximately 201 megatonnes and had two components: imports and recycled materials. Total imports accounted for 198 megatonnes (98 per cent) of total material input, while recycled materials made up the remaining 2 per cent. Since domestic extraction is considered negligible for this study, 'flows crossing through' are not a factor in Figure 5.3 on page 116. Outflows from the economy in 2008 totalled 149 megatonnes. Of this figure, approximately 100 megatonnes (67 per cent) were exports, 6 megatonnes (4 per cent) constituted municipal solid waste, 36 megatonnes (24 per cent) was carbon dioxide and 7 megatonnes (5 per cent) consisted of other emissions such as sulfur dioxide, nitrous oxides, carbon monoxide, non-methane volatile organic compounds, fluorinated gases and methane.

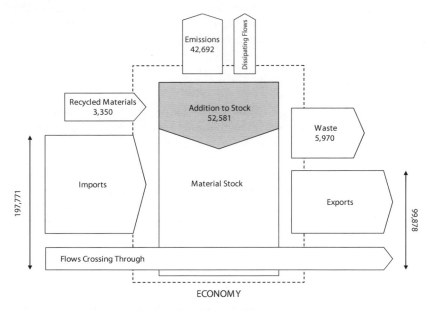

Figure 5.3: Singapore MFA Diagram 2008 (units in kilotonnes)

DMC was calculated from these numbers. According to Eurostat (2001) methodology, DMC equals materials domestically extracted plus imports minus exports. With negligible domestic extraction, DMC for Singapore was found by calculating the net weight of materials imported into the country. Based on this calculation, Singapore's DMC for 2008 was 98 megatonnes, corresponding to roughly 20 tonnes per capita. This figure compares favourably to the 2004 DMC per capita of 20 tonnes, as calculated by Niza *et al.* (2009) for Lisbon. Net addition to stock was found to be approximately 53 megatonnes in 2008 and was calculated by subtracting total outflows from total inflows to the economy. The 2009 SITC data showed that four of Singapore's top five import commodities were also four of its top five export commodities, a reflection of the island's role as a highly connected trans-shipment port that specializes in the storage, processing and reshipping of goods moving between other countries.

To get a sense of the material trends over time, we conducted the MFA for 2000 and 2004 as well. An interval of four years was chosen to increase the likelihood that meaningful changes would be captured. For waste, however, only landfilling and incineration data were available for 2000 and 2004, so a detailed picture of waste generated could not be established for those years. Also, since not all emissions data were available for these years, only import and export figures were recorded. DMC calculated for 2000 and 2004 was found to be 222 and 58 megatonnes respectively. On a per capita basis, these numbers correspond to 55 and 14 tonnes per person respectively. It is worth noting that Schulz's 2007 study of the inputs to Singapore presented a DMC per capita number of 54 tonnes per person in 2000.

In the absence of complete waste and emissions data, we estimated waste and emissions using corresponding numbers from 2008. By applying the ratio of recycled waste to incinerated plus landfill waste for 2008, we were able to estimate figures for total waste generated in 2000 and 2004. Where unavailable, emissions data for these years were approximated by finding the per-unit gross domestic product (GDP) mass of emissions for 2008 and multiplying this by 2000 and 2004 GDP figures. The implied assumption implicit in performing this estimation is that emissions per unit of GDP and the recycling rate were both unchanged for the years 2000, 2004 and 2008. Though the rates were likely to have changed over this time, the assumption allows us to estimate total outflows and additions to stock numbers for 2000 and 2004. These have been summarized in Table 5.1, together with the data for 2008. For the year 2000, total outflows came to 105 megatonnes, with an addition to the stock of 175 megatonnes. For the year 2004, total outflows came to 132 megatonnes, with an addition to the stock of 11 megatonnes.

Table 5.1: Detailed Singapore MFA overview

(units in kt unless otherwise stated)	2000	2004	2008
Inflows:			
Net Imports (DMC)	222,182	57,587	97,893
Recycled Material	3,571	3,169	3,350
Total Inflows	280,282	143,191	201,121
Outflows:			
Waste Generated	6,363	5,648	5,970
CO2	38,684	39,212	35,945
Other	5,361	4,564	6,747
Total Outflows	104,938	131,859	148,540
Addition to Stock	175,344	11,332	52,581
Net Imports (DMC)	222,182	57,587	97,893
Less: HS Code 25	175,445	16,999	42,333
Adjusted Net Imports (DMC)	46,737	40,588	55,560
DMC per Capita (tons)	55.2	13.8	20.2
Singapore (2000)	53.8	-	-
Lisbon (2004)	-	20.1	-
Adjusted DMC per Capita (tons)	11.6	9.7	11.5

Table 5.2 on page 118 shows the breakdown of waste in 2008 according to material category, and the recycling rates for each of these categories. Waste that is not recycled is either sent to landfills or incinerated. Paper- and cardboard-based waste accounted for 21 per cent of the total, while construction debris and ferrous metals followed at 15 per cent and 13 per cent of the total respectively. Plastic waste was fourth-largest, at 11 per cent of the total waste generated. The overall recycling rate for Singapore is 56 per cent, but this tends to vary greatly amongst waste categories. For example, the recycling rates of paper and cardboard

as well as plastics were only 48 per cent and 9 per cent respectively, which seem low, given the waste minimization and recycling programmes the government has in place. On the other hand, the recycling rates for construction debris and ferrous metals were over 95 per cent, as reported by MEWR, although additional details were not available.

Table 5.2: Singapore 2008 waste breakdown

Waste Stream	Generated (kt)	% of Total	Recycled (kt)	Implied Rate
Paper / Cardboard	1,260	21.1%	610	48.4%
Construction Debris	920	15.4%	900	97.8%
Ferrous Metals	780	13.1%	740	94.9%
Plastics	680	11.4%	60	8.8%
Used Slag	570	9.5%	560	98.2%
Food	570	9.5%	70	12.3%
Others (e.g. waste, ceramics, silt etc.)	310	5.2%	10	3.2%
Wood / Timber	270	4.5%	190	70.4%
Horticultural Waste	230	3.9%	100	43.5%
Sludge	110	1.8%	0	0.0%
Non-ferrous Metals	90	1.5%	70	77.8%
Textile / Leather	90	1.5%	10	11.1%
Glass	60	1.0%	10	16.7%
Scrap Tyres	30	0.5%	20	66.7%
Total	5,970	100.0%	3,350	56.1%

Source: MEWR, Key Environmental Statistics 2010

VI. Discussion

A sector-by-sector breakdown is presented in Table 5.3. The most apparent trend in the data is the disparity between DMCs for the years 2000, 2004 and 2008, which swings from 222 to 58 megatonnes and then back up to 98 megatonnes over these three years respectively. In 2000, 97 per cent of net material inflow was accounted for by HS codes 25, 26 and 27 – generic mineral products, within which 82 per cent of the net material inflow was classified under construction minerals (HS code 25). This sector refers generally to construction-related materials such as sand, cement and granite, in addition to other non-fossil-fuel and ore-based mineral products, and accounts for most of the changes in DMC for the years 2000, 2004 and 2008. In 2000, net material inflow for Construction Minerals was 175 megatonnes. In 2004, this figure dropped to 17 megatonnes before recovering to 42 megatonnes in 2008.

Table 5.3: Sector breakdown of flows

Category	Net Inflow / (Outflow) (kt)		
	2000	2004	2008
Animals and Animal Products	349	478	492
Vegetable Products	1,580	1,580	1,664
Foodstuffs	281	310	426
Mineral Products	216,272	55,529	93,088
Chemical and Allied Industries	139	(2,070)	(1,278)
Plastics and Rubbers	(1,068)	(2,557)	(2,774)
Rawhides, Skins, Leathers and Furs	14	21	21
Wood and Wood Products	301	532	338
Textiles	110	72	43
Footwear and Headwear	16	24	13
Stone and Glass	1,274	1,036	2,112
Metals	2,742	2,192	3,579
Machinery and Electrical	85	88	7
Transportation	(84)	226	(7)
Miscellaneous	171	128	168
Total	222,182	57,587	97,893

Source: UN Comtrade

The magnitude of these fluctuations is a particularly interesting finding. To explore the possibility of data error, we cross-checked the import data reported by Singapore against the export data reported to the UN Comtrade database by its partner countries. This revealed large discrepancies for commodities within the construction minerals category, particularly between Singapore's and Malaysia's data, supporting the notion that there was an error in reporting. A view of Singapore's construction activity at the time, however, supports the accuracy of the finding and highlights a conflict in the island's recent history.

First, we consider trends in Singapore's construction sector between 2000 and 2005. Figure 5.4, based on the data from Singapore's Building and Construction Authority (2005), shows a local maximum in construction growth rates in 2000 before a precipitous decline in the first quarter of 2003, bottoming out at around -17 per cent. Growth became positive in the first quarter of 2004 before resuming a negative trend in the rest of 2004 and the first half of 2005. Examining the trend for sand, one of the major constituents of the construction minerals category, for all years between 2000 and 2008, reveals a matching trend that shows imports of sand bottoming out between 2004 and 2005 before resuming growth in 2006 (Table 5.4). Hence, there is reason to believe that the net import data for the construction minerals category reflect the construction activities taking place in Singapore at the time.

Second, the cause of this observed upswing in sand imports is directly linked to the large amount of land reclamation activity that the Singapore government has undertaken. Since the 1960s, Singapore's land area has grown by over

20 per cent, owing to land reclamation (Henderson, 2010). As a result, whenever land reclamation projects take place, the country's demand for sand increases significantly. The most notable land reclamation project to date has been Jurong Island, the 30-hectare industrial complex located southwest of Singapore (Kelly, 2009) and the origin of 80 per cent of the manufacturing output of the chemical and energy sectors in the country. According to an article in the *Asia Times* (Guerin, 2003), 'Singapore's sand needs boomed in 1999' for the reclamation of more land for the airport and two other coastal areas, Jurong and Pasir Panjang. Part of this boom almost certainly originated from the commencement of the Jurong Island project at the beginning of the 21st century and would explain the 169 megatonnes of sand imported in 2000.

Table 5.4: Sand imports, 2000–2008

Year	2000	2001	2002	2003	2004	2005	2006	2007	2008
Imports (kt)	168,504	69,905	41,164	86	27	26	40	3,530	11,560

Source: UN Comtrade

As most of the sand used has historically been exported from Malaysia and Indonesia (Guerin, 2003), tensions have arisen between these countries and Singapore, owing to the environmental degradation caused by sand extraction activities. These tensions resulted in several bans and restrictions on sand exports to Singapore from both Indonesia (Arnold & Fuller, 2007) and Malaysia (Guerin, 2003; Cheong, 2009). In 2003, a temporary ban supposedly brought land reclamation and construction projects in Singapore to a standstill (Guerin, 2003). If the reports of these bans and restrictions are accurate, this would explain the sudden decline in sand imports in 2003.

Additionally, sources report that representatives in Malaysia and Indonesia have permitted suppliers of sand in those countries to sell sand illegally in Singapore despite supposed restrictions (Henderson, 2010). The illegal nature of these exports may also explain the discrepancy in the data reported by Malaysia and Singapore. Rising sand import figures for 2007 and 2008 reflect the emergence of Cambodia and Vietnam as increasingly important sources of sand for Singapore (UN Comtrade) and what was likely to be the recovery of the construction sector.

The next most substantial component of Singapore's DMC was HS code 27, which is largely comprised of fossil fuels such as coal, petroleum and natural gas and their basic derivatives. In 2000, 40 megatonnes of fossil-fuel products were imported on a net basis, accounting for 18 per cent of DMC, while the corresponding figures for 2004 and 2008 were 38 megatonnes (66 per cent) and 50 megatonnes (50 per cent) respectively. These statistics speak of the importance of Singapore as a petroleum-refining hub. By the 1990s, the country had become 'the third-largest oil-refining center in the world after Houston and Rotterdam, the third-largest oil-trading center after New York and London, and the largest fuel oil bunker market in volume terms' (Lee, 2000).

The net imports of fossil-fuel products are primarily used as raw materials for the refining and petrochemical industries, which manufacture petrochemical products and plastics for export. The UN Comtrade data corroborate this theory. Singapore is a net exporter of plastics, rubber products and chemicals for all three years in the analysis, exporting between 1 and 5 megatonnes on a net basis each year. The remainder of the material under HS code 27 is likely sold as fuel to container ships and oil tankers passing through Singapore's trans-shipment port. The relative stability of this data over time may be explained by the maturity of a sector that has been a staple of Singapore's economy for over a decade.

Outside of the mineral products category, the vegetable products category was the next-largest component of DMC, with roughly 1.6 megatonnes imported each year on a net basis. Edible vegetables, edible fruits and cereals were the three largest components of this category. Unlike mineral products and most of the non-biomass-related categories, DMC of vegetable products does not fluctuate as much over the three years in our analysis. What is curious, however, is that on a per capita basis, there is actually a slight decline in the DMC vegetable products. Parallel to this is an increase in the per capita DMC of animal products and prepared foodstuffs. The relative stability of all three sectors on an aggregate basis indicates the inelastic nature of their demand, which is unsurprising given their status as daily necessities. Because meat consumption tends to increase as per capita incomes rise (Schroeder et al., 1996), the shift in DMC away from vegetable products towards animal products and prepared foodstuffs may be indicative of changes in Singaporean lifestyles and eating habits that have been funded by strong per capita income growth over the last decade.

While meat production is inherently more water consumptive than vegetable production due to the high level of crop inputs required, the water drawdowns are not happening in Singapore, but in the originating countries where meat is produced and then embedded into exports. Hence what is called in the literature as 'virtual water' is imported to Singapore, keeping Singaporeans from needing to deplete their stocks of water to be able to eat meat. According to Hoekstra and Chapagain (2007), the global average virtual water contents of chicken meat, pork and beef are 3,900, 4,900 and 15,500 cubic metres/tonnes respectively, whereas the virtual water contents for maize, wheat and rice only amount to 900, 1,300 and 3,000 cubic metres/tonnes respectively. Singapore's growth in meat consumption therefore highlights an important aspect of the country's environmental impact, which may need to be addressed in the future through government policy, as has already been suggested by scientists (Vidal, 2004).

The magnitude of the changes in construction minerals cited above was such that other trends in DMC per capita were obscured. Controlling for this sector allows us to analyze those materials more closely, and reveals a much less volatile trend for 2000, 2004 and 2008. DMC per capita, adjusted for the construction minerals category, was 12 tonnes per capita in 2000, 10 tonnes per capita in 2004 and 12 tonnes per capita in 2008. This reflects a highly stable trend in DMC per capita, which could be evidence of decoupling in the economy, given the corresponding rises in GDP per capita that take place. Reasons for this could include the slowdown of Singapore's manufacturing sector and a shift towards the service sector, or an improvement in resource use productivity. The dip in DMC for 2004 is caused primarily by a decrease in net imports of fossil-fuel products, chemicals and plastics, which showed net declines of 2.6, 2.2 and 1.5 megatonnes respectively, although various other sectors, including metals, stone and glass, also showed net declines. An increase in the material weight of exports was responsible for these net declines, which is a potential indicator of several things happening in the economy.

The first and simplest explanation is that there was an increase in the amount of re-exports relative to exports of materials manufactured in Singapore. This could be proven once data on re-exports are available. It should be noted that exports increased without a corresponding increase in imports for these sectors or their raw materials, an observation with two possible implications that represent our second and third possibilities. The second is that there was a significant improvement in resource use productivity, permitting more materials to be manufactured from the same amount of imported raw materials. The third possibility is that Singaporeans themselves consumed less – a decrease in imports that was then compensated for by an increase either in raw materials imported or in the availability of goods manufactured in Singapore for export. The 2004 declines in these sectors were offset by moderate increases in net imports for animal products, wood and wood products, and transportation goods.

It is unclear what caused this dip in DMC per capita. In general, sectors experiencing declines in net imports between 2000 and 2004 exhibited a reversed trend between 2004 and 2008 – a change that was of a greater magnitude but in the opposite direction. The cause of the trend over the three different years of our analysis may potentially be explained by world economic activity. Basic economic theory dictates that, as incomes rise, consumer spending rises accordingly. Hence, for most countries, economic booms and busts result in corresponding increases and decreases in consumer spending that could potentially be reflected in changes in DMC per capita. Singapore's dependence on exports as a key component of its GDP, however, means that, as the world economy fluctuates – particularly in developed nations and China – so will Singapore's export levels. Therefore, even though there were recessions in developed nations in 2000 and 2008, Singapore's DMC per capita was high in these years due to depressed export levels. The converse is true in 2004, when the global economy was experiencing expansion.

VII.   Limitations and Future Research

Challenges and limitations in this study arose in the following forms: (a) Singapore's re-import and re-export data were unavailable. As a result, we were unable to separate the quantity of materials that only pass through Singapore from imports for domestic consumption and goods produced domestically for export. (b) As waste data were classified according to type of material, and import and export data were classified by product category, changes in the stock of each material category could not be calculated. (c) Data from UN Comtrade were obtained through voluntary reporting from individual countries. Since countries were not required to report the data they considered confidential, the material weights of traded goods might not be comprehensive. (d) Dissipative flows for Singapore, a part of the material outflows of the economy, were not included in this study. Niza *et al.* (2009) and Matthews *et al.* (2000) state that fertilizer and pesticide usage is the primary source of dissipative flows.

Because of the time-consuming nature of cataloguing and organizing UN Comtrade data, we were limited to conducting this MFA for just three years. We believe that further work needs to take into account re-exports and re-imports as well as to attempt to quantify inflows and outflows for specific material categories so that more details are available concerning from which sectors the flows entering and leaving Singapore originate.

VIII.   Conclusion

This study illustrates the utility of MFA in evaluating the material consumption trends of a dynamic economic system by assessing a series of material flow 'snapshots' over multiple years. Singapore, as a highly urbanized island city-state, provides a particularly interesting study site for both urban metabolism and island sustainability. One of the more distinctive findings of the study speaks to that point directly: as an island with limited land area, Singapore needs to import land itself, in the form of sand, in order to support its economic growth. That one figure alone caused great volatility across the study years. Using an MFA approach also uncovered a trend of increased impact from 'virtual water' imports, as, increasingly, Singaporeans appear to favour meat over vegetable products in their diets measured on a per capita basis. Finally, fluctuations in Singapore's import and export activity were shown to affect its DMC inversely, compared to global economic activity. Thus, this study has found that there are many lessons from an industrial ecology approach using MFA to measure the urban metabolism of Singapore.

Acknowledgements

The authors gratefully acknowledge a grant from the Ministry of National Development of Singapore to the National University of Singapore, enabling the Yale team to work with Prof. KUA, Harn Wei and Prof. Malone LEE, Lai Choo of the Center for Sustainable Asian Cities on urban metabolism in the Jurong Lakes District related to the overall theme of 'Planning and Development for Sustainable High Density Living'. Advice was also received from Prof Thomas GRAEDEL, Barbara RECK and BinBin JIANG at Yale at various stages of the preparation.

References

Abou-Abdo T, Davis NR, *et al.* (2011) *Dynamic Modeling of Singapore's Urban Resource Flows: Historical Trends and Sustainable Scenario Development.* International Symposium on Sustainable Systems and Technology (ISSST), Chicago, Illinois

Arnold W, Fuller T. Neighbor leaves Singapore short of sand. *The New York Times.* [Article on the Internet] March 16, 2007. [Cited 28 Apr 2010]. Available from: http://www.nytimes.com/2007/03/16/world/asia/16singapore.html.

Ayers R, Ayres L (eds) (2002) *Handbook of Industrial Ecology.* Cheltenham, U.K and Northampton, Massacusetts, USA: Edward Elgar Publishers.

Brunner, P H, Rechberger, H (2004) *Practical Handbook of Material Flow Analysis,* Boca Raton, FL, USA: CRC Press, 5–8.

Chan WW, Chang G, Low M (2005) *Second Quarter 2005 Review.* Report. Building and Construction Authority, Singapore.

Cheong A. Corporate: Searching for Sand. *The Edge.* [Article on the Internet] November 2, 2009. [Cited 28 Apr 2010]. Available from: http://www.theedgemalaysia.com/features/154464-searching-for-sand.html.

Deschenes PJ, Chertow M (2004) An island approach to industrial ecology. *Journal of Environmental Planning and Management* **47(2),** 201–17.

Eurostat (2001) *Economy-wide Material Flow Accounts and Derived Indicators. A Methodological Guide.* European Community, Luxembourg.

Globalis. An interactive world map based onUN Common Database from UN Population Division [Article on the Internet] 2010. [Cited 27 Apr 2010]. Available from: http://globalis.gvu.unu.edu/indicator_detail.cfm?Country=SG&IndicatorID=30#row.

Graedel T (1999) Industrial ecology and the ecocity. *Bridge* **29(4),** 4–9.

Graedel T, Allenby B (2009) *Industrial Ecology and Sustainable Engineering.* Prentice Hall, Saddle River NJ.

Guerin B. The shifting sands of time – and Singapore. *Asia Times.* [Article on the Internet] July 31, 2003. [Cited 28 Apr 2010]. Available from: http://www.atimes.com/atimes/Southeast_Asia/EG31Ae01.html.

Henderson B. Singapore accused of launching 'Sand Wars'. *The Telegraph.* [Article on the Internet] February 12, 2010. [Cited 28 Apr 2010]. Available from: http://www.telegraph.co.uk/news/worldnews/asia/singapore/7221987/Singapore-accused-of-launching-Sand-Wars.html.

Hendriks C, Obernosterer R, Müller D, Kytzia S, Baccini P, Brunner PH (2000) Material flow analysis: A tool to support environmental policy decision making. Case-studies on the city of Vienna and the Swiss lowlands. *Local Environment* **5(3),** 311–28.

Hoekstra AY, Chapagain AK (2007) Water footprints of nations: Water use by people as a function of their consumption pattern. *Water Resources Management* **21**, 35–48.

International Trade Centre. Difference between the Standard International Trade Classification (SITC) and the Harmonized System (HS) [Article on the Internet] 2010. [Cited 28 Apr 2010]. Available from: www.intracen.org/mas/sitchs.htm.

Kannan R, Leong KC, Osman R, Ho HK (2004) LCA-LCCA of oil fired steam turbine power plant in Singapore. *Energy Conversion and Management* **45(18–19)**, 3093–107.

Kannan R, Leong KC, Osman R, Ho HK (2005) Life cycle energy, emissions and cost inventory of power generation technologies in Singapore. *Renewable and Sustainable Energy Reviews* **11(4)**, 702–15.

Kannan R, Leong KC, Osman R, Ho HK (2004) LCA-LCCA of oil fired steam turbine power plant in Singapore. *Energy Conversion and Management* **45(18–19)**, 3093–107.

Kannan R, Leong KC, Osman R, Ho HK, Tso CP (2006) Life cycle assessment study of solar PV systems: An example of a 2.7 kWp distributed solar PV system in Singapore. *Solar Energy* **80(5)**, 555–63.

Kelly R. Land reclamation completed for Jurong Island. *Channel Newsasia*. [Article on the Internet] 25 Sept 2009. [Cited 28 Apr 2010]. Available from: http://www.channelnewsasia.com/stories/singaporebusinessnews/view/1007282/1/.html.

Kennedy C, Cuddihy J, Engel-Yan J (2007) The changing metabolism of cities. *Journal of Industrial Ecology* **11(2)**, 43–59.

HWLee KY (2000) *From Third World to First*. HarperCollins, New York.

Matthews E, Bringezu S, Fischer-Kowalski M, *et al.* (2000) *The Weight of Nations: Material Outflows from Industrial Economies*. World Resources Institute, Washington, DC.

Niza S, Rosado L, Ferrão P (2009) Urban metabolism: Methodological advances in urban material flow accounting based on the Lisbon case study. *Journal of Industrial Ecology* **13(3)**, 384–405.

Organisation for Economic Co-operation and Development (2008) *Measuring Material Flows and Resource Productivity: Synthesis Report*. OECD, Paris, France.

Olds K, Yeung HWC (2004) Pathways to global city formation: A view from the developmental city-state of Singapore. *Review of International Political Economy* **11(3)**, 489–521.

Schroeder TC, Barkley PB, Schroeder KC (1996) Income growth and international meat consumption. *Journal of International Food and Agribusiness Marketing* **7(3)**, 15–30.

Schulz NB (2007) The direct material inputs into Singapore's development. *Journal of Industrial Ecology* **11(2)**, 117–31.

United Nations Department of Economic and Social Affairs (2002) *World Population Prospects: The 2001 Revision, Data Tables and Highlights.* UNDESA publication sales no. (ESA/P/WP.173). UN, New York.

Vidal J. Meat-eaters soak up the world's water. *The Guardian.* [Article on the Internet] XxxAugust, 23 2004. [Cited 3 May 2010]. Available from: http://www.guardian.co.uk/environment/2004/aug/23/water.famine.

Wernick IK, Irwin FH (2005) *Material Flow Accounts: A Tool for Making Environmental Policy.* World Resources Institute, Washington, D. C.

White R (1994) Preface. In: Allenby B, Richards D (eds) *The Greening of Industrial Ecosystems*, National Academy Press, Washington, DC, v–vi.

Wolman A (1965) The metabolism of cities. *Scientific American* **213(3)**, 179–90.

World Resources Institute WRI.. Climate and atmosphere: Searchable database [Article on the Internet] 2007. [Cited 28 Apr 2010]. Available from: http://earthtrends.wri.org/searchable_db/index.php?action=select_variable&theme=3.

Yang PP, Ong BL (2004) Applying ecosystem concepts to the planning of industrial areas: A case study of Singapore's Jurong Island. *Journal of Cleaner Production* **12(8–10)**, 1011–23.

Biodiversity Challenges: Nature Conservation

CHAPTER SIX

Challenges to Southeast Asia's Status as the Global Coral Reef Hotspot

CHOU Loke-Ming
Department of Biological Sciences, National University of Singapore

"Regional response to the protection of its rich coral reef heritage has been slow, a pace which confirms an ignorance of its high economic value and what it means to possess the richest and most productive of coral reefs in the world. The threats have so far been allowed to operate and management capacity, while strong in some localities remains weak for the region. The opportunities lie with the economic potential of the ecosystem services that coral reefs provide but protection priorities are not given the required urgency."

– *CHOU Loke-Ming*

Challenges to Southeast Asia's Status as the Global Coral Reef Hotspot

CHOU Loke-Ming[1]

Department of Biological Sciences, National University of Singapore

Abstract

Southeast Asia is recognized as the global hotspot for coral reefs. An estimated 28 per cent of the world's reefs are contained in its seas, which cover only 2.5 per cent of Earth's ocean surface. Species richness of scleractinian corals and other reef-associated flora and fauna is the highest globally. Rapid economic expansion since the mid-1950s and a fast-growing coastal human population added tremendous pressures to the region's reef ecosystem, resulting in a rich natural heritage severely threatened by continued degradation and habitat loss. Further impacts from climate change are expected to significantly exacerbate the situation. The present management challenges have to be adequately addressed if the region is to retain its reef biodiversity hotspot position. An analysis of the strengths, weaknesses, threats and opportunities of Southeast Asia's coral reefs is presented to give a better understanding and appreciation of the significance and value of this reef hotspot status to the region, and the consequences of its loss.

Key words: coral reef, global hotspot, degradation, biodiversity, management challenges, Southeast Asia

I. Introduction

Southeast Asia is mostly tropical and situated across the equator between latitudes 21°N to 12°S and longitudes 93°E to 141°E. The world's two largest archipelagos, Indonesia and the Philippines, have approximately 25,000 islands. Almost all of the region's nations on the Asian continent possess extensive coastlines and numerous offshore islands, many of which are coral or volcanic. The combined coastline length of about 92,500 kilometres is almost 16 per cent of the world's total. This extensive coastline with different geomorphological features, such as cliffs, coves, beaches (rocky, sandy, muddy), deltas, spits, dunes and lagoons, results in a huge diversity of coastal and marine habitats.

The region's seas separate the continents of Asia and Australia but link the Pacific and Indian Oceans. The marine environment is characterized by the extensive shallow Sunda and Sahul continental shelves and also by deep basins, trenches,

---

[1] Department of Biological Sciences, National University of Singapore. Blk S3, 14 Science Drive 4, Singapore 117543. Email: dbsclm@nus.edu.sg; Tel: (+65) 6516 2696; Fax: (+65) 6779 2486.

troughs and continental slopes in between. Tides and currents are influenced by the seasonal monsoons, and heavy precipitation transports vast amounts of terrestrial nutrients to the sea. The elevated marine-nutrient level is further facilitated by the large number of islands distributed throughout, and only the offshore areas of some of the larger seas have a low nutrient input.

The great diversity of coastal and marine habitats combines with the warm tropical climate to favour and support very high species richness. The coral reef ecosystem is extensive, and species richness of corals and other reef-associated flora and fauna is the highest throughout the world, making Southeast Asia the global hotspot for coral reefs (Kelleher et al., 1995).

The reefs, however, face enormous pressures from the rapidly expanding human population and strong economic growth, especially in coastal areas. Economic development remained the clear focus of Southeast Asian states, with the pace intensifying from the 1960s (UNEP/COBSEA, 2010). Economic growth generated mainly by industrialization and international trade remained high during the 1980s and early 1990s (JEC, 2000). The threats continue to increase reef habitat loss through destruction and degradation through pollution and over-exploitation. Climate change is expected to have a huge impact on coral reefs, particularly by warming seas that are also becoming more acidic. The region's coral reef biodiversity is already under high threat from anthropogenic impacts that is compromising the habitat's generation of high levels of ecosystem services.

The strengths, weaknesses, threats and opportunities of this rich, natural heritage are analyzed to highlight its importance and emphasize the need for greater attention and stronger management.

II. Strengths

Southeast Asia's varied coastal geomorphology and oceanographic features, coupled with the warm climate, contribute to highly productive seas that support rich and extensive coral reefs. The total marine area of 9 million square kilometres occupies only 2.5 per cent of Earth's ocean surface but holds 28 per cent of the world's coral reef ecosystem (Burke et al., 2011). All reef types are represented, with fringing reefs dominating. The region is recognized as the faunistic centre of the entire Indo-Pacific (IUCN/UNEP, 1985). It has about 80 per cent of the world's hard coral species (Spalding et al., 2001), and this coral diversity remains the highest throughout the world (Veron, 1995). Almost all known living hard corals throughout the Indo-Pacific are represented in the region. The eastern half from Borneo lies in what is known as the Coral Triangle, which has the highest known diversity of corals (Veron et al., 2009), fish (Carpenter & Springer, 2005) and many other reef-associated groups throughout the world. The reef ecosystem supports a high diversity of associated species of plants and animals, contributing further to the region's status as the global centre of diversity for marine invertebrates, such as molluscs and crustaceans (Briggs, 1974).

Many species are endemic to the region in spite of marine connectivity. The highest species concentration of a common reef gastropod *Strombus* occurs in an area spanning Okinawa, the Philippines and Indonesia, with a gradient of declining diversity to the east and west (Abbott, 1960). The pattern of declining species diversity away from the region is repeated for many groups of reef invertebrates and fish. The species number of a common family of reef fish, the pomacentrids, diminishes east of Southeast Asia (Allen, 1985). The situation is repeated for inshore fishes, many of which are associated with coral reefs, and species decline east of the region suggests that Southeast Asia is the source of larval recruits to the Pacific Ocean (Myers, 1991).

Southeast Asia's position as a global hotspot for reef biodiversity is unparalleled, and numerous species remain undiscovered. Species new to science continue to be described, such as the hard corals, *Acropora suharsonoi* from Lombok, Indonesia, and *Acropora kosurini* from Surin Islands, Thailand (Wallace 1994). The 1998 discovery of the coelacanth from the reefs of Manado, Indonesia, (Erdmann *et al.*, 1998) is probably one of the most significant finds in recent time. It highlights the wealth of diversity in the region's reefs, much of which we still do not know. There is little doubt that the region's reefs hold a rich genetic resource with an enormous potential for the discovery of novel bioactive compounds.

As a global marine biodiversity hotspot, the region has significance globally as an evolutionary source (Briggs, 2005). Many species originated from and are present only in the region. Their loss translates simply into extinction on a global scale. This species uniqueness has to be guarded against further loss. Investigations into population genetics of reef organisms such as corals (Knittweis *et al.*, 2009), fish (Lourie & Vincent, 2004; Timm & Kochzius, 2008), crustaceans (Barber *et al.*, 2006), molluscs (Kochzius & Nuryanto, 2008) and echinoderms (Kochzius *et al.*, 2009) indicate high levels of genetic structuring with distinct signatures from the region.

III. Weaknesses

Both management and monitoring capacities are identified as major weaknesses that have resulted in the loss and degradation of the region's coral reef habitat. Strong pressures are exerted on marine resources, as economic development intensifies and management efforts have, on the whole, not been able to counter the widespread degradation. Although many national, bilateral and regional projects have been implemented, they have yet to make a sustained and coordinated contribution to the protection of coral reefs. Most projects are initiatives over fixed terms of less than a decade, more often less than five years, and all have not succeeded in maintaining the developed capacity beyond the projects' conclusions. Every new regional initiative goes through the process of rebuilding the momentum lost between projects and readjusting to new directions and agenda of different funding agencies.

A number of intergovernmental bodies focusing on marine environment management operate in the region, but membership is not common amongst Southeast Asian states. Some are not represented in a few of these regional arrangements, and regional targets are not easily achievable. Three of the longer-established intergovernmental bodies are the Association of Southeast Asian Nations (ASEAN), the Coordinating Body on the Seas of East Asia (COBSEA) and the Partnerships in Environmental Management for the Seas of East Asia (PEMSEA). All are concerned with the broader aspects of managing the marine environment and have a wider agenda. The protection of coral reefs is sometimes included, depending on opportunities, but does not always stay in full focus. The lack of a concerted and sustained regional effort to conserve coral reefs is identified as one of the main weaknesses.

Despite having almost 30 per cent of the world's coral reefs, only three areas with coral reefs are listed as UNESCO World Heritage Sites (Indonesia's Ujung Kulon National Park and Komodo National Park, and Philippines' Tubbataha Reef National Marine Park), showing a lack of international recognition. Many more reef areas, such as the Spratlys in the South China Sea (under territorial dispute between six states), North Borneo/Balabac Strait/Turtle Islands (Malaysia and the Philippines), Cagayan Ridge (the Philippines), Berau Islands (Indonesia), Semporna/Tawitawi (Malaysia), Raja Ampat (Indonesia), Banda and Lucipara (Indonesia), all fulfil the criteria for being listed as World Heritage sites, and their consideration will correct the conspicuous under-representation in the region. At the regional level, ASEAN Heritage Sites with coral reefs include the Philippines' Tubbataha Reef National Park, Myanmar's Lampi Marine National Park and Thailand's Tarutao National Park, Mu Ko Surin-Similan Marine National Park and Ao Phang-nga Marine National Park.

Marine Protected Areas (MPAs) are a common tool for the protection of marine biodiversity. Kelleher *et al.* (1995) indicated the existence of 233 MPAs, large and small, throughout the region. Of those for which data are available, about 13 per cent had a 'high' management level, 31 per cent had a 'moderate' management level, and the remaining 56 per cent had a 'low' management level (with 'high' indicating that the MPA generally met management objectives, and 'low' meaning the management objectives are not met). The number of MPAs increased to about 310 (declared) and 132 (proposed), as estimated by UP-MSI *et al.* (2002), and 443, as estimated by Burke *et al.* (2002), but both groups of authors assessed that only about 10 per cent of the region's MPAs were effectively managed. The most recent assessment indicates 599 MPAs, which cover 17 per cent of the region's reefs (Burke *et al.*, 2011). But more depressing is that only 2 per cent of the MPAs are rated as effective in terms of management, while 69 per cent are not. An earlier analysis by Burke *et al.* showed that only 8 per cent of the region's reefs lie within MPAs and that only 1 per cent were within MPAs that are effectively managed. The increasing number of MPAs is likely a response to the Convention of Biological Diversity's Global 2010 target of establishing a Marine Global Network and effective management of all protected areas by 2012, but the management effectiveness shows little improvement.

Coral reef monitoring throughout the region is still very much left to an informal network of coral reef researchers that was first established under the ASEAN–Australia Living Coastal Resources Project (1986–1994). The network made it possible for the region to contribute to the 'Status of Coral Reefs of the World' reports compiled for the Global Coral Reef Monitoring Network, which started in 1998 (Chou 1998, 2000). The status of Southeast Asia's coral reefs has been reported since then through an informal arrangement by the network (Chou et al., 2002; Tun et al., 2004, 2008). This informal process has advantages and disadvantages (Chou, 2011). Through goodwill, members contribute time and effort to compile national status reports to feed into the regional report. However, members find it increasingly difficult to continue doing this voluntarily because of other commitments and because gaps in regional reporting are inevitable. The sustainability of this informal arrangement also comes into question.

IV. Threats

The level of ecosystem services that support human well-being and economic development has declined from anthropogenic impacts, a trend that is common globally (UNEP, 2001; MA, 2005; ASEAN, 2006). Coastal population is the highest compared with other reef regions worldwide, with half (140 million) of the world's population living within 30 kilometres of coral reefs present in Southeast Asia, making the region's reefs the world's most threatened (Burke et al., 2011). Compared with other reef regions of the world, local threats in Southeast Asia are the most severe. As a result, reef biodiversity has suffered significant loss and degradation (UNEP/COBSEA, 2010).

Overfishing and destructive fishing are the largest threats to Southeast Asia's reefs (Burke et al., 2011). Coastal development and marine pollution also pose serious threats as well. Towards the last decade of the previous century, decimation of the region's rich reef system caused the collapse of an estimated 11 per cent, with a further 48 per cent under high threat of collapse predictably within two decades in the absence of management intervention (Wilkinson et al., 1993). Reef degradation continued into this century, with 88 per cent under high risk, (Burke et al., 2002), with the most recent assessment indicating that 95 per cent are now threatened, with 50 per cent in the 'very high' and 'high' threat categories (Burke et al., 2011). Future projections, taking into account climate change impacts and assuming that local pressure remains unchanged, place 82 per cent of reefs under 'high', 'very high' and 'critical' categories by 2030 and increasing to 95 per cent by 2050.

Many coral, fish, gastropod and lobster species have restricted geographic ranges (Roberts et al., 2002) and remain at high risk of extinction from localized reef degradation. Biodiversity hotspots throughout the Indo-Australian Archipelago are arbitrarily defined, as studies have been geographically patchy and unevenly distributed, with information gaps that imply that these 'recognized' hotspots may not necessarily support high numbers of endemics or serve as a speciation source.

Conservation of areas beyond these hotspots is therefore important (Bellwood & Meyer, 2009). Weeks *et al.* (2010) also recognized from their assessment of community-based MPAs in the Philippines that these small, protected areas should be supplemented by larger no-take reserves for biodiversity conservation targets to be more effectively met.

The link between climate change and biodiversity is well-established (IPCC, 2007). Marine biodiversity is exposed to a wide range of climate change impacts that include sea level rise, elevated sea temperature, increased frequency of extreme weather, intensification of water column stratification and ocean acidification (Chou, 1994). These impacts generate accompanying environmental effects on coastal and marine systems through coastal erosion, sudden salinity fluctuation, increased sedimentation, nutrient loading, saltwater intrusion, coastal inundation, and changes in coastal geomorphology and circulation patterns (Chou, 1992). Coral reefs are particularly vulnerable to these impacts (SCBD, 2009), and the region stands to lose much as a global coral reef hotspot. Although it is not easy to separate climate change impacts from local anthropogenic impacts, some idea of the former can be derived from current events that simulate climate change scenarios (Chou, 2010).

Yap (1994) reviewed the implications of expected climate change impacts on natural coastal ecosystems in the region. Climate change impacts on individual species vary, and impacts on biodiversity occur at species and community levels. For species, physiological constraints limit their tolerance to environmental fluctuations. For communities, an altered community structure affects ecosystem functioning. Motile species can migrate to new areas with more suitable conditions, while sessile and sedentary ones either adapt or perish. However, many are not expected to adapt to the rate and intensity of projected climate change scenarios and risk extinction (SCBD, 2010). A likely outcome is a possible decrease in equatorial biodiversity that compromises ecosystem resilience and services.

Rising sea levels inundating low-lying coastal plains can overwhelm adaptation response of coastal biomes, and temperature elevation is expected to elicit physiological and behavioural responses from species that could be detrimental to the entire biological communities and ecosystem integrity. Increased warming of the seas makes them inhospitable to corals and reef-associated species, while sudden increased precipitation adds further challenges to the tolerance limits of these species.

Increasing acidity from greater dissolution of carbon dioxide is detrimental particularly to corals, shellfish and plankton, as it interferes with the efficient production of their protective calcium carbonate skeletons. Coral reefs in particular are highly vulnerable to lowered ocean pH (Hoegh-Guldberg *et al.*, 2007), and the region with its high proportion of the world's reefs will suffer a greater loss. Should atmospheric CO_2 concentrations reach 500 parts per million after 2050, as projected by Rogeli *et al.* (2009), coral growth will be arrested as dissolution takes place (Silverman *et al.*, 2009). Coral reefs exposed to low-pH conditions over the long term at cool, shallow-water carbon dioxide seeps in Papua New Guinea had lower coral diversity, less structural complexity and reduced recruitment (Fabricius *et al.*, 2011). But the study showed that live-coral cover remained the same between sites with pH 8.1 (present condition with atmospheric carbon dioxide at 390 parts per million) and 7.8 (expected when atmospheric carbon dioxide reaches 750 parts per million by the end of this century). This suggests that some species can tolerate increased acidity, but the overall reduced diversity and complexity will compromise reef resiliency. Reef development, however, is arrested when pH drops below 7.7, and will push coral reefs well over the tipping point when coupled with other climate change impacts like warming. The death of coral reefs will be significant regions where millions of people depend on them for subsistence.

Two impacts related to climate change that have manifested since the late 1990s give some indication of its consequences to coral reefs. These are elevated sea surface temperatures and the prolonged lowering of salinity from sudden increased precipitation. Again, these give some idea of a linear cause–effect relationship and do not take into account the synergistic effects of multiple impacts.

Elevated Sea-Surface Temperature

The 1998 El Niño phenomenon affected coral reefs worldwide at an unprecedented scale. Warming sea surface temperature from mid-1997 to the late 1998 triggered the mass bleaching of corals throughout the world and drew attention to the urgency of protecting reef resiliency. Mortality of shallow water corals was as high as 95 per cent in some parts of the world, while no mortality was recorded in some reefs (Wilkinson & Hodgson, 1999). Management is needed to prevent degradation of reef system integrity.

Moderate to extensive bleaching occurred throughout Southeast Asia at a level previously unknown (Chou *et al.*, 2000). Indonesian reefs started bleaching in early 1998 in West Sumatra, resulting in over 90 per cent mortality. It spread to other reefs throughout the country, causing the loss of live-coral cover that ranged from 30 to 90 per cent. Recovery was variable after a few years, with some reefs showing overall live-coral cover decline between 10 and 40 per cent. Mean live-coral cover in the Philippines decreased by 19 per cent after the 1998 bleaching in Tubbataha, with no further loss or recovery two years later. At Danjugan Island in Negros Occidental, where coral mortality from the bleaching was high, recovery was observed over the next two years. In the Gulf of Thailand, widespread bleaching of shallow reefs occurred, while corals on pinnacles extending from deeper water were spared. The impact resulted in local extinctions of some *Acropora* species. Recovery in the inner Gulf was lengthened because of the low coral recruitment, but, along the east and west coasts, the higher availability of coral recruits facilitated recovery. Recovery was slow even after two years in Vietnam's Con Dao islands, where 37 per cent of coral colonies bleached. In Singapore, the bleaching affected all hard coral species and some species of soft coral and colonial sea anemones. Recovery commenced with the lowering of sea surface temperature to normal, and mortality was restricted to 20 per cent.

The 1997/1998 mass bleaching was the most severe ever observed, and raised the question of whether it was an isolated event or if similar occurrences would follow at greater frequency with the manifestation of global warming (Wilkinson & Hodgson, 1999). In early 2010, the sea surface temperature increased to levels higher than those in 1997/1998 and triggered mass bleaching once again. The region experienced widespread bleaching at a severity similar to or greater than the 1997/1998 event (Tun *et al.*, 2011). Initial mortality estimates of between 10 and 90 per cent were reported, but reefs in the oceanic eastern part of Indonesia mostly escaped the bleaching. Malaysia and Thailand both responded by closing off popular dive sites to minimize visitor pressure and to allow severely bleached reefs to recover. In Singapore, recovery was faster than the previous bleaching event, and mortality was less than 10 per cent. Monitoring is still in progress, and the full impact of the 2010 bleaching will become clearer over the next few years.

Sudden Salinity Decrease

Extreme weather events are expected to occur more frequently as global warming intensifies. Periods of drought alternating with intense rainfall will result in rapid and wide salinity fluctuations that are detrimental to intertidal reef-associated species and corals. Such an event occurred at Chek Jawa of Pulau Ubin, one of Singapore's offshore islands (Chou, 2010). Unusually heavy and sustained rainfall at the end of 2006 and beginning of 2007 caused a sudden drop of salinity that killed many soft-bodied invertebrates, such as carpet anemones, sea stars and sea cucumbers, and induced the bleaching of hard corals on the intertidal flat.

Recovery occurred as weather conditions returned to normal and reef-associated species that suffered the mass kill re-established and increased in abundance. However, the impact allowed the establishment of other species not previously known from the site. The Asian mussel *Musculista senhousia*, a native of northern Southeast Asia, spread to Singapore as an introduced species. Its presence was previously unrecorded at Chek Jawa, and the salinity loss that temporarily killed many species presented a chance for this introduced species to be established within a few months after the environmental disturbance (Loh, 2008). The opportunistic colonization of Chek Jawa by this mussel demonstrated the potential of an invasive species to dominate a biodiversity-rich natural habitat following an abrupt environmental change and to disrupt recovery of the original community structure.

V. Opportunities

Coral reefs are recognized for their valuable ecosystem services and, if well-managed, are indeed a significant asset to Southeast Asia. The socio-economic significance is high, as coastal population is greatest in the region. Indonesia and the Philippines alone have 100 million people living within 30 kilometres of a reef (Burke *et al.*, 2011). Well-established ecosystem services provided by coral reefs include fisheries, tourism and coastal protection.

Coral reefs can potentially supply 12 per cent of the world's fish catch (Munro & Williams, 1985), and their contribution throughout the region is significant, as fish forms a major protein source for the region's population. Each square kilometre of a healthy reef can provide up to 15 metric tonnes of seafood a year (Jennings & Polunin, 1995; Newton *et al.*, 2007). The Philippine reefs in excellent condition generate 18 metric tonnes per square kilometre annually (McAllister, 1998). In Sabah, East Malaysia, reef fishes make up 25 per cent of the total fish catch (Mathias & Langham, 1978). In Terengganu, West Malaysia, the figure reaches 30 per cent in certain months (De Silva & Rahman, 1982). Well-managed reefs have a valuable role in the provision of food security.

Effectively managed reefs attract dive tourists, and the diversity of reefs and their wealthy reef life make Southeast Asia a popular dive destination. The revenue generated can be substantial. Annual tourism incomes in US dollars from some protected coral reef sites have been estimated (CI, 2008): Indonesia's Pulau Weh MPA – $230,000, Bunaken National Park – $110,000, Wakatobi National Park – $286,000; Malaysia's Pulau Payar – $390,000, Pulau Redang – $545,000; Vietnam's Hon Mun MPA – $243,190; the Philippines' Lingayan Gulf – $4.7 million, Bohol Marine Triangle – $5.06 million. All these values emphasize that good reefs have both the drawing and earning power that can support management costs and benefit local communities.

It is estimated that 150,000 kilometres of shoreline throughout the world receive some protection from coral reefs (Burke *et al.*, 2011), but coastal protection is an important ecosystem service that is not fully appreciated. Reef degradation exposes shores to the full energy of strong waves, which result in beach erosion. This connection became apparent after the 2004 Asian tsunami, where the high waves swept further inland behind degraded reefs (Wilkinson *et al.*, 2006). In Bali, USD1 million was spent to protect a 500-metre shore, a function that could be provided for free if the reef was not degraded (Cesar, 1996). Hodgson and Dixon (1988) showed that the logging industry at El Nido in Palawan, which could generate $8.6-million gross revenue over a decade, would, in the same period, cause a loss of $6.2 million from fisheries and $13.9 million from tourism because of the impact, mainly sedimentation, to the coral reefs.

It is evident that coral reefs do provide economic, food and social security to coastal communities and that the high economic valuation does not come as a surprise. For the Philippines and Indonesia, the two countries in the region with the most reefs, the annual benefits are estimated to be $258 million from tourism, $2.2 million from reef fisheries and $782 million from shoreline protection (Burke *et al.*, 2002).

VI. Conclusion

Regional response to the protection of its rich coral reef heritage has been slow, a pace that confirms an ignorance of its high economic value and what it means to possess the richest and most productive of coral reefs in the world. The threats have, so far, been allowed to operate, and management capacity, while strong in some localities, remains weak for the region. The opportunities lie with the economic potential of the ecosystem services that coral reefs provide, but protection priorities are not given the required urgency. A number of countries have responded to the call by the Convention on Biological Diversity to establish more MPAs as a means of providing better protection of coral reefs, but the real issue is that of management effectiveness.

Southeast Asia's coral reefs can boost ecotourism and provide good economic returns, and stronger management is needed to prevent further degradation. Various reef management modes ranging from government- to community-level efforts have emerged. Strong legal and institutional structures to promote compliance are required to manage reefs at a national or state level. MPAs lying within the larger management umbrella of an Integrated Coastal Management (ICM) framework have positive effects on coral reefs, but this may not necessarily be so for community-managed no-take zones (Christie, 2005). Chou *et al.* (2010) highlighted the need for integrated management to slow the pace of reef degradation along the northern coast of Bintan Island (Indonesia). The Regional Programme PEMSEA under the Global Environment Facility (GEF), United Nations Development Programme (UNDP) and International Maritime Organisation (IMO) has been active in promoting the adoption of ICM as a holistic management approach that should give coastal ecosystems, including coral reefs, more effective protection.

There are, however, some signs of the reef degradation rate slowing down, accompanied by recovery and improvement of reef conditions (Tun *et al.*, 2008; UNEP/COBSEA, 2010). Community-based management has worked effectively at local levels (Christie *et al.*, 2002; Alcala & Russ, 2006) and is widely replicated throughout the Philippines.

Climate change impacts pose great threats to the region's reefs, and a sensible response is to strengthen management in a sustained effort to nurse degraded reefs back to health and simultaneously prevent further degradation of healthy reefs. This will improve and toughen reef ecosystem resilience, which can minimize serious damage from climate change forces (SCBD, 2010). Although some argue that climate change impacts will eliminate all coral reefs because the anthropogenic carbon dioxide loading of the atmosphere is too rapid and causes environmental shifts that are too sudden, doing nothing is not really an option. The immediate task is to reduce present anthropogenic pressure and improve management. This is a great challenge for the region, which has relied much on reef ecosystem services, but, at the same time, compromises its ability to continue doing so.

Reef restoration is an option that cannot be ignored in view of the widespread degradation that has affected many reefs. Various techniques and approaches have been investigated and tested (Edwards, 2010) and implemented (Chou *et al.*, 2009). These include low-cost approaches to aid reef recovery from blast fishing, where rubble stabilization and rock piles were found to encourage better coral recruitment and growth compared with scattered rubble (Fox *et al.*, 2005; Raymundo *et al.*, 2007).

Various regional projects have been initiated or concluded recently, such as the UNEP/GEF South China Sea project (UNEP, 2004), which has established reef demonstration sites for reversing degradation trends. The Coral Triangle Initiative is an ongoing regional project, with partners seeking to protect coral reefs in the very centre of reef biodiversity (stretching from eastern Indonesia to Solomon Islands). These, as well as past regional projects, have strong lessons on what works and what doesn't, and factors contributing to the successful management or rehabilitation need to be replicated to slow down coral reef loss, reverse the declining trend and allow the region to benefit from the status of being the global hotspot for coral reefs.

References

Abbott RT (1960) The genus *Strombus* in the Indo-Pacific. *Indo-Pacific Mollusca* **1**, 33–146.

Alcala AC, Russ GR (2006) No-take marine reserves and reef fisheries management in the Philippines: A new people power revolution. *Ambio* 35, 245–54.

Allen GR (1985) *Butterfly and Angelfishes of the World, Vol. 2.* Mergus Publishers, Melle.

ASEAN (2006) *Third ASEAN State of the Marine Environment Report 2006.* ASEAN Secretariat, Jakarta.

Barber PH, Erdmann MV, Palumbi SR (2006) Comparative phylogeography of three codistributed stomatopods: Origins and timing of regional lineage diversification in the coral triangle. *Evolution* 60, 1825–39.

Bellwood DR, Meyer CP (2009) Searching for heat in a marine biodiversity hotspot. *Journal of Biogeography* **36**, 569–76.

Briggs JC (1974) *Marine Zoogeography.* McGraw-Hill, New York.

Briggs JC (2005) Coral reefs: Conserving the evolutionary sources. *Biological Conservation* **126**, 297–305.

Burke L, Selig E, Spalding M (2002) *Reefs at Risk in Southeast Asia.* World Resources Institute, Washington, DC.

Burke L, Reytar K, Spalding M, Perry A (2011) *Reefs at Risk Revisited.* World Resources Institute, Washington, DC.

Carpenter KE, Springer VG (2005) The center of the center of marine shore fish biodiversity: The Philippine Islands. *Environmental Biology of Fishes* **72**, 467–80.

Cesar HSJ (1996) *Economic Analysis of Indonesian Coral Reefs.* Working Paper Series 'Work in Progress', World Bank, Washington, DC.

Chou LM (1992) Potential impacts of climate change and sea-level rise on the East Asian Seas Region. *Journal of Southeast Asian Earth Sciences* **7**, 61–4.

Chou LM (ed) (1994) *Implications of Expected Climate Changes in the East Asian Seas Region: An Overview.* RCU/EAS Technical Report Series No. 2. Regional Coordinating Unit, East Asian Seas, UNEP, Bangkok.

Chou LM (1998) Status of Southeast Asian coral reefs. In: Wilkinson CR (ed) *Status of Coral Reefs of the World: 1998,* pp. 79–87. Australian Institute of Marine Science, Townsville.

Chou LM (2000) Southeast Asian reefs – Status update: Cambodia, Indonesia, Malaysia, Philippines, Singapore, Thailand and Vietnam. In: Wilkinson CR (ed) *Status of Coral Reefs of the World: 2000,* pp. 117–29. Australian Institute of Marine Science, Townsville.

Chou LM (2010) Anticipated impacts of climate change on marine biodiversity: Using field situations that simulate climate change in Singapore. In: Sajise PE, Ticsay MV, Saguiguit Jr. GC (eds) *Moving Forward, Southeast Asian Perspectives on Climate Change and Biodiversity,* pp. 131–8. Southeast Asian Regional Center for Graduate Study and Research in Agriculture and Institute of Southeast Asian Studies, Singapore.

Chou LM (2011) Regional coral reef monitoring and reporting through an informal mechanism – the experience of Southeast Asia. In: *Status of Coral Reefs in East Asian Seas Region: 2010,* pp. 117–8. Ministry of the Environment, Tokyo.

Chou LM, Huang DW, Tun KPP, Kwik JTB, Tay YC, Seow AL (2010) Temporal changes in reef community structure at Bintan Island (Indonesia) suggest need for integrated management. *Pacific Science* **64**, 99–111.

Chou LM, Vo ST, PhilReefs, *et al.* (2002) Status of Southeast Asia coral reefs. In: Wilkinson C (ed) *Status of Coral Reefs of the World: 2002,* pp. 123–52. Australian Institute of Marine Science, Townsville.

Chou LM, Yeemin T, Abdul Rahim GY, Vo ST, Alino P, Suharsono S (2009) Coral reef restoration in the South China Sea. *Galaxea* **11**, 67–74.

Christie P (2005) Observed and perceived environmental impacts of marine protected areas in two Southeast Asia sites. *Ocean and Coastal Management* **48**, 252–70.

Christie P, White A, Deguit E (2002) Starting point or solution? Community-based marine protected areas in the Philippines. *Journal of Environmental Management* **66**, 441–54.

Conservation International (2008) *Economic Values of Coral Reefs, Mangroves and Seagrasses: A Global Compilation.* Center for Applied Biodiversity Science, Conservation International, Arlington.

De Silva MWRN, Rahman RA (1982) *Management Plan for the Coral Reefs of Pulau Paya/Segantang Group of Islands.* Faculty of Fisheries and Marine Science, University Pertanian Malaysia, and World Worldlife Fund, Malaysia.

Edwards AJ (ed) (2010) *Reef Rehabilitation Manual.* Coral Reef Targeted Research & Capacity Building for Management Program, St. Lucia, Australia.

Erdmann MV, Caldwell RL, Moosa MK (1998) Indonesian 'king of the sea' discovered. *Nature* **395,** 335.

Fabricius KE, Langdon C, Uthicke S, *et al.* (2011) Losers and winners in coral reefs acclimatized to elevated carbon dioxide concentrations. *Nature Climate Change* **1,** 165–9.

Fox HE, Mous PJ, Pet JS, Muljadi AH, Caldwell RL (2005) Experimental assessment of coral reef rehabilitation following blast fishing. *Conservation Biology* **19,** 98–107.

Hodgson G, Dixon LA (1988) Logging versus fisheries and tourism in Palawan. Occasional Papers No. 7. East-West Environment and Policy Institute, Honolulu.

Hoegh-Guldberg O, Mumby PJ, Hooten AJ, *et al.* (2007) Coral reefs under rapid climate change and ocean acidification. *Science* **318,** 1737–42.

IPCC (2007) *Climate Change 2007: Impacts, Adaptation and Vulnerability. Contribution of Working Group II to the Fourth Assessment Report of the Intergovernmental Panel on Climate Change.* Parry ML, Canziani OF, Palutikof JP, van der Linden PJ, Hanson CE (eds). Cambridge University Press, Cambridge.

IUCN/UNEP (1985) *Management and Conservation of Renewable Marine Resources in the East Asian Seas Region.* UNEP Regional Seas Reports and Studies No. 65. United Nations Environment Programme, Nairobi.

JEC (2000) *The State of the Environment in Asia 1999/2000.* Japan Environmental Council, Springer-Verlag, Tokyo.

Jennings S, Polunin NVC (1995) Comparative size and composition of yield from six Fijian reef fisheries. *Journal of Fish Biology* **46,** 28–46.

Kelleher G, Bleakley C, Wells C (1995) *A Global Representative System of Marine Protected Areas,* Vol. 3. The World Bank, Washington, DC.

Knittweis L, Kraemer WE, Timm J, Kochzius M (2009) Genetic structure of *Heliofungia actiniformis* (Scleractinia: Fungiidae) populations in the Indo-Malay Archipelago: Implications for live coral trade management efforts. *Conservation Genetics* **10,** 241–9.

Kochzius M, Nuryanto A (2008) Strong genetic population structure in the boring giant clam, *Tridacna crocea*, across the Indo-Malay Archipelago: Implications related to evolutionary processes and connectivity. *Molecular Ecology* **17**, 3775–87.

Kochzius M, Seidel C, Hauschild J, *et al.* (2009) Genetic population structures of the blue starfish *Linckia laevigata* and its gastropod ectoparasite *Thyca crystallina. Marine Ecology Progress Series* **396**, 211–9.

Loh KS (2008) Life and death at Chek Jawa. *Nature Watch* **16**, 18–23.

Lourie SA, Vincent ACJ (2004) A marine fish follows Wallace's Line: The pylogeography of the three-spot seahorse (*Hippocampus trimaculatus*, Syngnathidae, Teleostei) in Southeast Asia. *Journal of Biogeography* **31**, 1975–85.

MA (2005). Ecosystem and human well-being: Current status and trends assessment. In: *Coastal Systems*, pp. 515–49. Millennium Ecosystem Assessment, Washington, DC.

Mathias JA, Langham NPE (1978) Coral reefs. In: Chua TE, Mathias JA (eds) *Coastal Resources of West Sabah*, pp. 117–51. University Sains Malaysia, Penang.

McAllister DE (1988) Environmental, economic and social costs of coral reef destruction in the Philippines. *Galaxea* **7**, 161–8.

Munro JL, Williams DM (1985) Assessment and management of coral reef fisheries: Biological, environmental and socioeconomic aspects. In: Gabrie C & Salvat B (ed) *Proceedings of the Fifth International Coral Reef Congress 4*, pp. 543–81. Antenne Museum-EPHE, Moorea, French Polynesia.

Myers RF (1991) *Micronesian Reef Fishes: A Practical Guide to the Identification of the Inshore Marine Fishes of the Tropical Central and Western Pacific.* Second Edition. Coral Graphics, Guam.

Newton K, Côté IM, Pilling GM, Jennings S, Dulvy NK (2007) Current and future sustainability of island coral reef fisheries. *Current Biology* **17**, 655–8.

Raymundo LJ, Maypa AP, Gomez ED, Cadiz P (2007) Can dynamite-blasted reefs recover? A novel, low-tech approach to stimulating natural recovery in fish and coral populations. *Marine Pollution Bulletin* **54**, 1009–19.

Roberts CM, McClean CJ, Veron JEN, *et al.* (2002) Marine biodiversity hotspots and conservation priorities for tropical reefs. *Science* **295**, 1280–4.

Rogeli R, Hare B, Nabel J, *et al.* M (2009). Halfway to Copenhagen, no way to 2°C. *Nature Reports Climate Change* **3**, 81–3.

SCBD (2009) *Review of the Literature on the Links between Biodiversity and Climate Change.* Secretariat of the Convention on Biological Diversity, Montreal.

SCBD (2010) *Global Biodiversity Outlook 3.* Secretariat of the Convention on Biological Biversity, Montreal.

Silverman J, Lazar B, Cao L, Caldeira K, Erez J (2009). Coral reefs may start dissolving when atmospheric CO_2 doubles. *Geophysical Research Letters* **36**, L05606.

Spalding M, Ravilious C, Green EP (2001) *World Atlas of Coral Reefs*. World Conservation Monitoring Centre, University of California Press, California.

Timm J, Kochzius M (2008) Geological history and oceanography of the Indo-Malay Archipelago shape the genetic population structure in the false clown anemonefish (*Amphiprion ocellaris*). *Molecular Ecology* **17**, 3999–4014.

Tun K, Chou LM, Cabanban A, *et al.* (2004) Status of coral reefs, coral reef monitoring and management in Southeast Asia, 2004. In: Wilkinson CR (ed) *Status of Coral Reefs of the World: 2004*, pp. 235–75. Australian Institute of Marine Science, Townsville.

Tun K, Chou LM, Low KYJ, *et al.* (2011) A regional review on the 2010 coral bleaching event in Southeast Asia. In: *Status of Coral Reefs in East Asian Region: 2010*, pp. 9–27. Ministry of the Environment, Tokyo.

Tun K, Chou LM, Thamasak Y, *et al.* (2008) Status of coral reefs in Southeast Asia. In: Wilkinson CR (ed) *Status of Coral Reefs of the World: 2008*, pp. 131–44. Australian Institute of Marine Science, Townsville.

UNEP (2001) *Asia-Pacific Environmental Outlook*. United Nations Environment Programme, Bangkok.

UNEP (2004) *Coral Reefs in the South China Sea*. UNEP/GEF/SCS Technical Publication No. 2. UNEP/GEF Project Co-ordinating Unit, United Nations Environment Programme, Bangkok.

UNEP/COBSEA (2010) *State of the Marine Environment Report for the East Asian Seas 2009*. Chou LM (ed). COBSEA Secretariat, Bangkok.

UP-MSI, ABC, ARCBC, DENR, ASEAN (2002) *Marine Protected Areas in Southeast Asia*. ASEAN Regional Centre for Biodiversity Conservation, Department of Environment and Natural Resources, Los Baños.

Veron JEN (1995) *Corals in Space and Time: The Biogeography and Evolution of the Scleractinia*. UNSW Press, Sydney.

Veron JEN, Devantier LM, Turak E, *et al.* (2009) Delineating the coral triangle. *Galaxea* **11**, 91–100.

Wallace CC (1994) New species and a new species-group of the coral genus *Acropora* (Scleractinia: Astrocoeniina: Acroporidae) from Indo-Pacific locations. *Invertebrate Taxonomy* **8**, 961–88.

Weeks R, Russ GR, Alcala AC, White AT (2010) Effectiveness of marine protected areas in the Philippines for biodiversity conservation. *Conservation Biology* **24**, 531–40.

Wilkinson C, Hodgson G (1999) Coral reefs and the 1997–1998 mass bleaching and mortality. *Nature & Resources* **35**, 16–25.

Wilkinson C, Souter D, Goldberg J (2006) *Status of Coral Reefs in Tsunami Affected Countries: 2005*. Australian Institute of Marine Science, Townsville.

Wilkinson CR, Chou LM, Gomez E, Ridzwan AR, Soekarno S, Sudara S (1993) Status of coral reefs in southeast Asia: Threats and responses. In: Ginsburg RN (ed) *Proceedings of the Colloquium on Global Aspects of Coral Reefs: Health Hazards and History*, pp. 311–7. University of Miami, Florida.

Yap HT (1994) Implications of expected climatic changes on natural coastal ecosystems in the East Asian Seas region. In: Chou LM (ed) *Implications of Expected Climate Change in the East Asian Seas Region: An Overview*, pp. 105–22. RCU/EAS Technical Report Series No. 2. Regional Coordinating Unit, East Asian Seas, UNEP, Bangkok.

CHAPTER SEVEN

Nature Conservation for Sustainable Development in Singapore

HO, Hua Chew
Nature Society, Singapore

"Sustainable development only makes sense if our natural capital is sustained and not decreased no matter how much human capital we may have. Natural capital complements human capital but cannot be substituted by it, as the increase in the fleet of trawlers (human capital) is no substitute for the loss of a fish population (natural capital) that is being harvested."

– *HO, Hua Chew*

Nature Conservation for Sustainable Development in Singapore

HO, Hua Chew[1]

Nature Society, Singapore

Abstract

The goal of achieving a 'Sustainable Singapore' in terms of forging ahead with our economic development in an environmentally sustainable way has been underscored by the establishment of the Inter-Ministerial Committee for a Sustainable Singapore in 2009, resulting in the blueprint, *A Lively and Liveable Singapore: Strategies for Sustainable Development.* This goal has also been the main focus of the Urban Redevelopment (URA)'s Concept Plan 2011. However, it must be said that the vision of a 'Sustainable Singapore' envisaged in these planned programs, so far, tends to concentrate on the brown issues such as energy efficiency, recycling, green buildings and green transport. In terms of nature conservation, with its focus on the protection of natural habitats and their wildlife, there is a trend to accept the status quo and to work merely towards their enhancement, restoration and consolidation as well as accessibility for all with respect to the areas already under official protection. There is a serious failure to break into new grounds and forge a greater and more comprehensive network of protection of what remain of our nature areas and their biodiversity. In Singapore, there remain nature areas that are significant in terms of natural habitats and their wildlife are urgently crying out for genuine, lasting protection. This article argues that nature conservation should not be neglected, to make the vision and goal of a 'Sustainable Singapore' meaningful and viable for a lively, liveable and well-loved home.

Key words: biodiversity, conservation, development, ecology, nature, sustainability

I. Introduction

The wake-up calls for governments to take serious and urgent substantive actions to solve the environmental crisis caused by global warming have surged like a tidal wave all over the world in the past decade. This may be too late to avert the

---

[1] This article is personal in nature and cannot be attributed completely to the standpoint of the Nature Society. My personal email is: hohc@starhub.net.sg.
Nature Society (Singapore), 510 Geylang Road, #02-05, The Sunflower, Singapore 389466.
Email: contact@nss.org.sg; Tel: 6741 2036.

forecasted scenario of environmental catastrophes. At a time when the scientific evidence is massive and when unity and consensus is the key towards ameliorating this nightmarish situation, it is indeed most disheartening to see at the United Nations (UN) Copenhagen Conference the chaotic and conflicting responses coming from the governments of both developed and developing nations – from the pointing of accusing fingers to the outright rejection of a compromised solution.

What is the situation here in Singapore? Have we done enough towards achieving the goal of sustainable development? This is not to say that our government is not doing anything serious to ameliorate global warming. The government will be focusing on raising energy efficiency, given that we do not have, in a big way, the choice of clean-energy alternatives; it is 'aiming for a 35 per cent improvement in energy from 2005 levels by 2030', as emphasized in its sustainable blueprint (Ministry of Environment and Water Resources, 2009: 13). For the Copenhagen Conference, it has set a target of 7–11 per cent for emission reduction by 2020 and a promise to set it to 16 per cent from 'business as usual levels' if there is a binding agreement to the climate change program. Since there was no accord achieved at Copenhagen, the commitment to 16 per cent need not be implemented (*The Straits Times*, 2010). However, could we not also support and participate in the compensation program called Reducing Emission from Deforestation and Forest Degradation (REDD)? Although there is no legally binding outcome, it is the most promising program to have been discussed and worked out at Copenhagen for a binding accord in the immediate future. In this light, it is pertinent to ask: Do we have a serious policy to preserve the existing greenery in our own backyards so that our contribution to reduce carbon emission is not cancelled out by its further release through the destruction of this existing greenery?

Singapore is a small city-state in terms of land size and population. Size, however, should not dictate what we can and should be doing to reduce or mitigate environmental catastrophes that will affect everyone to the point of making life on Earth an arduous struggle for survival, and even unliveable, for most earthly creatures, including humans.

II. Singapore's Ecological Footprint

It is most disappointing for our government to keep emphasizing that, being a small nation, we cannot be expected to do proportionately as much as big nations do, citing the reason that our national percentage of carbon emission is only a tiny fraction of that for the globe, or even for China. It is correct, but to look at the problem in this way is to diffuse the responsibility of each and every single citizen, who enjoys the high standard of living we have achieved, to play his individual part in reducing carbon emission. One of our concerned citizens vehemently declared in the mass media: 'As a Singaporean, I am not sure if I can continue to hold my head high when I know my individual carbon footprint is mistakenly condoned ... just because my country says its national contribution to climate change is insignificant ... In terms of long-term sustainability, it is apt to remind ourselves that we are an island state, surrounded by seas whose levels are predicted to rise with increasing global temperatures' (Lau, 2009). Surely, every individual's effort counts, given the gloomy predictions of the UN's Intergovernmental Panel on Climate Change (IPCC). To have a clear view of the situation, we should look at our carbon emission in terms of per capita emission. Given this approach, the carbon emission contribution of each person residing here is obviously very high, as the material standard of living here is closer to that of developed and industrialized nations such as the USA and the UK, than to nations in the developing world, such as states in Africa. In failing to recognize this officially, the government is signalling, intentionally or not, to each and every Singaporean that he can go on with business as usual, or even that he can even be entitled to emit more than usual.

Here, it would be enlightening to consider the eco-footprint statistics from the results of the study for 153 out of the world's 185 countries by the Ecological Footprint Network (2010). Singapore's eco-footprint per person comes to 5.34 global hectares (gha), the 21st-largest in the list of 153 nations covered. This is larger than the eco-footprint of highly developed and industrialized nations such as Germany (5.08), UK (4.89), South Korea (4.87), Japan (4.73) and so on. The world's average is 2.27 gha as of 2007, with a world average biocapacity of 1.8 gha per person, which entails a world ecological deficit of 0.9 gha. Given that our biocapacity per person is 0.02 gha, this indicates that there is a national ecological deficit of 5.32 gha per person, making an ecological deficit that is five to six times greater than the world's average.

As a small city-state, we are able to sustain our economic progress and achieve a First World status because we are exploiting productive land and natural resources outside Singapore, while contributing to environmental deterioration and destruction elsewhere. We are able to have more industrial, business and housing spaces because we have extended a tremendous amount of space for food and other resources for growth to other countries. In other words, our eco-footprint transcends the limit of the productive space that we have and is comparatively gargantuan for such a small state (710 square kilometres in size). Obviously, we could not have sustained our current population of 5.18 million people at the current standard of living if we had used our own land to sustain, for example, the food to feed the population. While we could achieve further economic progress by increasing our population to 6.5 million at the current standard of living, this will only be possible by increasing the use of productive land from other countries to supply our needs and thus also increasing the world's ecological deficit, as indicated the study by the Ecological Footprint Network.

The larger the footprint, the larger the impact we will have on productive land spaces, the health of ecosystems and natural processes that have brought about materially-rich civilizations. Thus, creating reservoirs abroad (Johor) or expanding our Central Catchment Reservoirs (e.g. Peirce and Seletar) entails drowning rainforests, and creating reservoirs by damming up rivers entails destroying our mangroves (e.g. Sungei Kranji). Increasing our food consumption entails having other countries to clear forests for farms. Our importation of sands for land reclamation, and granite and timber for construction entails the degradation or destruction of natural ecosystems, marine and terrestrial, at home and abroad. Singaporeans generally do not see these degradations or destructions because these occur far away but nevertheless contribute to the general deterioration of the global natural ecosystem and biodiversity. This does not mean that we cannot import resources, but one has to be fully cognizant of the extent of our contribution to the global ecological crisis and make an enlightened and committed effort to reduce its impact both at home and abroad. For example, effort should be made at the international level, as was proposed by Dr Chris Hails, Chairman of WWF Singapore and Director (Network Relations) of WWF International, by importing resources from the right sources, such as 'buying timber from sustainably certified sources and by putting pressure on other countries to farm sustainably' (*Today*, 2010: Online, 14 October).

III. Lacunae in the Singapore Green Plan

It must be said that the thrust of the government's sustainable development programs are focused on brown issues, presuming that, when these are taken care of adequately, economic growth can carry on without limit, almost forgetting that Singapore is not all concrete jungle. What about biodiversity conservation, the protection of our natural habitats?

A collaborative study done by the Centre for Remote Imaging, Sensing & Processing, National University of Singapore and National Parks has shown that 47 per cent of our total land area is still undeveloped greenery, a fact unknown to many Singaporeans (*The Straits Times*, 25 June 2008). While this greenery includes public parks, it also includes scattered patches of remnant secondary forests, woodlands, scrublands, etc., in our remaining countryside. Forty-seven per cent is almost half of Singapore. The study also reveals that Singapore is 'getting greener'. The evidence for this is that the area of the island covered by greenery has gone up from 36 per cent in 1986 to 47 per cent in 2007, despite the country's population shooting up from 2.7 million to 4.6 million during this period. In *The Straits Times*' report, National Parks states that '10 per cent of the land here is set aside for nature reserves and parks, allowing for biodiversity in habitats, including lowland rainforests, freshwater swamp forests and coastal forests to be conserved'. What will happen to the rest, that is, the 37 per cent that is left unprotected?

If economic development is allowed to run rampant for the sake of growth, this precious greenery that has taken decades to establish itself will not survive for long. Given that there are many existing green areas, should we be content to allow only 10 per cent of it to be protected for public parks and biodiversity or nature conservation? Given that there are substantial undeveloped green areas still existing in Singapore, more could and should be done to rope in and protect other areas that are rich in biodiversity within the framework of the Singapore Green Plan (SGP). Also, more can be done to strengthen the Green Plan's framework itself so that what comes under its protection can be made to have long-term survival value rather than be subjected to short-term economic interest.

As mentioned above, about 10 per cent of the green, undeveloped areas are designated as public parks as well as Nature Reserves and Nature Areas in the SGP. Out of this 10 per cent, however, only 4.5 per cent constitutes the Nature Reserves in the Green Plan (ENV 2009). The other areas in the Green Plan are called 'Nature Areas', but 'Nature Areas' are designated as areas that, although having biodiversity value, can nevertheless be used for development when there is a need. This logic can be discerned in the decision to use the main part of the Kranji Marshes, a 'Nature Area' as defined in the SGP, for a golf course without even consulting the stakeholders prior to the decision and despite protests from the Nature Society (*The Straits Times*, 12 February 1999). Despite what happened at Kranji, further eroding of the Green Plan followed. In URA's last *Master Plan* (2004), the four Coral Zones (Semakau, Hantu, St John and Sudong) and three Nature Areas (Mandai Mangrove, Pulau Semakau and Khatib Bongsu) of the original Green Plan (1993) had even been delisted as 'Nature Areas', and their future is now left subjected to the vagaries of economic forces. However, these areas are still intact. There is a serious need to review their current status with all the stakeholders involved to consider whether deleting them from the SGP is justified or necessary (Nature Society, 2002, unpublished data).

Since independence (1965), only one nature area, Sungei Buloh, has been designated as a Nature Reserve. Two Reserves, however, Kranji and Pandan, out of the five established by the British Colonial government, have been degazetted. The three others that still remain are Bukit Timah, Central Catchment and Labrador. The degazetting of Pandan for industrial development and the Kranji Reserve to create a reservoir in the early decades of Independence is, naturally essential for economic survival or sustainability. After Singapore has become a First World nation, however, the Green Plan should have been be seriously directed to compensating these losses in some way. This can be done by establishing more nature reserves, given that there are 'Nature Areas', such as Pulau Ubin and Pulau Tekong, that are extremely rich in biodiversity and resident to some rare or endangered species, such as the greater mousedeer, barred eagle owl, brown wood owl in Ubin, and the black-naped monarch, the mangrove blue flycatcher and the Malayan porcupine in Tekong.

IV.   Green Areas for Biodiversity Conservation

Most of these green and undeveloped areas are in the remaining countryside north of the Pan Island Expressway from the east to the west, and are covered by substantial patches of woodlands and secondary forests. They can be clearly seen along such roads as Upper Thomson, Bukit Batok, Mandai, Jalan Bahar,

Choa Chu Kang and Yio Chu Kang (at the western section), or by trawling through any satellite map for Singapore. From what Nature Society can gather, such patches (e.g. at Khatib Bongsu, Bukit Brown, Lorong Halus and Springleaf) serve as havens or refugee sites for wildlife that are generally considered to be rare or endangered in Singapore, such as the leopard cat, Sunda Pangolin, the large Indian civet, the red jungle fowl, the Oriental pied hornbill and the straw-headed bulbul (Nature Society, 2009, unpublished data). The Oriental pied hornbill, supposedly unsighted on the main island of Singapore for decades, is making a comeback and has been spotted in some of these patches (e.g. the Clementi Forest in suburbia). The jungle fowl and the straw-headed bulbul are spreading in many of these wild woodlands outside Pulau Ubin, which is, at one time, the only area in Singapore where they can be found. These patches have also become extended foraging grounds for many species of wildlife in our forested Nature Reserves, some of which are rare or endangered, such as the grey-headed fish eagle, changeable hawk eagle and white-bellied woodpecker. The grey-headed fish eagle, the changeable hawk eagle and the uncommon white-bellied sea eagle have even resorted to these wild woodlands for nesting sites. These areas are also important refuelling stations and havens for hordes of birds seeking warmer climes from the winter in the temperate zone, such as the migratory flycatchers, warblers, pittas, cuckoos and raptors.

The existing forested Nature Reserves (Bukit Timah and Central Catchment) are clustered in the heart of the main island and are becoming increasingly like island fortresses surrounded by a sea of inhospitable landscape. Such natural habitats are called 'habitat-islands' by ecologists. They are comparatively very small (about 2,000 plus hectares combined, minus the reservoirs), and many of their typical wildlife have small population sizes, especially the uncommon and the rare species. The gene pool for these typical forest wildlife, such as the red-crowned barbet, the Asian drongo cuckoo and the banded leaf monkey, will tend to have low genetic variability. It is a well-established fact that 'genetic variation allows species to adapt to a changing environment. Rare species usually have less genetic variation than widespread species do and, consequently, are more vulnerable to extinction when environmental conditions change', with the exception of some plant species (Primack, 1998: 33). Thus, there is an urgent need to preserve the existing patches of woodlands or forests outside this Nature Reserves to act as stepping stones or green corridors for species from the Malay Peninsula in search of new habitats due to their expanding population or the destruction of their existing habitats.

This remaining natural greenery also has great potential to serve the therapeutic, aesthetic, recreational and eco-educational needs of all Singaporeans if their existence is widely publicized and their accessibility is promoted. More and more people – citizens as well as foreign residents – are gravitating to these areas. This is manifested in the hordes of cyclists roaming all over our countryside looking for challenging, wild or rough terrains and tracks; big groups of hikers, joggers and cross-country runners traversing through such areas; crowds of nature photographers bashing in the wild places looking for rarities; and so on. A very vivid and strong indicator of the value and attraction of this natural greenery amongst Singaporeans is the proliferation of lyrical advertisements by condo developers (e.g. The Meadows at Upper Thomson) showing their buildings rising amidst a sea of greenery that is depicted almost like a jungle. As the existing Nature Reserves have increasingly become rather limited in recreational space and under tremendous pressure from the visiting human crowd, it would be a tremendous relief from such pressure if this wild greenery outside their boundaries can be promoted for such recreational uses as well.

The government has done a great deal in the creation of public parks in line with the officially established model of Singapore as a garden-city, or a city in a garden, to raise the quality of life for its citizens. This was a great success in the context of the widespread and relentless urbanization of Singapore over the past decades since Independence. However, as observed by the former President of Nature Society (Singapore), the 'garden city was criticized as being sterile, artificial and superficial. Initially led by nature lovers and conservationists whom the government regarded as a narrow interest group, their voices grew in strength and gained increasing public support, as more and more nature areas and even nature reserves succumbed to bulldozers' (Geh, 2010: 389). This was evident in the tremendous support shown by members of the public in the petition-campaign for the preservation of Senoko bird sanctuary mounted by Friends of Senoko in 1994, when a total of 25,000 signatures were collected, not via the Internet but on the ground, arguably the largest petition mounted for any causes in Singapore (*The Straits Times*, 21 October 1994).

V.   Greenery Against Global and Local Warning

It is now widely accepted by scientists, as represented by IPCC (2007), that global warming due to carbon dioxide emission is a certainty over the past half century and that 90 per cent of this is derived from human activities such as (1) transportation and manufacturing, which involve the burning of fossil fuels (coal, oil and natural gas); (2) burning of firewood; and (3) the destruction of plants through the destruction of forests and other plant habitats (grasslands, mangroves, etc.). The destruction of plants through burning or decomposition releases carbon dioxide into the air, while dead plants can no longer carry out the photosynthesis that will sequester the carbon dioxide in the air. At this stage of global carbon emission, it is so prodigious and massive that, 'even if we plant as many forests and grasslands as possible, they will not (by themselves) remove all of the additional carbon we are pouring into the atmosphere' so as to save the world from excessive greenhouse effects (Rice, 2009: 76). Nevertheless, it must be emphasized that 'we can hardly expect success without saving the trees we have and planting more. Green plants are essential allies in the effort to prevent a disastrous greenhouse future' (Rice, 2009: 79–80).

Another highly important point is that greenery serves as natural air conditioners through the process of transpiration, when water from the soil is exuded into the air, which produces a cooling effect. It has been scientifically established that, as a city becomes more and more urbanized and more forests and woodlands are cleared, the predominance of concrete causes a 'heat island' effect, where the ambient temperature is higher than the surrounding forested areas or the green countryside. Thus, urbanites become heavily dependent on artificial air conditioners to cool down, but this is extremely costly and adds to the pollution in the air. To reduce such costly expenditure and pollution, we should let plants do the air-conditioning for us. As an example, such economic benefits can be shown by New York City parks where, using a computer program developed at the University of California, Davis, it is 'estimate[d] that the 592,130 trees in New York public areas generate $5.60 in savings for each $1.00 spent to plant and maintain them' (Randall, 2007).

Singapore has done a tremendous amount of tree planting. More tree-planting programs are in the works, initiated by the government to soften up the concrete landscape in the city and public housing estates, in the ongoing process of creating a garden-city. It has recently promoted rooftop greenery and gardening as well, as can be seen in the new Khoo Teck Puat Hospital. All these planting programs are highly laudable although extremely costly. Unfortunately, this has misled many pro-development proponents and decision-makers into thinking or excusing themselves that a forested area can be developed if you plant trees after the development, by way of compensation. Tree planting in the form of 'lollipop' trees in public parks, gardens, housing estates and roadsides by itself cannot compensate for the loss of natural forest and woodlands as carbon sinks, let alone as an ecosystem and habitat for wildlife. It is ecologically better and more economical to leave existing forests and woodlands, whether in the city or suburbs or the remaining countryside, as they are, rather than destroy them, as a first principle of land-use planning in this age of global warming. Forests and woodlands have a greater cooling capacity, as they have a greater abundance of leaves, which are involved in the transpiration process – what is known amongst botanists as the 'leaf area index' (Rice, 2009: 84). Moreover, if we plant trees, we must wait for a long time or 'we must wait decades for their shades ... This is another reason that we should preserve existing trees rather than just plant new ones' (Rice, 2009: 91). This becomes more urgent, as global warming intensifies as the years roll on.

Together with the forested Nature Reserves, our remaining undeveloped greenery have been serving us as carbon sinks as well as natural air conditioners without any recognition on the part of authorities and unbeknownst to most of the general populace. This situation should be rectified. There should be a study done to determine the carbon-sink weightage of this greenery. Even though the magnitude is relatively minute on the global scale, our proportional contribution to the retention of carbon sinks relative to our territorial and/or population size can be large. In this light, we should be contributing to UN's REDD so that we can be a shining example of what this little red dot can do to ameliorate global warming, no matter how small it is. To do so will be Singapore's most challenging and forthright contribution to ameliorate the crisis of global warming.

Singaporeans have been complaining loudly against the clearing of forests by burning in our neighbouring countries, which has resulted in incoming haze, as well as the heavy carbon emissions. However, these complaints sound rather hollow when we do not make our utmost effort to nurture and protect our own remaining woodlands, forests and other greenery. The recent *URA Master Plan 2008* looks promising in this direction, providing a thrust towards the use of brownfield sites and existing reclaimed lands for development, as in Kallang, Geylang, Marina South and Keppel Harbour, the last of when its lease as a port expires in 16 years' time. Future development plans should be geared primarily in this direction.

VI. Framework for Sustainable Development

It is very soothing to say that economic development and environmental sustainability should go hand in hand. One should not be at the expense of the other, where economic growth is pursuable so long as it is sustainable. However, we would not be able to see where we are heading, nor contribute towards improving the situation if we do not have a clear vision of what constitutes sustainable development and the yardstick for measuring whether we have remained within or transgressed beyond the limit of sustainability. Of course, our economic development and current standard can be sustained and upgraded so long as we still have other productive lands from other countries to exploit and fall back on. This is economic sustainability if you like. But have we achieved a harmony with ecological or environmental sustainability?

A nation or the world cannot endlessly pursue economic growth measured as gross national product or gross domestic product (GDP) growth, which has been increasingly recognized in economic circles as an inadequate or unreliable measure of well-being, because the reality is that the total economic system is a subsystem of the closed, finite global ecosystem. The global ecosystem's source and sink functions are limited. Growth, going beyond such limits, will cause environmental deterioration and the ultimate collapse of the economic system itself. As stated by Goodland (1996), 'The best evidence that there is an absolute limit is the calculation by Vitousek and others in *Bioscience* (1986), that the human economy, directly or indirectly, uses about 40 per cent of the net primary product of terrestrial photosynthesis today. This means that, with a mere doubling of the world's population in, say, 35 years, we will use 80 per cent, and 100 per cent shortly after' (Mander & Goldsmith, 1996: 209). Thus, unlimited economic growth at the physical or material level is an impossibility. Sustainable development only makes sense if our natural capital is sustained and not decreased no matter how much human capital we may have. Natural capital complements human capital but cannot be substituted by it, as the increase in the fleet of trawlers (human capital) is no substitute for the loss of a fish population (natural capital) that is being harvested.

This does not mean that sustainable development is an impossibility, but there is a serious need to be clear about the meaning of the terms 'growth' and 'development'. As Daly states, '*To grow* means "to increase naturally in size by the addition of material through assimilation or accretion". *To develop* means "to expand or realize the potentialities of; to bring gradually to a fuller, greater, or better state". When something grows, it gets bigger. When something develops, it becomes different. The earth ecosystem develops (evolves) but does not grow. Its subsystem, the economy, must eventually stop growing, but can continue to develop. The term "sustainable development" therefore makes sense to the economy, but only if it is understood as "development without growth", that is the qualitative improvement of a physical economic base that is maintained in a steady state by a throughput of matter – energy that is within the regenerative and assimilative capacities of the ecosystem' (Daly, 1993: 267–8). For example, the economic growth of a nation (measured in GDP) can plateau off at a sustainable level but develop further in terms of a more equitable distribution of incomes or wealth. There is an urgent need for a big or even revolutionary shift in traditional or establishment-based economic thinking towards working out a different measure of human well-being that will embrace in a forthright way environmental/ecological factors and values as fundamental to the calculation, as was shown by Robert Costanza and others (Daly & Cobb, 1994).

VII. Issue of Land Scarcity

This brings me to the issue of the 6.5 million people that our government wants to attain for Singapore to promote economic growth such that Singapore will become another metropolis comparable to Tokyo, New York or Hong Kong. Minister Mentor Lee Kuan Yew has cast doubts on the viability of this objective, which would involve 'just solid buildings, one building blocking the sunlight of another' and losing, in consequence, the sense of not being crammed. He advocates that 5.5, or even 5 million, people will be enough for our comfort and well-being (*The Straits Times*, 2 February 2008). This appears to be a sensible compromise, but, given the extreme discomfort and economic burden bourn now by Singaporeans from all walks of life in terms of transport, housing, education and even recreational spaces when we are now at 5 million, it is indeed stretching things too far to go beyond this. Going beyond the five-million mark will certainly erode drastically the green spaces that are still available for us to *develop* a greater quality of life rather than to continue to *grow* in our economic standards, to use the ecological perspective defined by Daly (1993) for sustainable development.

Land scarcity should not be an excuse to overwhelm our remaining green areas for development when the so-called 'scarce land' is used for golf courses, which comes to 23 eighteen-hole and 15 nine-hole courses (National Sports Council, 2007). Golf courses cater to about 2 per cent of the population of Singapore from age 15 and above, while jogging, walking and cycling come to 34.5 per cent combined (National Sports Council, 2005). Based on a conservative calculation of about 65 hectares for an 18-hole and 25 hectares for a 9-hole course, we have a total of 1,870 hectares of land used up for golfers. Two of these 18-hole courses, the Kranji Sanctuary (2004) and the Marina Bay Golf Course (2006), were constructed within the last six years, in a decade when the mantra of 'land scarcity' was repeatedly invoked against nature conservationists. It is indeed ridiculous to place golf as a priority over preserving our natural greenery for its value as a carbon sink and air conditioning as well as other outdoor recreational purposes, whether or not Singapore is land-scarce.

It is stated by pro-development proponents that there is a shortage of land for residential development (as with their responses to advocates of the preservation of Bukit Brown cemetery for its cultural and biodiversity value). Of course, housing citizens should have top priority. This, however, holds no water when 20 per cent of the resale Housing Development Board flats are allowed for permanent residents (*The Straits Times*, 26 June 2011). This so-called land scarcity for housing citizens becomes more unconvincing when the development of private properties that also cater to foreign buyers and permanent residents is taken into account. A recent estimate for the second quarter of 2011 showed that this came up to 30 per cent of the buyers in this sector (*The Straits Times*, 19 August 2011). The pressure for land to develop housing that is available and affordable for Singapore's citizens will certainly ease off if this liberal and broad policy is drastically revised.

In this light, we Singaporeans should not be cowed by the issue of land scarcity. Even if we are land-scarce, why should this entail areas of natural greenery and wildlife habitats being sacrificed to economic growth when we are no longer a Third World nation? Land scarcity, in the context of a First World nation, rather entails that we should put under intense scrutiny the need for any economic projects, such as the Singapore Tourism Board's project at Mandai Lake Road (Nature Society, 2009, unpublished data), that would destroy or jeopardize the remaining nature areas outside of the Nature Reserves.

VIII. Conclusion

The remaining natural green areas outside of the Nature Reserves and the public parks should not be simply regarded as merely land banks for future economic growth. In the thrust towards a sustainable Singapore, we should and can afford to be 'distinctive' and 'cutting edge, breaking new ground as a city of tomorrow'. We should be bold and take this remaining almost-50 per cent undeveloped greenery as a 'wilderness playground or park' to 'provide room for fun and the imagination', to borrow terms employed by URA's Concept Plan 2011 Focus Groups in their feedback report. We should go beyond a garden-city or city-in-the-garden concept with its dull uniformity and manicured greenery to embrace this undeveloped greenery as our countryside in its own right with its charming naturalness and vital untidiness.

Any new economic development should be restricted to brownfield sites and existing reclaimed land areas. Let our creativity, inventiveness, imagination come up with the solutions to land-use planning within these brownfield areas, and leave this greenery to grow and bloom in all its natural glory. This is our invaluable natural heritage, which enhances Singapore's liveliness and 'liveableness' in a variety of ways as carbon sinks and natural air conditioners, softens our landscape with lovely greenery, bestows us with a rich diversity of non-human neighbours to commune with and provides a wider arena for our ecological education as well as outdoor pursuits and recreations. It is in this direction that our thrust towards a Sustainable Singapore as a well-loved home will have a greater and wider appeal amongst our citizens.

References

Daly HE, Cobb Jr JB (1994) *For the Common Good: Redirecting the Economy toward Community, the Environment, and a Sustainable Future*. Beacon Press, Boston.

Daly HE (1993) Sustainable growth: An impossibility theorem. In: Daly HE, Townsend KN (eds) *Valuing the Earth: Economics, Ecology, Ethics*, pp. 267–8. The MIT Press, Cambridge, Massachusetts.

Ecological Footprint Network (2010). *Ecological Footprint Atlas 2010*. Available from: http://www.footprintnetwork.org/en/index.php/GEN/page/ecological_footprint_atlas_2008/.

Geh Min (2010) The Greening of the Global City. In: Chong T (ed) *Management of Success: Singapore Revisited*, p. 389. Institute of Southeast Asian Studies, Singapore.

Goodland R (1996) Growth has reached its limit. In: Mander J, Goldsmith E (eds) *The Case Against the Global Economy*, p. 209. Sierra Club Books, San Francisco

Intergovernmental Panel on Climate (IPCC) (2007) Climate change 2007: Climate change impacts, adaptation, and vulnerability. Summary for policy makers. Available from: http://www.ipcc.ch/pdf/asssessment-report/ar4/wg2-spm.pdf.

Lau YS (2009) Are we doing enough? In: *The Straits Times: Forum* (printed in 10 December 2009).

Ministry of Environment and Water Resources (2009) *A Lively and Liveable Singapore: Strategies for Sustainable Singapore*. Ministry of Environment and Water Resources, Singapore.

National Sports Council (2005) *National Sports Participation Survey 2005*. National Sports Council Library, Singapore.

National Sports Council (2007) *A Census of Sports Facilities in Singapore 2006*. National Sports Council Library, Singapore.

Primack RB (1998) *Essentials of Conservation Biology.* Sinauer Associates, Sunderland, Massachusetts.

Randal DK (2007) Perhaps only God can make a tree, but only people could put a dollar value on it. *New York Times,* 18 April 2007.

Rice SA (2009*) Green Planet: How Plants Keep the Earth Alive.* Rutgers University Press, New Brunswick, New Jersey & London.

The Straits Times (2011) Chinese again top foreign home buyers (19 August).

The Straits Times (2011) Who the resale HDB flat buyers are. (26 June).

The Straits Times (2010) Singapore to go ahead with carbon emission cuts (12 January).

The Straits Times (2008) MM 'not quite sold' on idea of 6.5 population (2 February).

The Straits Times (2008) Singapore is Getting Greener (25 June).

The Straits Times (1999) Do we really need another golf course? (12 February).

The Straits Times (1994) 25,000 appeal for Senoko Bird Habitat to be saved (21 October).

Today (2010) One Singapore's resident's ecological footprint equals that of 33 Africans (Online: 14 October).

CHAPTER EIGHT

The Biodiversity Crisis in Asia: Singapore as a Model for Environmental Change and Threats to Terrestrial and Freshwater Ecosystems

Darren CJ YEO, Richard T CORLETT and Hugh TW TAN
Department of Biological Sciences, National University of Singapore

"The environmental changes that have resulted in species loss in freshwater and terrestrial systems in Singapore are not easily reversed. Indeed, the high proportion of endangered and critically endangered species among the native flora and fauna suggests that further extinctions are inevitable. Forests and forest streams can potentially be restored and nationally extinct species reintroduced from the surrounding countries, but, in reality, only a small proportion of Singapore can be dedicated to natural habitats. It may also be possible to make urban habitats more suitable for the more tolerant native species. Current projects aimed at bringing native species into urban areas include the rehabilitation of selected stretches of freshwater canals and the experimental planting of a wide variety of native forest plant species in parks and along roadsides."

— Darren CJ YEO, Richard T CORLETT and Hugh TW TAN

The Biodiversity Crisis in Asia: Singapore as a Model for Environmental Change and Threats to Terrestrial and Freshwater Ecosystems

Darren CJ YEO[1], Richard T CORLETT[1] and Hugh TW TAN[1]
Department of Biological Sciences, National University of Singapore

Abstract

Of the many global environmental changes brought about by human activity, the Biodiversity Crisis – the rapid decline in biological diversity – is arguably one of the biggest and most complicated of challenges. This is partly because the crisis is being driven by a combination of other similarly severe anthropogenic environmental changes, including habitat loss, biological invasion, pollution, over-exploitation and climate change. The problem is particularly acute in Asia, with many examples of freshwater and terrestrial ecosystems that are under serious threat. In this essay, we highlight examples and important drivers of biodiversity decline in these habitats in Singapore, which mirror what is happening (or will soon be happening) in other parts of Asia.

Key words: biodiversity, environment, freshwater, Singapore, terrestrial, threats

I. Introduction

The rapid and spectacular loss of species diversity is probably the most prominent manifestation of the Biodiversity Crisis. Other aspects of this global problem include the loss of genetic diversity (e.g. through the wholesale loss of populations, through biotic homogenization, hybridization and introgression) as well as the loss of ecosystem diversity (e.g. loss of entire habitats and their associated communities of interacting species). There has been debate about the precise causes of this latest anthropogenic 'mass extinction', but, generally, habitat loss or modification, alien species introduction, pollution, disease, over-exploitation and climate change are often regarded as the primary drivers of species loss and are consequently seen as the most important threats to species diversity.

Because of its small size (712.4 square kilometres) and high population density (7,126 people/square kilometres), Singapore has experienced exceptionally high levels of species loss, with around 30 per cent of all species recorded here since

---

[1] Department of Biological Sciences, National University of Singapore. 14 Science Drive 4, Singapore 117543. Correspondence: Darren CJ Yeo; Email: dbsyeod@nus.edu.sg.

the 19th century now nationally extinct (Brook *et al.*, 2003). Half of those that remain are threatened with extinction. Singapore has suffered the same human impacts as the rest of the region, but up to a century earlier. Forest conversion to crop plantations and timber extraction, which are now major threats to tropical forests worldwide, had already transformed Singapore's landscape by the 1880s. Today, only 0.3 per cent of the original primary forest remains, surrounded by a larger area of secondary forests of various ages. Singapore can thus act as an early-warning system for the region as a whole, as the rest of the region undergoes a similar transformation.

II. Habitat Loss and Modification

Habitat loss and modification are prevalent throughout Asia as its nations develop in response to growing populations and consumption. In Singapore, this is very clearly seen in the loss and modification of both freshwater and terrestrial habitats.

Several of the country's southern river drainages, most notably the Singapore and Kallang Rivers, were intensively used for commerce from 1819 and, by the late 20th century, had already lost all the original riparian habitats and associated biodiversity along the lower reaches. These and other river drainages around Singapore have also been subjected to 'drowning' by the construction of dams to create inland as well as coastal or estuarine reservoirs. In addition, most inland watercourses in Singapore have lost their original ecological structure through widespread and intensive canalization (modification involving straightening, deepening and cementing of the waterways), done in the name of controlling both floods and disease vectors (e.g. mosquitoes). More recently, acidification of freshwater streams in the Bukit Timah Nature Reserve (BTNR), possibly from acid rain, has been suggested as a form of habitat modification contributing to the decline of an endemic and critically endangered freshwater crab species, the Singapore stream crab (*Johora singaporensis*), found there.

Losses of terrestrial habitats (accompanied by the loss of the plants and animals that inhabited them) began not long after Singapore's founding as a maritime trading port in 1819 by Sir Stamford Raffles. Gambier (*Uncaria gambir*) and pepper (*Piper nigrum*) were already being grown as cash crops when Raffles arrived, and their cultivation expanded rapidly into the interior as the new settlement grew. Gambier was cultivated on a large scale on land newly cleared of forest where it grew best. Once the soil was depleted of its nutrients by the gambier crop, the area would be abandoned for another newly cleared patch of forest, leaving behind only scrub and grassland that was capable of thriving on the degraded soil. A succession of other crop plants followed, with the last major impact on Singapore's native forests coming from the expansion of rubber tree (*Hevea brasiliensis*) cultivation, which, at its peak in the 1930s, covered almost 40 per cent of Singapore's land area.

Land-intensive agriculture is no longer practised in Singapore. Instead, urbanization, including construction for housing, industry and infrastructure, has now taken over as the prime cause of habitat loss. However, this is on a much smaller scale compared with agriculture in the past, and, in most cases, urban areas expand into land that had previously been cleared for agriculture. That said, many patches of abandoned cultivated land have 'recovered' to varying degrees to become occupied by secondary forests, although the most recent of these tend to be dominated by alien (non-native) trees. While the structure and species composition of secondary forests undoubtedly differ from that of the original primary forest, the older, native-dominated secondary forests in the Central Catchment Nature Reserve (CCNR) have nevertheless become important for supporting what remains of Singapore's natural flora and fauna, and as buffer areas for the more sensitive remaining patches of primary forest. It is to these secondary forests (and not to the well-protected primary forest) that habitat loss or modification from urbanization is a major threat.

Despite the huge majority of Singapore's natural habitats having been destroyed or modified, a surprising amount of natural biodiversity still clings on to existence in the remnant pockets of original natural habitat as well as in secondary forests and some of the other less heavily modified habitats. For instance, 50 per cent of all of the island's native flora and fauna can be found in less than 1 per cent of Singapore's original rainforest – Bukit Timah Nature Reserve.

III. Over-Exploitation

Singapore's forests have been heavily exploited for timber, firewood and other forest products. At the height of this exploitation, in the 19th century, valuable plant species, such as gutta percha (*Palaquium gutta*) and kranji (*Dialium laurinum*), became scarce as a result of destructive harvesting for their latex and timber, respectively.

Hunting led to the extirpation of various large mammals, including sambar deer (*Rusa unicolor*) and barking deer (*Muntiacus muntjak*), as well as some of the larger bird species, such as the rhinoceros hornbill (*Buceros rhinoceros*). Laws were passed to protect these species, but they were either too late or insufficiently enforced to save these animals. By the 1960s, large species were either very rare or altogether extirpated in Singapore. Some species, such as tigers, were killed as vermin or for trophies and disappeared well before World War II, while others were hunted for their purported medicinal uses (see Yang *et al.*, 1990; Goh, 1994; Wee & Ng, 1995; Davison *et al.*, 2008; Tan *et al.*, 2010).

Poaching around the edges of BTNR and CCNR persisted even after World War II. Traps such as mist-nets, sticky traps and small cages were used to catch a range of small and medium-sized mammals, including leaf monkeys, flying foxes, pangolins and giant squirrels, for personal consumption and for the live-food trade, and tree shrews and small squirrels for pets and the pet trade.

Birds and fish were also caught for food and for the pet trade, while snakes, turtles, lizards and frogs were collected for food. Even invertebrates were not spared, with various insects and spiders being collected as curios. Conservation and enforcement efforts were severely lacking in Singapore as late as the mid-1970s, when locally collected live animals (together with imported animals) were still commonly seen being sold in Chinatown.

Today, although increased awareness and better enforcement (and, it must be said, a shortage of animals) have greatly reduced illegal hunting activities, poaching still continues on a small scale in Singapore. National Parks Board staff and researchers in the nature reserves have occasionally encountered poachers but more often come across traps set for both terrestrial and aquatic animals. Targeted species include ornamental or songbird species and terrestrial animals such as slow lorises, colugos, pangolins and water monitors, as well as various ornamental or game fish species.

IV. Alien Species

Alien species dominate most artificial or modified freshwater and non-forest terrestrial habitats in Singapore, as well as many of the younger secondary forests. This pattern is repeated near urban areas throughout Asia, and indeed globally, and is likely to spread into areas transformed by intensive agriculture. This may demonstrate a meltdown of sorts in which multiple environmental change drivers (in this case, habitat degradation together with alien species invasion) enhance each other's negative effects on native-species diversity. Others argue, however, that alien species are merely 'passengers' making the most of loss or modification of original habitats that have already caused species decline in the first place. This appears to be supported, in Singapore at least, by the fact that the remaining native forests seem relatively resistant to invasion so far, presumably because of the great diversity of locally adapted native species that are able to make the most of the available resources. Despite this, an increasing number of alien species have been reported from even the most natural remaining areas of primary forest, including a small South American shrub, Koster's curse (*Clidemia hirta*), and the yellow crazy ant (*Anoplolepis gracilipes*), believed to have originated from Africa.

Globalization and Singapore's role as an international trade and travel hub mean that alien species introduction is a particularly serious biodiversity threat, with alien or non-indigenous species being introduced either intentionally or unintentionally or both. Threats to biodiversity by alien species can be direct (e.g. through herbivory, predation on or direct competition with native species) and/or indirect (e.g. alien plant species can alter soil water content, nutrient cycling and light conditions, and ultimately modify the habitat, making it less suitable for native species).

The horticultural, pet, aquarium and live-food trades are probably the biggest pathways for alien species introduction into Singapore. These have brought in various potential threats to Singapore's native biodiversity, including the Chinese soft-shell turtle (*Pelodiscus sinensis*), a possible competitive threat to the native turtle species such as the Asian soft-shell turtle (*Amyda cartilaginea*), which appears to have declined in recent years; the South American golden apple snail (*Pomacea canaliculata*), which is a rice-field pest in surrounding countries, but has been implicated in the decline of a resident apple snail (*Pila scutata*) in Singapore; the red-claw freshwater crayfish (*Cherax quadricarinatus*), which, as an ecological equivalent to native freshwater crabs, could potentially threaten the place of Singapore's endemic freshwater crab species; and the South American pumpwood tree (*Cecropia pachystachya*), which may compete with the very similar native pioneer trees in the genus *Macaranga*.

Travel-related pathways include introductions via 'hitchhiking' on various transportation modes or in packaging materials and so on. The introduced changeable lizard (*Calotes versicolor*) is thought to be one such hitchhiker, which may have been translocated to Singapore from its native range in northern Malaysia and Thailand via the railway system that runs from Singapore up to the Thai–Malaysian peninsula. This species appears to have successfully outcompeted and excluded a native equivalent, the green-crested lizard (*Bronchocela cristatella*), from many habitats, restricting the latter now to forested areas, when it was once also commonly found in open rural to suburban areas.

Freshwater ecosystems in Singapore have experienced some of the most notorious alien species invasions. The invasive aquatic weeds, water hyacinth (*Eichhornia crassipes*), water spangle (*Salvinia molesta*) and hydrilla (*Hydrilla verticillata*), have invaded many of Singapore's artificial waterways and water bodies. In the 1970s, water hyacinth and water spangle were so abundant that they covered the water surface of some reservoirs (Wee & Corlett, 1986). Other exotic aquatic plants that have been introduced to the reservoirs or park ponds include burhead (*Echinodorus* species), hornwort (*Ceratophyllum* species), lotus (*Nelumbo nucifera*), mayaca (*Mayaca fluviatilis*) and water lily (*Nymphaea* species). Many of these exotic aquatic plants are available in aquarium shops as ornamentals. Today, many remain established in reservoirs and are kept in check only by very costly physical removal measures.

Singapore's reservoirs, ponds and other artificial bodies of fresh water and waterways also abound with released (or rather abandoned) pets. Some of these are illegally traded species (Kaur, 2003; Goh & O'Riordan, 2007), which are unlikely to establish here in part because of the low frequency or small numbers that actually pass through the trade and are released or escape. Legally imported pets, however, are a different matter. Despite a potentially unsuitable climate (different from their native conditions) for reproduction, the large volume of imported animals poses bigger problems because of the sheer numbers that find their way into the wild. The American bullfrog (*Lithobates catesbeianus*), for example, is a large species imported for the live-food trade (but sometimes also sold in the aquarium trade as food for large predacious fish) that, while not yet established in the wild in Singapore, nevertheless poses a potentially serious predatory and competitive threat to the native frog species as well as to other freshwater organisms.

The most well-known example and possibly the most serious pest, however, is the American red-eared slider or terrapin (*Trachemys scripta elegans*). This hardy species is a popular pet when small (below 5 centimetres in size), when their delicate features and bright-green colour appeal to children. However, once they reach adulthood at sizes of 15–20 centimetres, they become darker, lose their bright colour and can cause injury by biting. This is when many pet owners 'outgrow' their interest and the animals are then released into the wild. Cultural practices involving the release of live animals for spiritual merit – sometimes as part of various religious festivals and ceremonies – have also resulted in the release of numerous red-eared sliders. Thousands of these animals are released into reservoirs, public ponds and waterways, sometimes resulting in mass die-offs from overpopulation and starvation, posing serious environmental and water-quality problems. There is also the threat that these turtles may displace local species, deprive them of food and/or space, and even spread diseases (Sulaiman, 2002; Ramsay *et al.*, 2007). Like the American bullfrog, the red-eared slider is not established in the wild (i.e. it does not have self-sustaining populations) in Singapore possibly because of the environmental limitations. That said, several hundred thousand terrapins transiting every year in a small island like Singapore will nevertheless pose a serious problem, given the frequent releases of large numbers of individuals and the longevity of these animals.

V.   Pollution

Pollution is a major environmental threat in freshwater systems globally as well as in Singapore. Effluents from industrial as well as domestic sources combine with surface run-off from urban areas and cultivated land to deliver potentially lethal cocktails of contaminants including heavy metals, detergents and biologically active compounds (e.g. endocrine disrupters) into freshwater habitats, posing a serious threat to freshwater biodiversity. Cultural eutrophication (anthropogenic nutrient enrichment) from various sources, for example, fertilizer run-off from golf courses and agricultural areas, is also a major threat to aquatic biodiversity through the encouragement of algal blooms and associated effects, such as hypoxic conditions or cyanobacterial toxicity, which in turn may result in the decline of species abundance and diversity. The threat of pollution to terrestrial habitats has received less attention in Singapore because the relatively high air quality.

VI.   Other Threats

Singapore's BTNR demonstrates two other threats to biodiversity that will certainly become more prevalent as cities develop and populations mature. The first is habitat fragmentation. BTNR is separated from the much larger CCNR by the Bukit Timah Expressway (BKE), and this isolation could lead to loss of genetic or even species diversity from BTNR because of the lack of gene flow and recruitment from CCNR. Even before BKE was completed, there was no forested connection between BTNR and CCNR, and species unwilling or unable to cross open areas were effectively isolated. An illustration of this threat was the gradual dwindling of the isolated BTNR population of banded leaf monkeys (*Presbytis femoralis*), a nationally endangered species in Singapore, to a single female, before the completion of the six-lane BKE in 1986 (the animal was killed a year later by dogs when she descended from a tree).

The second threat alludes to BTNR being literally 'loved to death' by thousands of visitors pounding the trails of the reserve in order to experience nature or perhaps just to get away from the concrete jungle. This is paradoxically related to the public's growing interest in Singapore's natural heritage and awareness of its nature areas. The challenge here is to maintain that interest and meet the needs of the growing population by allowing continued access while balancing the need to minimize visitor impact on the forest. The number of forest trails has been reduced in recent years, but this has increased pressure on the trails that remain.

As more land is developed for housing and recreation in land-scarce Singapore, and more natural environments abut human habitations, biodiversity more often comes into contact and conflict with human society. Conflict often stems from cultural attitudes to wild animals. For instance, urban commensal animals, such as the common palm civet (*Paradoxurus hermaphroditus*), which frequents parks and gardens and is generally harmless but rarely seen and poorly understood, are sometimes viewed with suspicion or even open hostility by uninformed members of the public. At the other extreme are long-tailed macaques (*Macaca fascicularis*), which have become a real nuisance and caused problems in some areas simply because humans fail to understand the consequences of their feeding of these animals. Although it is illegal to feed macaques, it is exasperatingly difficult to stop people from doing so, with the laws (and fines) being difficult to enforce. Despite the best efforts of the National Parks Board, many people still do not realize that feeding macaques will make them less fearful of people (or they perhaps do not see that as a bad thing), develop unhealthy food preferences, and also increase populations beyond what the forest ecosystems can support. Unsurprisingly then, there has been a rise in the number of complaints of aggressive behaviour or even attacks by rogue monkeys on people over the years. Wild pigs (*Sus scrofa*) have recently re-invaded Singapore Island from Johor, following their earlier extirpation by hunters, and the population is expanding exponentially. Pigs are bigger and more aggressive than macaques and more willing to cross open areas, so future conflicts with people are almost inevitable.

VII. Climate Change

Most of the threats mentioned above can be significantly reduced by action in Singapore. Human-caused climate change, in contrast, is a global issue and one that Singapore can only deal with in consort with the global community. Singapore's National Climate Change Secretariat is expecting a rise in temperature of 2.7–4.2°C by 2100, along with possible changes in rainfall and a sea level rise of at least 60 centimetres. The implications of this for Singapore's biodiversity are still unclear but are almost certain to be negative (Corlett, 2011).

VIII. Conclusion

The environmental changes that have resulted in species loss in freshwater and terrestrial systems in Singapore are not easily reversed. Indeed, the high proportion of endangered and critically endangered species amongst the native flora and fauna suggests that further extinctions are inevitable. Forests and forest streams can potentially be restored and nationally extinct species re-introduced from the surrounding countries, but, in reality, only a small proportion of Singapore can be dedicated to natural habitats. It may also be possible to make urban habitats more suitable for the more tolerant native species. Current projects aimed at bringing native species into urban areas include the rehabilitation of selected stretches of freshwater canals and the experimental planting of a wide variety of native forest plant species in parks and along roadsides. There are also ongoing research projects looking at the potential for the control of invasive species. Further progress in slowing, and then reversing, biodiversity loss in Singapore is likely to involve a range of conservation tools, from better protection of existing habitats, through to habitat restoration and species re-introductions. Singapore has been a model for biodiversity decline under intense human pressure. Hopefully, it can also be a model for its recovery.

Acknowledgements

We thank Peter Ng for his helpful comments on this manuscript. This article is an update based on an essay titled 'Threats to Biodiversity in Singapore' published in *Singapore Biodiversity: An Encyclopedia of the Natural Environment and Sustainability* (Yeo *et al.*, 2011).

References

Brook BW, Sodhi NS, Ng PKL (2003) Catastrophic extinctions follow deforestation in Singapore. *Nature* **424,** 420–3.

Corlett RT (2011) Impacts of warming on tropical lowland rainforests. *Trends in Ecology & Evolution* **26,** 606–13.

Davison GWH, Ng PKL, HC Ho (eds) (2008) *The Singapore Red Data Book: Threatened Plants and Animals of Singapore*, 2nd Edition. Nature Society, Singapore.

Goh SN (1994) The last mammals. *The Straits Times,* 16 May 1994, pp. 2–3.

Goh TY, O'Riordan RM (2007) Are tortoises and freshwater turtles still traded illegally as pets in Singapore? *Oryx* **41,** 97–100.

Kaur S (2003) Singapore a wildlife 'centre' for illegal trade? *The Straits Times,* 7 October 2003.

Ramsay NF, Ng PKA, O'Riordan RM, Chou LM (2007) The red-eared slider (*Trachemys scripta elegans*) in Asia: A review. In: Gherardi F (ed) *Biological Invaders in Inland Waters: Profiles, Distribution, and Threats,* Invading Nature – Springer Series in Invasion Ecology 2, pp.161–74. Springer, Dordrecht, the Netherlands.

Sulaiman S (2002) American ex-pets pushing out locals. *The Sunday Times,* 29 September 2002, p. 30.

Tan HTW, Chou LM, Yeo DCJ, Ng PKL (2010) *The Natural Heritage of Singapore,* 3rd Edition. Pearson Prentice Hall, Singapore.

Wee DPC, Ng PKL (1995) Swimming crabs of the genera *Charybdis* De Haan, 1833, and *Thalamita* Latreille, 1829 (Crustacea: Decapoda: Brachyura: Portunidae) from Peninsular Malaysia and Singapore. *Raffles Bulletin of Zoology* **Supplement 1,** 1–128.

Wee YC, Corlett, RT (1986) *The City and the Forest: Plant Life in Urban Singapore.* Singapore University Press, Singapore.

Yang CM, Yong K, Lim KKP (1990) Wild mammals of Singapore. In: Chou LM, Ng PKL (eds) *Essays in Zoology: Papers Commemorating the 40th Anniversary of the Department of Zoology,* pp. 2–24. National University of Singapore, Singapore.

Yeo DCJ, Ng PKL, Corlett RT, Tan HTW (2011) Threats to Singapore Biodiversity. In: Ng PKL, Corlett RT, Tan HTW (eds) *Singapore Biodiversity: An Encyclopedia of the Natural Environment and Sustainable Development,* pp. 96–105. Editions Didier Millet, Singapore.

Preparing for Climate Change: Adaptations

Chapter 9: Building Climate Change Resilience for the
Asia-Pacific Transport Sector: Linking Climate Change
Adaptation and Environmental Safeguards – Case Studies
Carsten M HÜTTCHE

CHAPTER NINE

Building Climate Change Resilience for the Asia-Pacific Transport Sector: Linking Climate Change Adaptation and Environmental Safeguards – Case Studies

Carsten M HÜTTCHE
Founder and Director
Environmental Professionals (Enviro Pro)/
Enviro Pro Green Innovations (S) Pte Ltd, Singapore

"Climate change adaptation and, likewise, environmental and social safeguards become useful instruments for decision-making, if applied early in the plan or project formulation process. Once a potential impact or risk has been identified, whether it is caused by climate change or a project activity, appropriate measures to avoid, abate or minimize this impact can be identified and described. For climate change impacts, the effects of climate change on natural and human systems, adaptation measures may be developed. Due to the similarities in activities, climate change adaptation can be integrated into environmental and social safeguard processes. This may lead to cost and time savings when performed concurrently, which is a crucial benefit for development planning."

— Carsten M HÜTTCHE

Building Climate Change Resilience for the Asia-Pacific Transport Sector: Linking Climate Change Adaptation and Environmental Safeguards – Case Studies

Carsten M HÜTTCHE[1]
Founder and Director
Environmental Professionals (Enviro Pro)/
Enviro Pro Green Innovations (S) Pte Ltd, Singapore

Abstract

Climate change is a global concern, and countries in Asia and the Pacific are likely to be amongst those hardest hit. Climate change impacts are expected to impose additional burdens on development in the emerging economies of the region. Poor and marginalized communities are most likely to suffer most from the challenges posed by climate change in the near future. This paper discusses recent trends that have seen the environmental assessment process in Asia and the Pacific encompassing impacts of climate change. Most recently, a new assessment tool described as Climate Change Adaptation has been developed and is being used to climate-proof essential infrastructure, such as road development projects. Climate Change Adaptation and Environmental Impacts Assessment are useful instruments for decision-making if applied early in the plan formulation process. Both processes can moderate environmental and social impacts, arising from a specific intervention or a climatic event, thus protecting investments made into a natural or human system. Climate Change Adaptation involves adjustments in natural or human systems in response to actual or expected climatic stimuli or their effects, moderating harm or exploiting beneficial opportunities. Mainstreaming climate change into development investments has only recently started and remains a work in progress. The paper discusses the practical experiences with environmental and climate change assessment applications on two case studies in the transportation sector. It presents some preliminary findings and lessons learned from the Road Network Development Project in Timor Leste and the Second Solomon Island Road Improvement Sector Project. For both projects, climate change adaptation assessments were undertaken by the Asian Development Bank.

---

[1] Enviro Pro Green Innovations (S) Pte Ltd, TradeHub 21, 18 Boon Lay Way, #10-114, Singapore 609966. Email: envpro@starhub.net.sg; Tel: +65 6465 1187, Fax: +65 6465 1186.

Key words: Climate Change Adaptation assessment, climate-proofing, environmental impact assessment and safeguards, vulnerability assessment, global climate models, downscaling, engineering and non-engineering adaptation options, economic cost–benefit analysis, mainstreaming climate change

I. Introduction

Climate change is a concern in Asia and the Pacific, and its impacts are projected to intensify in the decades to come, threatening the development and security of the region. The impacts of climate change are expected to impose additional burdens on development. A report by the Asian Development Bank (ADB, 2009) titled *The Economics of Climate Change in Southeast Asia: A Regional Review* estimates a potential loss equivalent to 6.7 per cent of combined gross domestic product (GDP) per year by 2100 for four countries in Southeast Asia. The projected impacts in countries such as Indonesia and Thailand will affect densely populated coastal lowlands and river basins/deltas, low-lying islands and fragile mountain ecosystems. The same may be safely stated for the South Pacific region with similar geographic-climatic characteristics.

The Environmental Impact Assessment (EIA) process and related legislative framework have been introduced to Asia and the Pacific since the late '70s. Many Asian and Pacific nations have developed EIA legislative frameworks by adopting and adapting those from developed countries in order to define and implement a formal process to assess environmental and social impacts of developmental sectors and projects. Historically, the EIA process targets large-scale industries like oil and gas, manufacturing, logging, agriculture, mining, tourism and infrastructure development. These sectors are driving the emerging economies of Asian countries and have also played a role in the Pacific region. The EIA legislative frameworks have been initially defined for projects in the context of a specific type and scale of development. As the EIA process is progressing further, governments have extended the principles of EIA to national or regional policies, plans and programmes (i.e. Strategic Environmental Assessment [SEA]).

For example, in the People's Republic of China, a new EIA law was passed in 2002 to incorporate Strategic Environmental Assessment (SEA) into the EIA legislative framework. According to the new EIA law, environmental impact assessment is mandatory for all developmental plans and construction projects. For developmental plans, EIA is to be conducted parallel to plan formulation, so that a plan will not be evaluated without the submission of an effective EIA Report. Therefore, the new EIA covers strategic planning as well as project design. As a result, more social and economic factors are taken into consideration in addition to the environmental assessment. However, Chinese EIA laws still do not cover policy aspects under the new legislative framework. Major flaws in the current EIA legislation are the lack of clearly stated public participation requirements and the low maximum penalty in case of non-compliance and corrupt practices.

Singapore has taken a different road with a planning-based approach rather than a legislative approach and currently has no mandatory EIA law per se. Environmental considerations are integrated in the spatial planning framework of the Southeast Asian city state, resulting in tight government planning and zoning laws. Pollution-control studies for industrial developments and mandatory environmental management practices for a wide range of industries and projects are incorporated through a range of acts and regulations (i.e. the Environmental Protection and Management Act). The Singapore Government, through its various agencies, has commissioned EIAs on an ad hoc basis, mainly for larger-scale industrial and infrastructure developments. More recently, the country's park and nature reserve authority has developed an environmental impact screening process that requires a classification of proposed development projects near ecologically sensitive areas early during the planning stage. Projects with potentially higher impact significance (i.e. category 3), as identified during the screening process, will require a Biodiversity Impact Assessment and/or EIA prior to project implementation. The drawback, however, is that a consistent EIA standard has yet to be established in Singapore. Current EIA practices, although improving, may not fully comply with best international EIA practices, for instance, in the areas of systematic public participation and disclosure of information. There is little environmental litigation in Singapore at present, partially as a result of the relatively stringent enforcement practices of applicable environmental laws and regulations by the Singapore Government, when compared to its Asian neighbours.

Internationally, the 'family' of environmental assessment frameworks is constantly evolving and growing. At the end of the last century, it had expanded into and linked together with Environmental Management Systems, Lifecycle Assessments for specific products and services as well as Social Impact and Poverty Assessment, amongst others.

This paper discusses recent trends that have seen the environmental assessment process in Asia and the Pacific, encompassing climate change impacts. Most recently, a new assessment tool described as Climate Change Adaptation (CCA) assessment has been developed or is in the process of being developed throughout the region. The author is observing that EIA methods and tools are being applied at least partially to address climate change impacts in a particular developmental project or sector.

For example, the assessment of the vulnerability of a location or region to climate change impacts is comparable in principle to the identification of environment-sensitive receivers to other environmental impacts in EIA, be it caused by industrial pollution or other development projects. In both cases, the common denominator is that of identifying and describing a potential future risk to a natural or human system or asset by an identified impact, resulting from a proposed activity or climate change.

Likewise, the aspect of public consultation with affected communities is featured prominently in EIA and in the CCA assessment during the baseline or vulnerability evaluation studies. For CCA, this will include investigations into recent extreme climatic events and natural disasters.

Once a potential impact or risk has been identified, whether it is caused by climate change or a project activity, appropriate measures to avoid, abate or minimize this impact can be identified and described. For climate change impacts, the effects of climate change on natural and human systems, adaptation measures may be developed. Adaptation involves adjustments in natural or human systems in response to actual or expected climatic stimuli or their effects, which moderate harm or exploit beneficial opportunities (ADB, 2011).

In the EIA process, mitigation measures are provided in the Environmental Management and Monitoring Plan (EMP or EMMP), which represents the quintessential output of EIA, whereas, for climate change impacts, adaptation options are then tested for their economic feasibility, applying the tool of cost–benefit analysis.

Overall, climate change adaptation and, likewise, EIA become useful instruments for decision-making, if applied early in the plan formulation process. Both processes can moderate environmental and social impacts, arising from a specific intervention or a climatic event, thus protecting investments made into a natural or human system.

International Financial Institutions (IFIs) such the World Bank Group, ADB and bilateral development agencies have started incorporating potential future climate change impacts in their investment programming and strategies, primarily in the infrastructure sectors of developing countries in Asia and the Pacific. These countries are amongst the most globally vulnerable to the adverse impacts of climate change. Poor and marginalized communities are likely to suffer most from the challenges posed by climate change in the near future. Recognizing these threats, ADB, for instance, has developed a long-term strategic framework to assist developing member countries in Asia and the Pacific to address climate change impacts on vulnerable sectors, such as transportation, agriculture, energy, water and health, and to build a climate-resilient region. By doing so, ADB attempts to provide help in 'climate-proofing' investments in these sectors to ensure that their intended outcomes are not compromised by climate change. This makes good economic sense, as failing infrastructure projects due to climate change impacts will adversely hinder developmental goals and the ability of developing countries to repay loans provided by these development agencies.

Environmental Safeguard Procedures and Climate Change Adaptation – Asian Development Bank

ADB's 2009 Safeguard Policy Statement (SPS) states the following three safeguard policies as included as part of the operational policies, the *Involuntary Resettlement Policy* (1995), the *Policy on Indigenous Peoples* (1998), and the *Environment Policy* (2002). Safeguard policies are generally understood to be operational policies that seek to avoid, minimize, or mitigate adverse environmental and social impacts, including protecting the rights of those likely to be affected or marginalized by the development process. All three safeguard policies involve a structured process of impact assessment, planning, and mitigation to address the adverse effects of projects throughout the project cycle.

Environmental safeguards are triggered if a project is likely to have potential environmental risks and impacts. ADB uses a classification system to reflect the significance of a project's potential environmental impacts. A project's category is determined by the category of its most environmentally sensitive component, including direct, indirect, cumulative, and induced impacts in the project's area of influence. Each proposed project is scrutinized as to its type, location, scale, and sensitivity and the magnitude of its potential environmental impacts.

Projects are assigned to one of the following main categories:

Category A – A proposed project is classified as category A if it is likely to have significant adverse environmental impacts that are irreversible, diverse, or unprecedented. These impacts may affect an area larger than the sites or facilities subject to physical works. An environmental impact assessment (EIA) is required.

Category B – A proposed project is classified as category B if its potential adverse environmental impacts are less adverse than those of category A projects. These impacts are site-specific, few if any of them are irreversible, and in most cases mitigation measures can be designed more readily than for category A projects. An initial environmental examination (IEE) is required.

Category C – A proposed project is classified as category C if it is likely to have minimal or no adverse environmental impacts. No environmental assessment is required although environmental implications need to be reviewed.

With an established environmental policy in place, ADB, in 2008/2009, introduced Climate Change Adaptation assessments for selected projects. For a transport sector developmental project by ADB in Timor Leste in 2009, the environmental and social baseline assessments necessary for IEE were extended to include the collection of data on climate change impact vulnerability of selected project locations. The data were then used for a Climate Change Adaptation assessment.

The ADB environmental and social safeguard frameworks provide a set of prescribed activities that can support the assessment of environmental and social risks related to existing or predicted climate changes. For example, the Rapid Environmental Assessment Checklists, prescribed in the ADB EIA process for early environmental impact screening and impact categorization now include climate change-related questions, though they remain optional at this stage. ADB is currently testing a project risk-screening tool for climate change and natural hazard risks. After it is tested, climate change and natural hazards may become a component of the environmental safeguard process. Most recently, ADB published *Guidelines for Climate Proofing Investment in the Transport Sector,* which is available as a free e-book on ADB's website.

Mainstreaming climate change into development investments and, in particular, into the environmental safeguard process remains a work in progress.

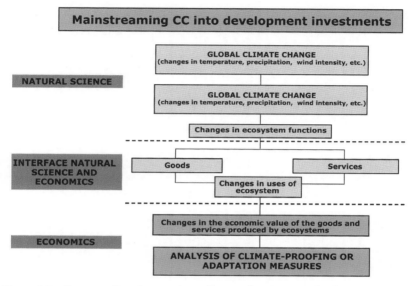

Figure 9.1: Process of mainstreaming climate change into development projects. The integrated, interdisciplinary analysis of technical and non-technical adaptation measures is followed by an economic cost–benefit analysis of all identified adaptation options. CC: climate change (Source: ADB)

II. Climate Change Adaptation – Challenges and Lessons Learnt from Development Investments

There are different definitions of adaptation to climate change. The latest Intergovernmental Panel on Climate Change (IPCC, 2007a: 869) assessment report, for instance, gives the following definition: 'Adjustment in natural or human systems in response to actual or expected climatic stimuli or their effects, which moderates harm or exploits beneficial opportunities'. In comparison, the definition of mitigation is simple. It is just the reduction of greenhouse gases (GHGs). Two strategies are necessary to reduce the risks of climate change:

1. *Mitigation* – The causes of climate change are removed by reducing GHG emissions ('avoid the unmanageable…').
2. *Adaptation* – The effects of climate change are dealt with by coping with their negative impacts ('… and manage the unavoidable').

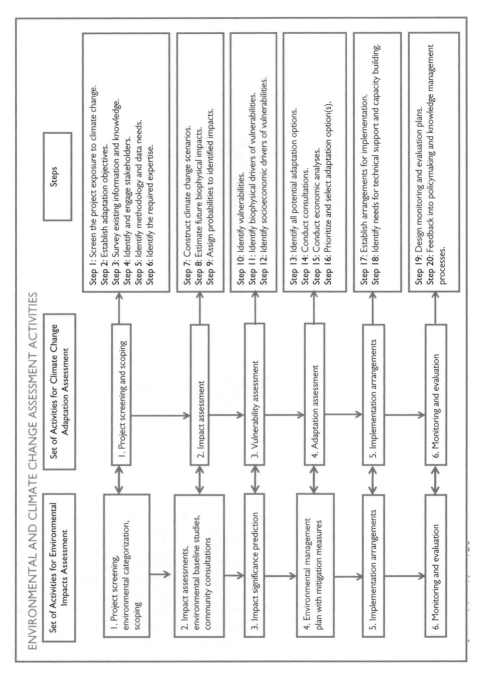

Figure 9.2: Activities and steps for environmental impact and CCA assessments. Due to the similarities in activities, CCA can be integrated into safeguard processes. This may lead to cost and time savings when performed concurrently. (Source: Adapted from ADB)

The two strategies are interlinked: the more successful the first strategy is, the less the second one is required.

Climate-Proofing and Adaptation Assessment

The term *climate-proofing* infrastructure generally refers to investing in measures that reduce risks to an acceptable level through long-lasting, environmentally sound, economically viable and socially acceptable activities (ADB, 2005). In practical terms, climate-proofing involves (Sveiven, 2010):

• identifying risks to a development project as a consequence of climate variability and change; and
• ensuring that those risks are reduced to acceptable levels through environmentally sound, economically viable and socially acceptable measures.

The process of climate-proofing includes, amongst others, an adaptation assessment (see Figure 9.2), in which the current and expected changes in hazards and vulnerability (risks) under climate change and variability are evaluated. This is followed by the identification of adaptation measures to reduce the risks to an acceptable level, and the selection of adaptation measures that reflect consideration of potential impacts on social, economic and environmental systems.

Adaptation options in the transportation sector can generally be divided into engineering (or structural) (such as subsurface conditions, material specifications, cross-section and standard dimensions, drainage and erosion, as well as protective engineering structures) and non-engineering options (such as maintenance planning and early warning, alignment, master planning and land use planning, and environmental management). In addition, it is important to recognize that, in a number of circumstances, a 'do nothing' response to climate change – for example, allowing an infrastructure to deteriorate and be decommissioned instead of climate-proofing the infrastructure – may be a preferred course of action.

The early applications of climate change adaptation assessment face many challenges. In order to make educated decisions about which adaptation options are best suited to moderate impacts by an actual or expected climatic stimuli, the risks need to be adequately identified and quantified. A range of Global Climate Models (GCMs) have been devised by leading scientific institutions around the globe, which make varying predictions for climate change on a global scale. However, for a project with limited geographical limits, such as infrastructure investments, these models may not provide sufficient resolution to predict and quantify climate change risks relating to this project. The larger the possible variations in climate change predictions, the harder it is to pinpoint and quantify suitable adaptation measures. This may have significant consequences for engineering and non-engineering responses to the identified climate change risks and their related investment costs, adding to the existing costs for proposed infrastructure projects. In order to provide large sums of investment dollars for proposed infrastructure projects or sectors, the decision will expect the project to be financially and economically viable. This is normally expressed by a defined internal rate of return (IRR) and net present value (NPV) for the project or programme. It has been shown that significant upfront investment costs for CCA measures may push individual projects or sector programmes below the required financial and economic benchmarks. As a result, the projects may be implemented with incomplete or downsized CCA programmes.

On the other side, the quantification and monetization of benefits derived from non-technical adaptation measures, such as the protection of water catchment areas for storm water infiltration and retention in order to reduce the risks of downstream flooding on a proposed infrastructure, are often not adequately considered. Economists still find it difficult to monetize the values of these important environmental services and assign relevant shares to specific projects. It appears to be, by far, simpler to work with human systems, such as the maintenance costs for a particular infrastructure project, and to consider the options of increasing or decreasing maintenance or operational costs with and without climate impacts. Based on a small number of case studies, the author finds that the cost–benefit analysis performed for investments, including climate change adaptation measures, emphasizes the human systems and fails to place in proper proportion the benefits of natural systems or environmental services delivered. Thus, non-technical adaptation measures, such as reforestation or rehabilitation of natural systems, have often been underestimated in their relevance to reduce harm from climate change impacts. This is aggravated by the fact that these non-technical measures are often spatially disconnected from proposed infrastructure intervention, pushing them outside the scope of the project.

In the recent climate-proofing guidelines, ADB concludes that greater attention needs to be given to the upstream decision-making processes such as transport master planning. Here, parallels can be drawn to the developments in the EIA sector, where SEA has been introduced to deal with environmental impacts at policy or master-plan level.

Climate Change Adaptation as a Development Challenge

For many developing nations in Asia and the Pacific, the issue of climate change is a fundamental development issue. The majority of the people are directly dependent on local forests, arable land and healthy rivers for their survival. Climate change affects a country's natural resource base, undermines existing agricultural systems, deepens food insecurity, compounds poverty, threatens infrastructure and water resources, and will likely overwhelm health-care systems. Since wealth offers adaptive capacity, the broadest and most effective climate mitigation and adaptation strategy is the eradication of poverty. Addressing poverty in ecologically sustainable ways will reduce the regions' vulnerability to climate change by increasing the ability of both social and environmental systems to adapt to climate change. The following lists a number of potential development policies that could help to reduce vulnerability to climate change:

(i) The development of adaptable farming systems

The development of adaptable farming systems may need to be prioritized for its beneficial impact on food security and economic development. Expected changes in temperatures, rainfall patterns and humidity have ominous implications for crop planting and harvesting. Historically, local agricultural patterns have evolved around the monsoon climate. Water is a critical resource and the most important environmental constraint on agricultural production. Variability in climate is significantly influenced by El Niño Southern Oscillation, which changes the timing and volume of rainfall. Global warming will exacerbate this problem. Characterized by seasonal climate forecasts, water resource infrastructure and irrigation systems, the development of adaptable farming systems is crucial in assisting in flood control and supplying water during dry periods, as well as the use of drought-tolerant species to maximize output, the production and sale of surplus food, insurance systems and government safety nets to guarantee welfare in times of drought and flood.

(ii) Forest preservation and restoration

Heavy commercial pressure on forest resources has existed in Asia and the Pacific for many decades. It threatens to deplete stores of natural resource, exacerbate the problem of soil erosion and interrupt the hydrological cycle with negative consequences for agriculture and infrastructure. Forest degradation will also lead to increased run-off during the rainy season and increased erosion will lead to landslides. Deforestation is linked to infrastructure damages to roads and bridges, which can be addressed by reforestation in upper catchment areas as a non-engineering adaption option.

(iii) Developing national meteorological services and climate early-warning systems

The first obstacle a developing country faces in dealing with a changing climate is the lack of reliable climate data collection and management systems. This lack encompasses temperature and rainfall data, and data on a range of climate-related processes like river run-offs, tides, floods and groundwater levels. An improved meteorological service, coupled with a system of disseminating seasonal forecasts on a country-wide scale, will provide local communities with information about predicted climate patterns, allowing them to better plan planting, harvesting, food storage or alternative food procurement. This would mean more efficient use of seeds, labour and other agricultural inputs.

(iv) Enhancing health care and access to potable water

Additionally, changing rainfall patterns combined with rising temperatures may alter the distribution and concentration of mosquitoes, which are vectors for diseases such as malaria and dengue fever. Declining water quality due to an increased volume of suspended sediments may encourage the spread of waterborne diseases, such as cholera and diarrhoea. It is suggested that, with disease-monitoring systems, potential communal health risks in the face of climate change will be better understood and managed. The availability of and access to good quality potable water will be major factors in any improved health-care system.

III. Case Study: A Road Network Development Project for Timor Leste

In 2009, ADB provided technical assistance to develop a sector programme for a road network development to the Government of Timor Leste. As the youngest nation of the 21st century, the country's existing road network has seen progressive degradation since Indonesia relinquished control of the territory in 1999. Being a highly mountainous country and a small island, Timor Leste is particularly vulnerable to the impacts of climate change. In addition, its low level of development and dependence on the agricultural sector mean that climate change impacts can be far reaching, while the capacity to adapt is low. Some of the changes are expected to result in increased drought in some regions and floods in others, reducing agricultural productivity and damaging coastal roads due to sea level rise. A large portion of Timor Leste's National Road Network is located in areas highly sensitive to climate change and is at risk of landslides, land degradation and sedimentation.

Figure 9.3: Road works in Timor Leste damaged by landslides. The young geological age and the high rate of tectonic uplift, combined with the presence of weak, poorly consolidated strata, produce intractable stability, slope failure and erosion problems in many areas of Timor Leste. Changes in rainfall patterns due to climate change may aggravate these problems. (Source: Feasibility Study of the Rehabilitation and Maintenance of District Roads in Timor Leste, EDF, 2011)

The initial technical assistance did not include CCA assessments, only general environmental and social safeguard procedures. However, due to an identified vulnerability to climate change during the field surveys, the scope of CCA assessment was added to the project. Two sample roads, a coastal and a mountainous road link, were subsequently chosen for a CCA assessment. The project adopted and implemented lessons learnt from other CCAs. Firstly, climate change parameters of concern to the projects need to be identified at the outset. For example, in the case of road projects, road engineers are more concerned with projected occurrences of peak rainfall rather than with annual averages so as to provide sufficient drainage facilities. Secondly, a projected climatic change on its own is not sufficient to make recommendations for project design. It should be interpreted (as with most modelling results) through complementary, field-based vulnerability assessment for adaptation options. Hence an integrated impact and vulnerability assessment was conducted.

The first step in the analysis was to assess projected climate change in Timor Leste. Based on these projections, the study team later assessed and developed adaption measures – both engineering and bio-engineering, including cost estimates. In turn, these cost estimates were used as input to the cost–benefit analysis under climate change scenarios.

The assessment of climate change was conducted by a climate change modelling expert, added as an external consultant to the study team. The climate change consultant prepared projections for Timor Leste based on leading global and regional mathematical models of climate change. The modeller zoomed in on Timor Leste and downscaled the global/regional models, concentrating on the two sample road areas. Amongst others, the methodology included verification, comparison with historical climatic data for Timor Leste, field study ('ground truthing') data collection by the project and results of 'focus group' analyses in villages along the sample roads. The assessment of climate change in Timor Leste included the following climate parameters:

- Changes in onset and intensity of seasonal rains;
- Changes in very hot days and heat waves;
- Expected sea level rise;
- Changes in intensity of and frequency of precipitation events, and flood patterns;
- Changes in seasonal precipitation and flooding patterns; and
- Changes in cyclone intensity, frequency and duration, and associated storm surges and wave actions.

Downscaling GCMs to a Regional Scale

The latest report from IPCC (2007b) brings evidence of climate change to be observed in the next few decades. With respect to small island countries, IPCC stated with 'very high confidence' that 'small islands, whether located in the tropics or higher latitudes, have characteristics that make them especially vulnerable to the effects of climate change, sea level rise and extreme events'.

There are inherent uncertainties in predicting climate changes, particularly for a 'data poor' country like Timor Leste. This makes it difficult to fully ensure that adaptation measures will be in line with future climate changes. Implementation risks also include a lack of institutional capacity to manage climate change risks, lack of quality assured data collection and co-ordination to monitor climate change. Timor Leste has just recently started to develop the National Adaptation Programme of Action for climate change.

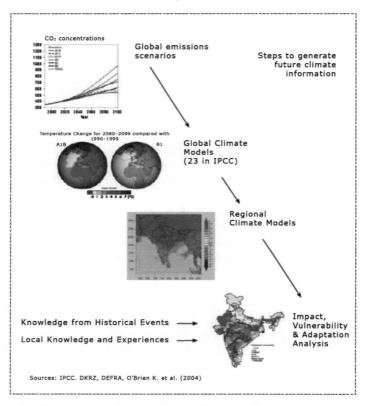

Figure 9.4: CCA process, showing the combination of 'top-down' Climate model downscaling and 'bottom-up' vulnerability analysis. During the latter, knowledge from historical events, and local knowledge and experiences with changing weather pattern as well as extreme climatic events are gathered and analyzed, together with regionalized climate change predictions. (Source: GTZ, 2009)

Table 9.1 on the next page lists key variables and their sources of information, and qualifies the level of confidence in the results.[2]

---

[2] According to IPCC (2007): "'Where uncertainty in specific outcomes is assessed using expert judgmdgement and statistical analysis of a body of evidence (e.g. observations or model results), then the following likelihood ranges are used to express the assessed probability of occurrence: virtually certain >99 per cent; extremely likely >95 per cent; very likely >90 per cent; likely >66 per cent; more likely than not > 50 per cent; about as likely as not 33 per cent to 66 per cent; unlikely <33 per cent; very unlikely <10 per cent; extremely unlikely <5 per cent; exceptionally unlikely <1 per cent'".

Table 9.1: Key variables, sources of information and level of confidence in the results

Variable	Source			Level of confidence
	GCMs	CCAM (CSIRO)	Literature	
Annual temperature	✓		✓	Extremely likely
Seasonal temperature	✓			Very likely
Monthly temperature	✓			Likely
Hot days and heat waves		✓	✓	Likely
Annual rainfall	✓			Very likely
Seasonal rainfall	✓			Likely
Monthly rainfall	✓		✓	Likely
Intensity and frequency of rainfall events		✓		Likely
Cyclone intensity, frequency and duration			✓	Likely
Droughts		✓		Likely
Flood patterns, intensity and frequency		✓		Likely
Sea level rise			✓	Very likely
Storm surge and wave action			✓	More likely than not

Source: Preparing the Road Network Development Project, ADB, 2009

Methodology Used to Project Changes of Climate and Limitations

In order to downscale climate predictions from GCMs to regional scales, various approaches can be applied. Downscaling methods increase both spatial resolution (e.g. from hundreds to tens of kilometres) and temporal resolution (e.g. from monthly to daily). ADB's 'Guidelines for Climate Proofing Investment in the Transport Sector' identifies two main approaches for downscaling: dynamical downscaling (using regional climate models [RCMs]) and statistical downscaling (using empirical relationships). Each downscaling method has its strengths and limitations, and the appropriate method will depend on the specific needs of the impact assessment, data availability and budget. However, it is important to note that, since downscaling is a transformation of GCM outputs, it cannot add skill or accuracy that is not present in GCMs. If GCMs do not accurately project changes in large-scale atmospheric circulation patterns, downscaling techniques cannot correct the errors.

Climate change projections can be useful in determining how climate variables, such as temperature and precipitation, may change in the future. However, projections based on climate model outputs are limited by the imperfect representation of the climate system within climate models, in addition to uncertainties associated with Greenhouse Gas Emissions (GHG). Therefore, climate projections are not forecasts or predictions, but provide alternative characterizations of possible future climate conditions. They are helpful in exploring 'what if' questions and do not aim to provide accurate predictions of what the climate will be in the future.

ADB finds that the best approach to use for a given project is chosen based on the adaptation decision context, availability of data, time frame and budget. The most common approach is to use existing GCMs or RCMs and apply a simple spatial downscaling technique using local historical climate data. This is a cost-effective option to obtain climate projections for the project site. In the developmental assistance context, given the time, technical expertise and resources required, building new RCMs or even running new simulations with existing RCMs is generally not advisable within the context of a single project, but may be justified for sector programs or plans.

In the following section, a short summary of a downscaling approach is given, which has been used to investigate the evolution of temperature and rainfall associated with climate change in Timor Leste (Roy *et al.*, 2008).

(i) Verification mode

- A database with observations of temperature and rainfall for the region of interest is organized.

- Then, centred on the region of interest, the most relevant climate simulations (GCM) for the reference period (simulations suggested by IPCC in their Fourth Assessment Report, 2007b) are downloaded.

- Based on their skill in reproducing the intra-annual pattern of observed temperature (see Figure 9.4) and rainfall of the recent decades, the most relevant climate models are then chosen.

(ii) Simulation mode

- Three sets of 30-year climate projections for time horizons 2020, 2050 and 2080 (projections selected by IPCC in their Fourth Assessment Report, 2007b) are downloaded for each combination of climate model and GHG scenario.

- For each time step (2020, 2050 and 2080), annual, seasonal and monthly temperature and precipitation projections for each and every climate projection are analyzed.

- The mean (average from all projections) expected changes of temperature and rainfall for the region of interest in 2020, 2050 and 2080 are estimated.

- Climate change signals of the future with respect to the present (reference period) are calculated. These changes are expressed as the difference (°C) for simulated temperatures and a ratio (per cent) for simulated precipitations.

Historical Data Comparison with Climate Models Simulation for the Reference Period

In order to verify the ability of the different available climate models in reproducing recent past climate conditions, the climate model simulations for the reference period are investigated. This information will be used to compare the climate simulation of the reference period with the observations over approximately the same region of the world.

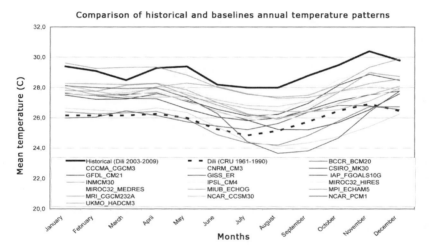

Figure 9.5: Observation and GCM simulations of mean annual cycle of temperature for the reference period 2003–2009 for Dili, Timor Leste. Note that each line represents the simulations of one GCM versus the historical data expressed by the blue bold line. By comparing actual past measurements with predictions by models for the same reference period, suitable GCMs for future predictions can be selected. (Source: Preparing the Road Network Development Project, ADB, 2009)

Climate Change Predictions for Timor Leste

From the analysis of climate projections and information gathered from recent literature; temperature in Timor Leste is expected to increase over the next decades. Despite a very wide range of possible climate conditions, mean annual precipitation is also expected to increase slightly (especially over the wet season).

- *Horizon 2020:*
  Mean annual temperature increase: +0.8°C (varying between +0.4 and +1.5°C)
  Mean annual precipitation difference: +2 per cent (varying between -12 and +15 per cent)

- *Horizon 2050:*
  Mean annual temperature increase: +1.5°C (varying between +0.7 and +2.8°C)
  Mean annual precipitation difference: +4 per cent (varying from -25 and +15 per cent)

- *Horizon 2080:*
  Mean annual temperature increase: +2.2°C (varying between +0.8 and +4.0°C)
  Mean annual precipitation difference: +6 per cent (varying from -21 and +32 per cent)

Future conditions may bring fewer extreme rainfall events, but their intensity would become more important. Tropical cyclone patterns and occurrences are not expected to change much in a warming world, while the intensity of these cyclones and their accompanied precipitations volume may increase over the next decades.

The combined effects of temperature increase, and no significant difference in rainfall over the dry season would have considerable effects on the run-off volume during the drought period (run-off volume could be significantly reduced). Similar to floods, both intensity and frequency are expected to increase significantly for the region of interest.

Projected sea level rise at the end of the 21st century relative to the reference 1980–1999 period for the six SRES scenarios ranges from 1.9 to 5.8 millimetre/year. Studies conducted worldwide have shown that the extreme wave height probability of occurrence and intensity associated with climate change along coastal regions are likely to increase.

Field Ground Truthing, Vulnerability and Risk Assessment

An inherent level of uncertainty exists for any such climate change impact assessment, which is model based, and even more so for a data-poor country such as Timor Leste. For this reason, the project used 'bottom-up' vulnerability and risk assessments in the field to complement the 'top-down' approach of the mathematical climate change models. Amongst others, the advantage of such 'bottom-up' assessments is to ensure that proposed measures are relevant to the existing vulnerabilities, and also to maximize co-benefits. This included field-level observations on currently observed climate changes and coping mechanisms at the two project sites.

The project is characterized by two regions facing different vulnerability characteristics: coastal roads along the northern coast are exposed to climate risks related to mean rising sea levels and cyclone activity, and roads passing through highly mountainous interior agricultural areas, which are exposed to landslides due to changes in the hydrological regime. Vulnerability varies widely across communities, sectors and regions. This diversity of the 'real world' is the starting place for the vulnerability or risk assessment. Trend analyses are often used at this point to assess the closeness of the model-generated climate change scenarios with the actual trends in climate change.

In this case, the project team produced climate vulnerability maps based on the field ground truthing and risk assessment. The maps were derived from geographic information system (GIS) analysis (i.e. digital elevation model) and field observation by engineers and environmental experts (i.e. existing landslide or flooding zones). During ground truthing surveys, the project team used 'Natural Hazard Location Mapping Forms' to record individual hazards that were subsequently digitized. A scoring system was used to produce vulnerability indices based on GIS natural hazard maps to further guide more detailed ground assessments. Climate-sensitive hotspots were identified and confirmed against the projected climate change scenarios. This process helped to identify priority areas for adaptation measures.

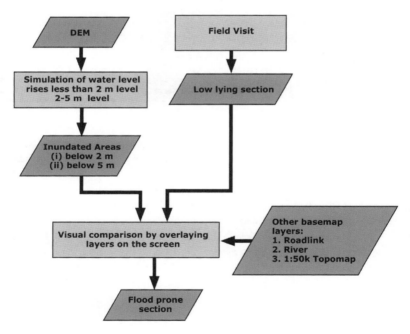

Figure 9.6: Sample process flow for vulnerability assessment as part of the CCA process. In this case, coastal areas along the sample road were plotted with a 2- to 5-metre sea level rise to identify and visualize flood-prone areas. During field ground truthing surveys, these areas were then visited to investigate and record signs of floods and inundation at current climate conditions. DEM: Digital elevation model (Source: Preparing the Road Network Development Project, ADB, 2009)

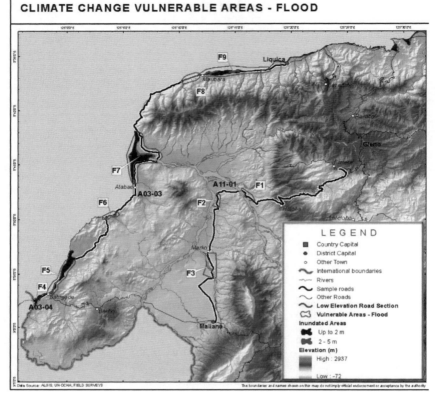

CLIMATE CHANGE VULNERABLE AREAS - FLOOD

Figure 9.7: Sample vulnerability map. Running parallel to the downscaling process of GCMs, ground assessment for selected sample roads was conducted to identify areas and sections of roads that are susceptible to various hazards associated with climate variability. At the desktop study phase, a GIS analysis was used to compile existing spatial data sets and to conduct a spatial analysis to identify areas along the selected road link that are susceptible to climate-related geo-hazards and to prepare detailed base maps for ground surveys. During the ground survey, GPS/GIS was used to assist with mobile tracking, field orientation and navigation as well as to manage the data captured by the project team. The collected data comprised road sections that were experiencing or are likely to experience natural hazards such as tidal floods, flash floods and various forms of earth movements. (Source: Preparing the Road Network Development Project, ADB, 2009)

In addition, temperature and precipitation records from seven weather stations nearest the two sample roads were collected and analyzed during the vulnerability assessment. Data quality is generally a problem. In this case, consistent, comparable weather data sets were only available for a period of four years between 1968 and 1971. With the assistance of GIS software, average annual rainfall, the average number of wet months per year and the average monthly rainfall of the three driest months per year, amongst other data, were spatially mapped.

The maps showed that certain low-lying areas displayed the lowest average monthly rainfall during the driest period of the year. With climate change, it is predicted that precipitation during the dry period is likely to decrease for the 2020 and 2050 time horizons, possibly further drying these areas as climate-sensitive hotspots.

Recent approaches to adaptation planning emphasized community-based participatory assessments of climate variability, including observed changes, current coping practices and enhanced adaptation needs. This last approach served as a basis for developing National Adaptation Programmes of Action (NAPAs) for climate change in Timor Leste. The project adopted this technique, which assisted in identifying adaptation options with considerable co-benefits in both project- and sector-wide planning. In this case study, a climate change field consultation was arranged in a total of seven sucos (villages) along the two sample roads with the help of a local non-governmental organization (NGO). The purpose of this consultation was to provide ground-level inputs of existing climate changes, such as the onset and intensity of seasonal rains and the increase in the intensity and frequency of flood, cyclone and storm surges, as well as to collect recommended adaptation options – practiced or planned by the communities – that can be further assessed.

Adaptation Assessment and Adaptation Options

Table 9.2 summarizes the most significant findings of CCA and the consequences for the sample coastal road, as well as the consequences and risks of impacts.

Table 9.2: Most significant findings of CCA and the consequences for the sample coastal road, as well as the consequences and risks of impacts

Climate change finding	Hazard/ Consequence	Risk of impact
Horizon 2020–2050: Cumulative seven-day rainfall highest percentile will increase in intensity by 15 per cent.	• More run-off water • Gully erosion • Flooding	• Significant increase in risk of flooding of pavement
Horizon 2020–2050: The combined effects of the increase in temperature and the decrease in rainfall would have considerable effects on the run-off volume during the drought period (run-off volume could be reduced by as much as 80 per cent).	• Lowering of water table • Settlement of subgrade and embankment • Reduced vegetation cover, leading to increased erosion	• Significant impact on where expansive clays are present • Significant impact on steep hillsides

Climate change finding	Hazard/ Consequence	Risk of impact
Horizon 2020–2050: For cumulative 30-day run-off sequences, the highest percentile is shown to increase in intensity by up to 20 per cent. The flood frequency above an arbitrary level (cumulative daily run-off of 200 millimetres over a seven-day sequence) is expected to increase by up to 35 per cent for the region of interest.	• More run-off water • Gully erosion • Flooding	• Significant increase of risk of flooding in the vicinity of river crossings
Projected sea level rise at the end of the 21st century relative to the reference 1980–1999 period for the six SRES scenarios ranges from 1.9 to 5.8 millimetres/year.	• Flooding and wave damage in low-lying coastal areas • Raising water table in low-lying coastal areas • More embankment failures • More landslides • Saturated pavement • Saturated subgrade	• Significant risk of flooding and wave damage in low-lying coastal areas • Low risk of embankment failures and landslides in low-lying coastal areas due to flat topography • Significant risk of saturated pavement and subgrade
Horizon 2020–2050: The probability of occurrence of an extreme wave height and the intensity associated with climate change along coastal regions are likely to increase.	• Flooding and wave damage in low-lying coastal areas	• Significant risk of flooding and wave damage in low-lying coastal areas

Engineering adaption strategies have been developed for each significant infrastructure risk, as listed in Table 9.2. These strategies focus on protecting the infrastructure from the impact of the environmental hazards due to climate change. The strategies involve a combination of capital and maintenance works to ensure that a reliable and safe transport link is provided.

The following adaptation strategies have been identified:

(i) Predicted climate change impact: Sea level changes and increased storm surge wave height

- *Realignment* – Where the elevation of the road is so low such that the sea will intrude on both sides of the road, the preferred strategy is to relocate the road.

- *Erosion protection* – Where the road will be subject to risk of erosion from wave action, the preferred strategy is to construct an earth levee bank with riprap protection against erosion by wave action.

- *Increased maintenance* – The quantity of maintenance increases in response to the faster rate of physical deterioration.

(ii) Predicted climate change impact: More intense short-duration precipitation

- *Increased capacity of transverse drainage system* – Where the intensity of short-duration precipitation events increases, the capacity of transverse drainage system will be increased by providing additional relief culverts.

- *Improved longitudinal drainage* – The ability of the longitudinal drainage systems to accommodate the higher quantity of run-off due to the higher precipitation rates will be improved by lining drains and providing larger ones.

- *Erosion protection* – Areas in the vicinity of the road at risk of erosion will be protected using bio-engineering techniques. In addition, steeply graded streams in the vicinity of the road will be provided with check dams to reduce sediment loads on the road drainage system.

- *Increased maintenance* – The quantity of maintenance increases in response to the faster rate of physical deterioration.

Non-Engineering Options

Traditionally, responses to hazards, such as flooding and landslides, would be designing and providing additional engineering structures, or reinforcing existing ones. Other climate change adaptation options include reforestation and bio-engineering measures along water courses and road slopes as well as upper catchment zones. These options are considered 'no regret' or 'low regret' adaptation activities (ideally a win–win situation for mitigation, adaptation and sustainability), by increasing land cover or coastal buffers and adding carbon storage capacity. The bio-engineering adaptations methods are complementary to the engineering methods. They can be implemented along with the road construction schedule, or ahead of it, as they may be located also outside the ROW (e.g. mangrove replanting and reforestation). These solutions are relatively inexpensive, flexible and reversible, and also increase adaptive capacities when facing uncertainty. Finally, they can be part of the community programmes, involving maintenance by the local communities.

(i) An economic cost–benefit analysis of adaptation measures

The goals of the economic analysis of adaptation options are to provide decision-makers with information on the expected costs and benefits of each of the technically viable options and to rank these options according to their net benefits.

A 2011 paper by Padma and Thuirarajah for IUCN highlights the economic analysis of projects as an important component of ADB's internal operations when selecting projects, using the criteria of economic efficiency, reflected in measures such as net economic benefits, NPVs, benefit–cost ratios (BCR), and economic IRRs, estimated using cost–benefit analysis. Each adaptation option would then be compared using such economic efficiency measures estimated using cost–benefit analysis. ADB's preferred measure is usually IRR, which represents the discount rate by which benefits and costs are equal. ADB's 'basic criteria for a project's acceptability' is an economic IRR of 12 per cent (Dole & Abeygunawardena, 2002). A return of as low as 8–10 per cent is also thought to be acceptable for projects where 'additional unvalued benefits can be demonstrated, and where they are expected to exceed unvalued costs', with the lower limit of 8 per cent accepted for weakly performing countries.

Climate change assessment and adaptation procedures are unique in terms of the duration of the assessment. Climate change projections are made for a relatively longer-term change, with a period of up to 50 years and involving high degrees of uncertainty. Thus, adaptation assessments and cost–benefit analyses are also applied to a period of 50 years. This is in comparison to the typical assessment of road projects (without climate change) conducted for a period of 20 years, with higher degree of certainty. The cost–benefit-assessment in Timor Leste was conducted with 3 per cent interest rate, instead of the 'standard' rate of 12 per cent given by ADB (for no-climate-change scenarios). Nevertheless, as a comparison, the analysis also showed economic evaluations at 12 per cent and for a 20-year period.

Based on a prevailing methodology of cost–benefit analysis for climate change, the project used the savings on annual maintenance costs as the major economic benefit of climate change adaptation. Other benefits were not quantified.

The engineering adaptation for climate change along the sample coastal road for both 'Precipitation Increase' and 'Mean Sea Level Rising', if it occurred in the first one to two years, showed negative NPV at 12 per cent and relatively low return at 3 per cent as well as relatively low IRR. However, if the investment is to ameliorate only 'Precipitation Increase' or to stage investment to ameliorate 'Mean Sea Level Rising' onto years 25–26, when sea levels will actually rise significantly, warranting investment, the results were significant and positive. The reason for these results is the large size of the investment required for ameliorating 'Mean Sea Level Rise', which was approximately USD22 million.

The economic cost–benefit analysis in the case of the coastal sample road concluded to: (a) apply relatively inexpensive adaptation to precipitation change upfront; and (b) postpone expensive adaptation measures to mean sea level rise as long as possible. Non-engineering and bio-engineering adaptations, in the case of the mountainous road, were relatively inexpensive and recommended as viable.

IV. Case Study: The Solomon Islands Coastal Roads under Threat by Climate Change

The Solomon Islands are a low-lying coastal country consisting of approximately 1,000 islands located east of Papua New Guinea between latitudes 5°S and 13°S, and longitudes 155°E and 169°E. The distance between the easternmost and westernmost islands is approximately 1,500 kilometres (930 mi). Its main islands are Choiseul (the Shortland Islands), the New Georgia Islands, Santa Isabel, the Russell Islands, Nggela (the Florida Islands), Malaita, Guadalcanal, Sikaiana, Maramasike, Ulawa, Uki, Makira (San Cristobal), Santa Ana, Rennell, Bellona, and the Santa Cruz Islands. Its three remote, tiny outliers are Tikopia, Anuta and Fatutaka.

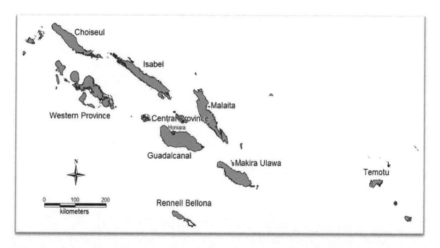

Figure 9.8: The Solomon Islands' geographical distribution (Source: Second Solomon Island Road Improvement Project, ADB, 2010)

As a **least developed country** (LDC) and a **small island developing state** (SIDS), the Solomon Islands face significant sustainable development challenges associated with impacts of climate change, variability and extreme events. LDCs and SIDS are recognized by the United Nations Framework Convention on Climate Change (UNFCCC) as countries most vulnerable to the adverse impacts of climate change.

Lal and Thuirarajah (2011) report that the Solomon Islands are exposed to a wide range of geological, hydrological and climatic hazards, including tropical cyclones, landslides, floods and droughts. Between 1980 and 2009, the country experienced 17 major disasters costing over USD20 million and affecting almost

300,000 people. Of these, there were six major natural disasters – two earthquakes and four tropical cyclones – and associated floods and storms, directly impacting over 100,000 people with over 100 deaths. Climate-related events, including floods, landslides and storms, dominate both in terms of the number of incidents as well as the number of people affected, damage and losses suffered.

Among the Solomon Islands, the rise in sea levels is already causing significant damage to rural infrastructure, including wharves and roads. Saltwater intrusion, as a result of king tides, is affecting food security and groundwater resources. Coastal erosion due to the rise in sea levels is threatening people's livelihood. Coral bleaching in some parts of the Solomon Islands has also been noted. Heavier, more frequent and less predictable rainfall (increasingly higher rainfall in wet months of summer and less rainfall in dry months) has caused flash floods, deaths, drought and ruined crops. The government is resettling populations away from low-lying coastal and riverside areas. The commercial logging of lowland rainforest is depleting world carbon stocks and destroying biodiversity. Climatic events such as more frequent and intense cyclones, storm surges, erosion and landslides are undermining agricultural systems and compounding poverty.

The Pacific Developing Member Countries, including the Solomon Islands, face several constraints in responding to climate change. One constraint is the limited capacity to plan for, or design ways to adapt to (or mitigate), climate change. Added to this is their inability to afford the technologies or financing requirements of climate change adaptation or mitigation interventions. Finally, there is the lack of capacity to handle the complex financing arrangements of existing global climate change funds.

ADB has responded with Pacific Climate Change Implementation Plan 2009 consisting of adaptation and mitigation, consultation with governments and scaling up of adaptation efforts. ADB has developed a Pacific Climate Change Program as a one stop-shop and a platform for multidonor technical and financial assistance.

Figure 9.9: Section of Malaita North Road in Malaita Province; Gabion baskets in place to protect the road from erosion due to sea water intrusion (photo taken at low tide). (Source: Second Solomon Island Road Improvement Project, ADB, 2010)

A three-pronged strategy is driving the Pacific Climate Change Program: fast-tracking and scaling up of climate change adaptation and mitigation investments; building capacity to strengthen the knowledge, skills and practices of agencies; and promoting more effective development partner responses. Five priority sectors are highlighted; namely water and sanitation, urban development, energy (energy efficiency and renewable energy), transportation, information and communication technology (ICT), as well as natural resources (coastal and marine resources management). Particular attention will be given to climate-proofing of infrastructure projects. In the transportation sector, support will be provided to reduce vulnerability to the rise in sea levels, floods and landslides through engineering, bio-engineering and land management solutions. The coastal infrastructure will be protected by implementing both engineering/armouring strategies and natural resource-based solutions (e.g. mangrove buffer zones).

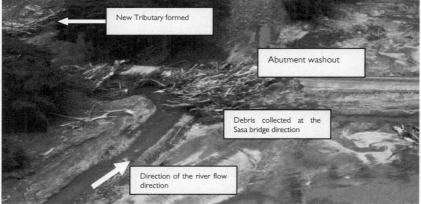

Figure 9.10 & 9.11: West Guadacanal, the Solomon Islands, Tamboko River during a flood (above) and Sasa River just after a flood in 2009. Heavy rains caused these extreme floods defined as by 1 to 50 years Average Recurrence Interval (ARI). (Source: Second Solomon Island Road Improvement Project, ADB, 2010)

Community members in Sasa Village in West Guadacanal have experienced extreme weather events (long rainy season, dry spells, strong winds, cyclones) with aftermaths such as soil erosion, water scarcity, crop failure and food shortage as well as health- and economic-related impacts for the community. Road infrastructure was also badly affected, with sediments blocking roads and bridges damaged by debris deposits. Most recently, the major floods of the last two years (January and February 2009 and 2010) have lead to a number of fatalities in the village and destroyed community assets, such as houses, cash crops and other food plants, drinking-water quality and other infrastructure. This, in turn, has led to escalating prices for essential food items, as fresh produce was not available after the floods and imported food items, such as rice and preserved food, were consumed at a higher cost. Prior to the 2009 flood, villagers were able to sell surplus produce at a marketplace at least once a week. Following the floods, the size of land area available for farming near the village was reduced. Access to markets and schools became difficult, as the flood had destroyed bridges and road sections.

A reduction of income has also affected payments of school fees for their children, as parents were unable to generate the cash from the sale of fresh produce. Older residents claim that they cannot recall any similar catastrophic events happening in the previous 60–80 years. After the major 2009 flood, the village relocated to higher grounds, away from the river, decreasing the damage caused by the 2010 flood to the village.

The name Sasa River means 'sweep' in local dialect and refers to the mobility of the river mouth where the community is located. In the last two years, the wet season stretched from November to April, with rainfall peaks in January to February. In 2010, prior to the large flood, it rained constantly for nearly three weeks. In contrast, the community found the dry season becoming hotter. Even river-water temperatures were reported to have increased during this period. Major logging operations in the upper catchment had taken place 25 years ago, with some logging concessions still reported operating in the upper catchment.

Project Approach

The National Transport Plan for the Solomon Islands prepared by the Ministry of Infrastructure Development, Transport Policy and Planning Unit in 2005, identifies more than 200 kilometres of provincial or rural roads as candidate roads for immediate rehabilitation. Under the Second Solomon Islands Road Improvement Project (SIRIP 2), with financial and technical assistance by ADB, sections of Malaita North Coast Road have been identified as a priority for restoration due to their vulnerability to coastal erosion and local flooding, as well as potentially higher exposure and risks to the impact of climate change issues.

The Malaita North Road study area is located at several separate coastal sites affected by localized foreshore erosion and degradation. It is understood that some of the initial damages occurred during the 1986 cyclone and that the site has further deteriorated over time during various storms and exceptionally high-tide events. The affected sections are located at low-lying areas and where high levels of community activities occur. Community activities, including settlement sites, market areas, mangrove timber collection for copra production and boat/canoe access locations, have all degraded the coastal environment, reducing its natural resistance to erosion and its defence against the impacts of storm surges and future climate changes.

The CCA assessment addressed the evolution of several climate parameters in relation to North Malaita, namely temperature, precipitation, tropical cyclones, storm surges and the rise in sea levels. It further incorporated future qualitative risks of freshwater flooding within the project's coastal river systems. The assessment of the climate and hydrological evolution for North Malaita was carried out between 2010 and 2011. In order for the potential climate change impact on the selected coastal sections to be assessed, the recent past and future climate and hydrological conditions were analyzed. Emphasis was placed on the expected lifespan of any proposed road up-grading works, which the engineers estimated to be 20 years (~year 2030). For the evolution of climate and hydrological conditions, the time horizons between 2020 and 2050 are thus more crucial for any proposed infrastructure project, though projections were also made until the end of the 21st century.

The author of this paper, together with a climate change specialist, conducted a study of climate evolution for temperature and precipitation, as well as a qualitative flood risk analysis for sections of the Malaita North coastal road. Further inputs on mean sea level rise and cyclone intensity were provided by climate change scientists from the UK Met Service–Hadley Centre, engineers and social safeguard specialists from SIRIP 2, which was also funded by ADB.

The study included gathering observation data of past extreme climate and hydrological events to identify recent trends over recent decades. Then, the most relevant climate simulations for the reference period (simulations suggested by IPCC in their Fourth Assessment Report, 2007b) were downloaded. By assessing their ability to reproduce the intra-annual pattern of observed temperature and rainfall of past decades in the region of interest, the most suitable GCMs were selected. In addition, physical data, such as weather, flow measurements in rivers (where available) and other hydrological parameters, were collected during field trips to Malaita and Honiara in 2010. Community consultations aiming to record the existing vulnerability to flooding of the coastal road, and settlements were conducted and are also reflected in the project's Economic Assessment.

Projected climate and hydrological evolution would put sections of Malaita North Road under risk of significant failure (i.e. by overtopping due to the rise in sea levels and river overbank flooding). As such, the subsequent task of this study was to identify CCA measures in an interdisciplinary approach, with the project's engineers, economists and social, environmental and climate change specialists evaluating engineering and non-engineering options. This aimed at developing and integrating measures to climate-proof the relevant road sections under risk for the lifespan of the project in an economically feasible manner. Climate-proofing is defined here as where 'the project can meet its outcome statement under future climatic conditions' (ADB, 2011).

Vulnerability and Risk Assessment with Community Consultation

Consultation was undertaken with stakeholders and communities within the catchment area of Malaita North Road (including villages up to a 30-minute walk from the coastline) through three primary methods: (a) village focus group discussions (FGDs) with men and women; (b) a household survey in 16 villages; and (c) semi-structured interviews with cocoa/copra buyers, truck operators and boat operators. The household survey was undertaken in villages that were located along the coast in the subproject road sections, including one inland village and three villages beyond the end of Section 5, but that relied on North Road. In total, 303 households were surveyed. FGDs were undertaken in six villages and three zones in the Malu'u subregional centre.

The participants of each of the FGDs stated that they had noticed changes in weather and tidal patterns over the past 10–15 years, including increased intensity of rain and shifts in rainfall season, increased numbers of very hot days (or heat waves), less rain during the dry season, a prolonged and hotter dry season, increased intensity and frequency of flooding, as well as increased storm surges. The women's groups also noted that the weather is simply more unpredictable, which made planning for crops and harvests very difficult. Traditionally, the elders used to be able to predict the weather patterns, but this was no longer the case. When asked about the negative impacts of the changes, the women's groups most frequently cited the following:

- Destruction of crops and food gardens by heavy and continual rainfall and flooding, followed by droughts and heat waves, leading to food shortages and additional hardship for households;
- Changes in the soil (harder, drier), making it difficult to grow some crops (leafy green vegetables);
- Decreases in harvest and quality of crops as a result of above, with some men and women estimating their coastal garden produce to be about 40 per cent less than that of 10 years ago (due to insects, crops rotting or destruction by rain or drought);
- Heavy rains (and for longer periods, up to three months) and strong winds affecting daily activities;
- Saltwater intrusion affecting crops and gardens along the coast; and
- Household water sources affected in two different ways: (i) saltwater intrusion affecting rivers/streams and shallow freshwater pools used for drinking water; and (ii) water shortages as a result of drought (a more pronounced and longer dry season).

The men's groups identified a number of the same issues as the women's groups, and also cited additional ones:

- Increased coastal erosion;
- Damage to or destruction of houses;
- Longer drier periods and more intense rains in wet season, bringing new insects that affected crops, as well as longer periods of heavy rainfall, making it difficult for water to drain away and resulting in increased breeding of mosquitoes in stagnant water; and
- Impacts on villages and communities, with some families relocating inland or to higher ground.

Of all the changes in weather and tidal patterns they had noticed, the groups were asked to prioritize the three most significant changes. While the men's and women's groups all identified the same changes when asked to identify the most significant of those changes, the groups came up with clearly different priorities. The women's groups stated that the most significant changes for them were rainfall, increased heat, and flooding and storm surges. Drought ranked fourth overall but was prioritized as the most significant change by three women's groups. The men's groups identified flooding, storm surges and drought, followed by coastal erosion as the three most significant changes. The rise in sea levels ranked fourth but was selected as the most significant change in three of the groups. The men's groups did not rank increased heat as amongst the most significant changes experienced.

Households are making adjustments to cope with the effects of changing weather and tidal patterns. The adaptive measures being put into place include the following:

- Planting a wider variety of crops, more root crops, and changing cropping seasons (i.e. Hongkong taro, swamp taro and banana in the wet season; sweet potato, yam and cassava in the dry season);
- Households becoming more dependent on copra, rather than garden vegetables for cash (to buy rice, etc.);
- Relocating houses and gardens further inland or to higher ground to reduce impacts of storm surges, flooding and coastal erosion;
- Women digging wells to access deeper, freshwater not affected by floods and storm surges (saltwater intrusion);
- Increased use of pesticides to deal with new insects and infestations;
- Buying more food than before, as they cannot rely on their own crops or gardens so much; and
- Education on and training in adapting cultivation types and techniques, from the Agriculture Department, to better cope with the impacts of climate change.

According to the Solomon Country Report (Roughan & Wara, 2010), vulnerability to extreme events is apparent from a long and growing list of adverse events resulting from climatic and tectonic incidents'. The authors reported some of the most significant recent disasters. From this document, the most recent climate events that affected the Solomon Islands were extracted.

Table 9.3: Summary of recent environmental disasters

Cyclone Namu 1986	Cyclone Namu was the most devastating and costly tropical cyclone to affect the Solomon Islands in history, resulting in 90,000 homeless and more than USD100 million in economy losses. It developed very quickly on 18 May 1986 whilst travelling from the north towards Sikaiana. The cyclone was very erratic in its path and with varying speeds (ranging from 5 to 20 knots). Many places reported wind speed estimates exceeding hurricane force (64 knots), approximately 100 knots. Widespread damage occurred to Sikaiana, South Malaita, Makira, Rennell, Bellona and Eastern Guadalcanal, with 103 dead and 33 missing.
Cyclone Nina 1993	A total of 30,000 people were affected and economic losses totalled about USD 20 million.
Cyclone Fergus 1996	A total of 30,000 people affected and relief cost amounted to USD 1.9 million.
Cyclone Zoe 2003	A category 5 cyclone reaching its peak on 28 December 2002 with extremely high sea levels, sustained winds of 245 kilometres/hour and gusts up to 340 kilometres/hour. The eye of the cyclone was located only 50 kilometres southeast of Anuta Island. Tikopia sustained more damage because Zoe was almost stationary near the island for a period of 16 hours. Despite the extent of damage, no casualties resulted from either Tikopia or Anuta Islands.
Tropical Cyclone Beni	Rennell and Bellona were affected, but the islands sustained little damage even though Beni occurred near the islands for a period of 10 hours.

Ontong Java High Swells 2008	For three days in December 2008, the two atoll island communities of Luaniua and Pelau experienced unusually high swells, which inundated swamp taro gardens. Near the atolls, the sea was generally calm, and there were no strong winds prior to that period as well. The swells came in from the north, probably as a result of build up of waters associated with the low areas to the far north near the equator at that time. These high swells resulted in saltwater intrusion into the communities' taro farms and gardens. The high swells also resulted in saltwater intrusion into the main freshwater lenses of the island atolls. Another issue was the increasing unusual high tides, resulting in continuous coastal erosion. Although no lives were lost, such phenomena underlined the challenges that such atolls might encounter.
Flooding of the North West Guadalcanal, Central Islands, Isabel, Malaita, Western Province and Makira, 2009	Heavy and continuous rainfall associated with a trough of low pressure south of the Rennell Bellona Province caused severe flooding in North West Guadalcanal. Further damage resulting from flooding and landslides was reported from eastern Guadalcanal, Central Province, Isabel, Western, Malaita and Makira Provinces. Flash floods swept through villages, destroying houses, gardens and water sources. The worst-affected area was Western Guadalcanal, though effects were also observed in eastern Guadalcanal, Central, Malaita, Isabel, Western and Makira Provinces. The disaster claimed 13 lives and affected approximately 270 villages and 20,000 people. Emergency response and relief were coordinated by the National Disaster Management Office.

Adapted from Roughan and Wara (2010).

The Solomon Islands were also affected by important droughts in the recent past. In 1997, the total rainfall during the first four months varied between 7 per cent below and 28 per cent above normal at the six recording stations. In May, it fell to between 29 and 89 per cent below normal at all stations. From May to December, a total rainfall amount of between 32 per cent and 58 per cent below normal was recorded. First reports that a serious drought was developing appeared to have reached Honiara in September and October. Reports on the effects of the drought indicated that springs and streams were drying up and that staple crops were small and underdeveloped.

Historic Weather Pattern Changes

For the purpose of this study, historic climate data for the area of concern to provide projections on climate evolution for the 21st century were evaluated. Analyses of the historical meteorological observations were based on the mean monthly minimum, maximum temperature and rainfall (1962–2010) for Auki

(Malaita Province) weather station and the mean monthly rainfall from Honiara weather (Guadacanal Province) station.

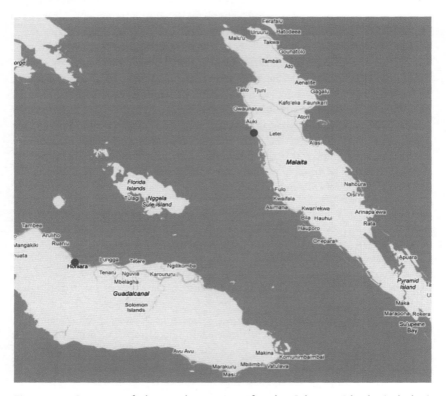

Figure 9.12: Location of the weather stations for the Solomon Islands (red dots) (AUKI: 8.14°S x 160.73°E; Honiara: 9.41°S x 159.97°E). (Source: Second Solomon Island Road Improvement Project, ADB, 2010)

Temperature

The diurnal differences between maximal and minimal temperatures remain approximately constant at 7°C. These patterns are typical of tropical areas, where temperatures are rather uniform throughout the year. Over the 48-year record period, it appears that T_{min} increased by about 1.1°C while the evolution of T_{max} over that period is even more important, with a difference of 1.65°C. Considering the past five decades, the rates of warming for both T_{min} and T_{max} are 0.23°C and 0.34°C respectively per decade.

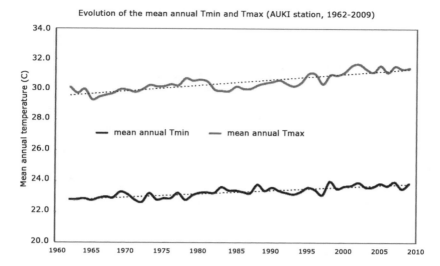

Evolution of the mean annual Tmin and Tmax (AUKI station, 1962-2009)

Figure 9.13: Evolution of the mean annual T_{min} and T_{max} at the AUKI station over the record period (1962–2009) (Source: Second Solomon Island Road Improvement Project, ADB, 2010)

Precipitation

The climate seasons definition (Winter: June, July and August; Spring: September, October and November; Summer: December, January and February; Fall: March, April and May) fits fairly well with wet (Summer and early Fall) and dry seasons (Winter and Spring) observed on the Solomon Islands. Therefore, it is suggested to consider Summer and Fall (except for May) the wet season and Winter and Spring the dry season.

Analyzing the intra-annual precipitation variations at Auki weather station, one can recognize the general intra-annual pattern with higher precipitations for the months of January, February and March. However, monthly precipitation deviation from the mean may be observed very often, as is the case over the wet period (December to March).

Looking at longer-term trends, Figure 9.14 illustrates the evolution of the annual rainfall at Auki station over the record period (1962–2009). In Figure 9.14, the total annual precipitation from 1962 to 2009 was plotted and superimposed as a linear trend line (dotted line). According to the trend line, rainfall has been decreasing since 1962. While annual rainfall was averaging 3,400 millimetres at the beginning of the record period, those values have declined in the range of 2,900 millimetres in the recent past. On average, it represents a decline of approximately 10 millimetres per year. Moreover, by looking closely at Figure 9.14, one may also consider three distinct periods. It appears that 1962–1977 was rather humid, while 1978–1995 could be considered dry. Finally, 1996–2009 is positioned in between the wet and dry periods and also characterized by a more important inter-annual variability.

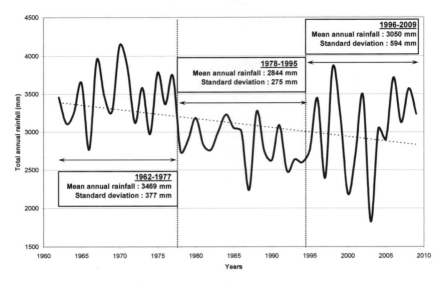

Figure 9.14: Evolution of the annual rainfall at the AUKI station (1962–2009)
(Source: Second Solomon Island Road Improvement Project, ADB, 2010)

The decline of total annual rainfall is also reported in the literature at Honiara station. A recent report published by the Secretariat of the Pacific Regional Environment Programme highlights that the negative trend in the total annual precipitation is translated in a decline of rainfall at an average rate of 8 millimetres per year. It is also suggested that this trend had been seen in several parts of the country (AUKI).

The authors raised the fact that 'climate variation poses a new threat to this scarce resource', while 'under this trend of historical records, the Solomon Islands are likely to face shortages of freshwater in certain parts and low fresh water in the rest of the country'. Manton *et al.* (2001) also analyzed trends in total annual rainfall from 1961 to 1998 in the South Pacific. They showed that both annual rainfall and the number of rain days (with at least 2 millimetres of rain) generally decreased during that period in the region. According to the authors, this decrease is associated with the predominance of El Niño events since the mid-1970s'.

For the specific case of the Solomon Islands, the authors have shown that the number of rainy days decreased at all stations. From the Honiara station, they have also shown that 'the proportion of annual rainfall from extreme rainfall has increased significantly' (Manton *et al.*, 2001).

Thus, maximum and minimum temperatures as well as annual rainfall show significant trends, suggesting climate evolution over the past five decades. These findings are also reported by Rasmussen *et al.* (2009) in a recent publication dealing with impacts, vulnerability and adaptation to climate change on three Polynesian outliers on the Solomon Islands. The authors mentioned that '… the analysis of temperature data at four synoptic stations on the the Solomon Islands reveals an increase in the near-surface temperature in recent years. The region around the Solomon Islands has experienced a long-term decrease in annual mean precipitation of 20 per cent to 40 per cent per century through the period 1901 to 2005' (Rasmussen *et al.*, 2009).

Projected Climate Changes

Based on the historical data gathered during the climate change vulnerability assessment of the Solomon Islands, a comparison was then made with the simulations by available climate models for the past reference period. This allowed researchers to select the most relevant climate models based on their skills in reproducing the intra-annual pattern of observed temperature and rainfall of the recent decades chosen. The selected climate models were later used for projections of future climate changes.

Following this method, the following predictions were made in this study. At the 2020 time horizons, expected changes of temperature will vary between +0.4 and +1.2°C and from -16 to +13 per cent for precipitation. Looking at the 2050 time horizon, the projections vary from +0.7 to +2.2°C for temperature and from -12 to +33 per cent for precipitation. Finally for 2080, the projections vary from +0.8 to +3.3°C for temperature and from -9 to +32 per cent for precipitation. Uncertainties associated with the evolution of precipitation are more important than the ones associated with temperature.

Without surprise, this specific information centred on the Solomon Islands is coherent with IPCC (2007) projections, which forecasted for the Southern Pacific region, a temperature increase of 0.45–0.82°C, 0.80–1.79°C and 0.99–3.11°C respectively for time horizons centred on 2020, 2050 and 2080.

As per future thermal conditions, the study has shown that temperature warming trends will persist until the end of the century. According to climate model projections, expected temperature could increase respectively by up to 0.8, 1.4 and 2.2°C for time horizons 2020, 2050 and 2080. Those changes are rather uniform within the year where no single monthly temperature would increase more or less than any other month.

Considering precipitation conditions, despite the recent observed decline of the total annual precipitation, rainfall could increase by up to 3, 8 and 10 per cent at time horizons 2020, 2050 and 2080 respectively. Winter- and summer-expected changes appear to be more important than the other two seasons. At horizon 2050, the months of February, July and August may record the most important monthly total precipitation volume increases. Average precipitation intensity may also increase by 2050, while the projected five-day maximum cumulative annual precipitation should not be significantly different from the actual climate conditions.

Based on temperature and rainfall projected by climate models, the total water volume flowing in the Malaita rivers could slightly increase by 2080. As per extreme hydrological events, neither floods nor droughts conditions should change much by 2050, which does not mean that severe droughts and floods, as observed in the recent past, would decrease in frequency or severity. The most important changes could affect the downstream river stretches near the ocean. The projected sea level rise, increased storm surges and wave heights suggested by other studies could significantly affect the river hydraulics in the coastal zones. Those important hydraulic constraints and the associated backwater effects could lead to frequent and intense river overbank flows, with damaging consequences to nearby infrastructure.

Mean sea level rise

Other studies were conducted on the Solomon Islands to assess the potential sea level rise, cyclone activities and storm surges. These studies used computer modelling to project sea level rise along Malaita North Road for 2030, 2040 and 2050. Mean relative sea level rise was found to pose a low hazard for this coastal road, despite one road subsection at risk of future inundation due to projected sea level rise. The time-mean relative sea level rise for 2030 is 0.15–0.23 metres. For 2040, this range is 0.19–0.31 metres, and for 2050, the range extends from 0.24 to 0.38 metres (UK Met Service–Hadley Centre, 2010).

Cyclone intensity

In the assessment of cyclone intensity in the Solomon Islands (UK Met Service–Hadley Centre, 2010), IPCC's Fourth Assessment Report (2007) was quoted stating that computer models suggest that tropical cyclones are likely to become more intense in the future, with larger peak wind speeds and more heavy precipitation. It is commonly assumed that increasing sea surface temperatures (SSTs) will lead to more tropical cyclones because the areas of formation are primarily determined by SSTs reaching a particular threshold. The 2010 study states that other factors need to be considered as well. Modelling studies indicate fewer tropical cyclones globally in the future because of changes in the atmospheric dynamics (e.g. wind shear, or variations in the wind at different heights) and thermodynamics (e.g. static stability, relative SSTs). In regions where wind shear increases and/or the relative SSTs (local-mean SSTs minus tropical-mean SSTs) decrease, there will be fewer tropical cyclones. This caused uncertainty in the prediction of cyclone frequencies.

The study that relates closest to the Solomon Islands is one by Walsh *et al.* (2004), who ran RCM predictions covering the north of Australia, Coral Sea and far western South Pacific. The results are consistent with global modelling studies, which show little change in overall storm frequency but an increase in stronger storms. Whilst there is uncertainty as to the magnitude of increases in tropical cyclone intensity as a result of climate change, projected rises in sea level could increase the vulnerability to storm surge flooding of low-lying areas, including some of the Solomon Islands.

Storm surges

Storm surges are weather-induced, short-term changes in sea level and can cause significant coastal flooding when they occur close to high tides. There are two key impacts of climate change-related sea level rise that pose an effect on storm surges:

- An increase in mean water level, which has a direct impact on the expected levels; and
- An increase in the likely depth over the reef and hence an increase in the likely wave run-up.

Sea level increases the maximum potential wave run-up that can occur across the reef. It is noted that climate change is estimated to more than double the period when the existing road is overtopped.

Due to a lack of reliable historic data for storm surges based on tidal gauge readings on the Solomon Islands, no conclusive projections on storm surges and surge return periods could be made for this study. More comprehensive and long-term input tidal, bathymetric and wind speed data would need to be researched in order to allow for some quantitative assessments of storm surges/tides and their heights as well as impacts on low-lying coastal road sections of North Malaita. Whilst there is uncertainty as to the magnitude of increases in tropical cyclone intensity due to climate change, projected rises in sea level could increase the vulnerability to storm surge flooding of low-lying areas, including some of the Solomon Islands.

Freshwater flooding

Important flooding was reported in 1986 and more recently in 2009 in the Guadalcanal plains, while, in 1995, drought severely affected most part of the country, causing severe food shortages. It is believed and expressed by the population that, recently, climate and weather extremes might be occurring more frequently due to climate changes and variations.

The rather limited consequences of climate change on the rainfall of the Solomon Islands will probably be echoed on the evolution of its hydrological regime, at least up to 2050. On average, the annual rainfall amount is bound to increase by up to 3, 8 and 10 per cent respectively for horizons 2020, 2050 and 2080, and the consequences on the annual run-off of the Solomon rivers might be in the same order of magnitude, perhaps a little less, given the increase of evapo-transpiration associated with the slight and uniformly increase of temperature throughout the year (by up to 2.2°C in 2080).

Despite the fact that, on average, the total annual water volume flowing in the rivers on the Solomon Islands would not increase that much, extreme rainfall events and longer dry spells should be considered as serious threats for the islanders in the future. Even if the evolution of the precipitation regime will not likely affect the extreme hydrological events of Malaita Island that much, one should seriously consider the potential effects of sea level rise, storm surge and wave heights on the hydraulics of floods. All these factors represent hydraulic constraints downstream the river, affecting considerably the river flow. For the study area, the sources of the river may reach up to 1,000 metres in altitude with a distance of 10 kilometres to the sea inferior. Upper river reaches are very steep, and the concave upward longitudinal profile of the streams reaches the low-lying portion of the watershed near the ocean.

The behaviour of water and its sediment load when approaching the sea is rather complex. As the rivers approach the coastal plains, fluvial dynamics changes as channel and valley slopes decline, while, generally, the river width increases. In the fluvial-estuarine transition zones, one can observe backwater effects associated with the river flow entering the mass of oceanic water and its associated tides. The dynamic equilibrium between the coastal and marine processes with their influence generally decreases inland, and the fluvial processes declining in strength downstream is a function of river discharge and tidal cycles. This precarious equilibrium reached over the past centuries is expected to be affected by sea level and wave height rise as well as potentially increased storm surge associated with climate change. This phenomenon could also be complicated by valley morphology, extensive water storage on floodplains and low-water tributaries that may act as distributaries in high-flow conditions.

In the lowermost reaches of the Malaita rivers, the power of the rivers may decrease even more if the hydraulic constraints associated with sea level and wave height rise and storm surge increases. In these conditions, the risk of observing the rivers flowing overbank increases in the future. These decreasing river power conditions would, consequently, limit the capacity of the river to carry its sediment load. Low-lying parts of the coast would then be inundated more often and fed by more sediments, and the resulting swampy areas could affect road integrity, growth of crops and general security of the people living in these areas.

Akamatsu *et al.* (2006) reported the results of their research work on the effects of sea level rise on rivers flowing into the ocean. Their case studies are located in Papua New Guinea in an environment similar to those observed on the Solomon Islands. Their research was motivated by the growing attention given to the potential effects of sea level rise associated with climate change on shore lines, river deltas and river-long profiles. To anticipate the effects of foreseen sea level on coastal fluvial geomorphology, they have investigated the effects on coastal morphology associated with Holocene sea level rise following the last glaciation. Their main findings are schematically illustrated in Figure 9.15 by using the downstream reach of the Kwaibala River in Malaita.

Figure 9.15(a) shows the expected delta shoreline evolution under constant base-level conditions at time t_1, t_2 and t_3. Under these normal conditions, assuming that the sediment source is constant, the delta should progress seawards with time. Figure 9.15(b) illustrates the potential effects of sea level rise on the delta shoreline position. As water level rises (also assuming storm surge increases), the shoreline moves landwards as the sea erodes the coastal zones, while the sediment load of the river is dropped because of the decreased river power associated with hydraulic constraints induced by the sea level rise. Consequently, the river base bed level could become higher. This scenario is valid only if the sediment availability persists in the future. It has been shown that any change in the sediment supply could have a drastic influence on the bed profile.

In summary, despite the limited impacts of climate change on the precipitation and thus hydrological regime at horizon 2050, sea level rise, storm surge and wave height evolution may significantly affect the river bed profile and the shoreline morphology.

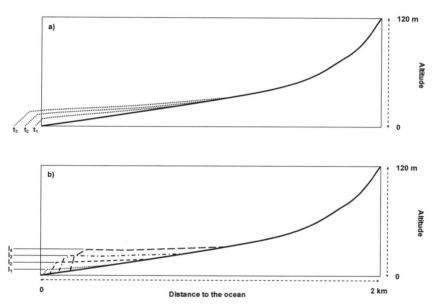

Figure 9.15: Potential evolution of the longitudinal profile of Kwaibala River under sea level rise conditions. (Source: Second Solomon Island Road Improvement Project, ADB, 2010)

Climate Change Adaptation Options for Malaita North Road and Lessons Learned

The findings of this climate change study were integrated with the engineering and economic assessments for Malaita North Road. In theory, six engineering options, including several adaptation measures, were identified and assessed for their economic feasibility. The economic feasibility was a major decision-making factor in determining climate change adaptation options.

Incorporating the effects of climate change over the next 20–30 years (lifespan of road works), the economic feasibility assessment has demonstrated that the option of resheeting the road plus coastal protection measures in low-lying areas would yield acceptable economic return rates as the only scenario.

Other adaptation options are not economically-viable, though technically possible. Resheeting works would include the installation of gabion protection barrier structure along the existing foreshore areas, backfilling any existing scour sections, raising road formation at low locations, protecting levy banks with suitable mangrove planting or by expanding existing mangroves.

The key benefit of the proposed adaptation response is increased access for the local population for both travel and transport of goods and services, resulting from the improvements in the infrastructure due to reduction in the periods of overtopping of the roads and bridges by coastal floods. More specifically, the beneficiaries include the following:

- Existing road users, whether vehicle drivers, passengers or non-motorized transport users;
- Households living in villages along and in the catchment area of the road who grow or sell a range of cash crops, including copra and cocoa;
- Passenger and goods transport service providers as well as commercial truck drivers;
- Schoolchildren, teachers and schools as well as hospitals that rely on direct access to roads and bridges;
- Small businesses and traders, including vendors at the local and or informal markets, trade store owners and produce buyers; and
- Provincial authorities and key social service providers such as the education and health sector.

V. Conclusion

The CCA assessment for the SIRIP 2 subproject shows that it is important to choose an approach based on the adaptation decision context, availability of data, time frame and budget. Cost-effectiveness of the adopted approach is an important selection criterion, especially in cases where actual development investments are limited by socioeconomic and physical constraints. Achieving climate adaptation results within reasonable timeframes is another important consideration, especially when development investments are targeting the rehabilitation needs of damaged infrastructures.

Current legislative framework and decision-making processes within the Solomon Islands may not explicitly include strategies to address climate change adaptation needs. There is also a limited capacity in making investments decisions. Integrating climate change adaptation within the existing legislative framework for environmental assessment and the available limited capacity in this sector may be a step in the right direction. Developmental assistance as in the case of the SIRIP 2 subproject could play a more active role in facilitating climate change adaptation by programming and providing technical assistance to multiple government counterparts, rather than the Ministry of Infrastructure alone.

Acknowledgements

The author expresses his appreciation to the following colleagues who have contributed to or provided materials for this article: Rene Roy, Canada; Yurdinus Lelean, Indonesia; Joseph Deng, China; Roni Adiv, Israel/USA. Gratitude is also due to his colleagues and staff at ADB, namely, Jay Roop, and the project teams of Cardno Acil and EGIS Bceom in Timor Leste and the Solomon Islands.

References

Akamatsu Y, Parker G, Muto T (2006) Effect of sea level rise on rivers flowing into the ocean: application to the Fly-Strickland river system, Papua New Guinea, River, Coastal and Estuarine Morphodynamics. In: Parker G, Garcia MH (eds) *Proceedings of the 4th IAHR Symposium on River, Coastal and Estuarine Morphodynamics*, RCEM 2005, pp. 686–95. Taylor & Francis, Urbana, Illinois.

Asian Development Bank (2011) Guidelines for climate proofing investment in the transport sector: Road Infrastructure Projects. Asian Development Bank, Manila, Philippines.

Asian Development Bank (2009) The Economics of Climate Change in Southeast Asia: A Regional Review. Asian Development Bank, Manila, Philippines.

Asian Development Bank (2005) Climate Proofing: A Risk-based Approach to Adaptation. *Pacific Studies Series*. Asian Development Bank, Manila, Philippines.

Climatic Research Unit of the University of East Anglia website (2010) http://cru.csi.cgiar.org/index.asp.

Cardno Acil (2009) Solomon Islands Road (Sector) Improvement Project: Guadalcanal Flood Damage Restoration Subproject, Guadalcanal Province-Engineering Assessment. Honiara, Solomon Islands Government, Ministry of Infrastructure Development.

Cardno Acil (2010) Solomon Islands Road (Sector) Improvement Project: North-West Guadalcanal Road, Poha to Naro Hill, Guadalcanal Province-Preliminary Climate Change Assessment. Honiara, Solomon Islands Government, Ministry of Infrastructure Development.

Dole D & Abeygunawardena P (2002) An analysis and case study of the role of environmental economics at the Asian Development Bank. Manila, Philippines, Asian Development Bank.

Hadley Centre (2010) Summary results – MORSE Projects of Solomon Islands (Draft), Meteorological Office, Hadley Centre, UK.

IPCC (2007b) Contribution of working Group I to the Fourth Assessment Report of the Intergovernmental Panel on Climate Change. In: Solomon S, Qin S, Manning M, *et al.* (eds) *Climate Change 2007: The Physical Science Basis*. Cambridge University Press, Cambridge and New York.

IPCC (2007a) Contribution of working Group II to the Fourth Assessment Report of the Intergovernmental Panel on Climate Change. In: Mimura N, Nurse L, McLean RF, *et al. Climate Change 2007: Impacts, Adaptation and Vulnerability*. Cambridge University Press, Cambridge.

Kropp J (2009): Climate Change Information for effective adaptation – A practioner's manual. *Deutsche Gesellschaft fuer Technische Zusammenarbeit (GTZ)*, Climate Protection Programme, Eschborn.

Manton MJ, Della-Marta PM, Haylock MR, *et al.* (2001) Treands in extreme daily rainfall and temperature in Southern Asia and the South Pacific: 1961–1998. *International Journal of Climatology* **21,** 269–84.

Ministry of Infrastructure Timor Leste (2009) Final Report, Preparing the Road Network Development Project – TA 7100 *Volume III Climate Change Assessment*, Prepared by: Cardno Acil in association with KWK Consulting (Unpublished).

PCIC website (2010) http://pacificclimate.org/.

Padma, NL & Thuirarajah V (2011) Social and economic assessment of climate proofing of road infrastructure in the Western Guadacanal, Solomon Islands (draft). (Unpublished).

Phillips JD, Slattery MC (2007) Downstream trends in discharge, slope, and stream power in a lower coastal plain river. *Journal of Hydrology* **334,** 290–303.

Rasmussen K, May W, Birk T, Mataki M, Mertz O, Yee D (2009) Climate change on three Polynesian outliers in the Solomon Islands: Impacts, vulnerability and adaptation. *Danish Journal of Geography* **109(1),** 1–13.

Roughan P, Wara S (2010) *The Solomon Islands Country Report for the 5 Year Review of the Mauritius Strategy for Further Implementation of the Barbados Programme of Action for Sustainable Development of SIDS (MSI+5).*

Roy R (2010) *Assessment of Climate Change and Potential Impacts in the Hydrological Regime of Solomon Islands at the horizons 2020, 2050, and 2080 (Second Draft)* (Unpublished).

Roy R (2009) *Assessment of anticipated climate change impacts on the proposed Road Network Development Project in Timor Leste* (Unpublished).

Roy R, Pacher G, Roy L, Adamson P, Silver R (2008) *Adaptive Management for Climate Change in Water Resources Planning and Operation.*

Secretariat of the Pacific Regional Environment Programme. http://www.sprep.org/.

World Bank Portal website (2010) http://sdwebx.worldbank.org/climateportal/. Sveiven S (2010) Are the European financial institution climate proofing their investments? Amsterdam, Netherlands, Institute of Environmental Studies, University of Amsterdam. *Report R-10/07.*

Green Buildings

CHAPTER TEN

Integrated Sustainability Policies For China's Cement Industry – A Case Study Approach

Harn Wei KUA
Department of Building, School of Design and Environment,
National University of Singapore

"Unintended and negative consequences occurred to sustainability indicators when: they are completely ignored by policymakers and/or policymakers fail to identify intrinsic but inconspicuous links between semmingly unrelated indicators."

– *Harn Wei KUA*

Integrated Sustainability Policies For China's Cement Industry – A Case Study Approach

KUA Harn Wei[1]

Department of Building, School of Design and Environment,
National University of Singapore

Abstract

This paper explores the reasons behind associated effects of policy tools employed to promote sustainable building materials. By comparing the original motivations and intended effects of these policies and their actual outcomes, and subsequently understanding the reasons behind any disparities between them, ways by which future policy planning can be improved have been suggested. The research is based on a series of international case studies on the applications of policy tools to promote sustainability issues related to concrete aggregates, cement and wood. Each of these cases is attended by negative, unanticipated outcomes. By analyzing these outcomes, an underlying pattern was observed – negative and unanticipated policy outcomes occur when either a sustainability indicator/issue is completely ignored by policymakers, or the policymakers fail to identify intrinsic but inconspicuous links between seemingly disparate indicators. These unexpected outcomes can be reduced, or avoided, if policymakers conceptualize policies more broadly, for which purpose the concept of integrated policymaking was proposed. This concept promotes the idea of co-addressing a wide range of sustainability indicators covering all the three bottom lines of sustainability: economy, environment and employment. Furthermore, in doing so, policymakers must promote interactions among the different levels of governmental agencies and between the governmental and non-governmental stakeholder groups. Finally, this concept of integrated policymaking has been applied to analyze how China's strategies on the production and consumption of cement can be improved to ensure better possibility of achieving sustainability for the industry.

Key words: building materials, integrated policies, sustainability, sustainable materials, unintended consequences, sustainable development

---

[1] Department of Building, School of Design and Environment, National University of Singapore. Blk SDE1, #05-07, 4 Architecture Drive, Singapore 117566. Email: bdgkuahw@nus.edu.sg; Tel: +65 65163428.

I. The Sustainability Impact of Building Materials and the Role of China's Cement Industry

Although the construction and use of buildings consume huge amounts of energy (Hwang & Tan, 2010), governments and non-governmental organizations (NGOs) around the world have also turned their attention to put in place a wide variety of policy tools to address the different sustainability challenges arising from the life cycle stages of various construction materials. Worldwide, the construction industry is, by far, one of the biggest consumers of extracted materials. Buildings are estimated to be responsible for two-fifths of the world's material and energy flows, one-sixth of its fresh water usage and one-quarter of its wood harvest (Augenbroe *et al.*, 1998). Due to its size and contribution to the economy, construction is not only one of the largest users of energy and resources but also a formidable polluter.

Every construction material presents its unique set of challenges to global sustainability. On a per-unit basis, cement is the most energy- and pollution-intensive component of concrete. The 1.45 billion tonnes of global cement production are accountable for about 2 per cent of the global primary energy, or close to 5 per cent of the total global industrial energy consumption; about 5 per cent of the global anthropogenic carbon dioxide (CO_2) emissions (Worrell *et al.*, 2001); and significant emissions of sulphur dioxide (SO_2), nitrogen oxides (NO_x), particulate matters and other pollutants (van Oss & Padovani, 2003). Since certain types of aggregates are also found to be highly localized geographically, substitution of 'problematic' materials (such as natural gravel and sand) by other similar materials (such as crushed stone) may end up increasing transportation needs. This in turn increases fuel use as well as operational costs of mining companies. The fact that the forest is a crucial element of the ecosystem and a cultural identity and livelihood for many civilizations means that any narrow perception of it as merely a source of structural timber is likely to trigger multiparty disputes. Even though CO_2 emissions from combusting wood fuel are not a major concern, given that trees extract CO_2 from the air and soil for photosynthesis, emissions of fine particulates have long received attention from both the workers' and users' health perspectives. The preservation of biodiversity and indigenous tribes' land rights are increasingly emphasized in the assessment of sustainable forestry.

China is presently the world's largest cement producer and consumer. Today, it produces roughly half of the total global output of cement, whereas the next three largest producers – India, Japan and the United States – together produce less than 20 per cent of the global volume. It is estimated that China's cement (much of which is produced in energy-inefficient, highly polluting kilns) consumes roughly 6 per cent of China's total energy demand. Eighty per cent of the energy supply comes directly from coal and other fossil fuels. The remaining 20 per cent comes from electricity, which is mainly derived from coal. Cement manufacturing is thus a major contributor to greenhouse gas (GHG) emissions in China. In late January 2011, the Chinese Government announced plans to double its annual investment in water projects to US$630 billion over the next 10 years. This increase in investment is expected to become a new driver of cement consumption. In fact, it is expected that China's demand for cement may grow 12 per cent year on year, to nearly 2.1 billion metric tonnes this year (in 2010, it produced 1.87 billion metric tonnes of cement, representing an increase of 15.5 per cent from the volume in 2009). For these reasons, strategies to improve the overall sustainability of the cement industry of China will have a profound and considerable effect on the global building material industry's contribution to climate change.

While researchers and policymakers are still in the process of fully understanding the full sustainability impacts of building materials and the industries and sectors related to their development and use, many sustainability policies have nonetheless resulted in negative and unanticipated outcomes. Why do these outcomes occur? What can we learn from these that would enable us to design and implement more effective policies in the future? How can we apply these to China so that its strategies for its cement industry can be improved?

II. Methodology

The overall research methodology of this article could be summarized with the flow chart in Figure 10.1. This research is based on a series of ex-post multiple comparative case studies related to the production and consumption of key building materials, namely cement, concrete aggregates and wood. These cases were chosen based on the criteria that policies were set to promote certain aspects of sustainable development and that some of the associated outcomes were observed to be unanticipated and negative. Besides literature review, data from multiple sources for each case were also sought in order to achieve triangulation. These sources of data included interviews, surveys and participation in academic focus group discussions. The multiple comparative case studies were carried out as follows:

1. Identification of original policy strategies and goals in each case;
2. Identification of the sustainability indicators that were related to these goals;
3. Identification of the policy anticipated outcomes (supported by data collected); and
4. Comparison between goals and actual outcomes and identification of unanticipated outcomes and neglected sustainability indicators.

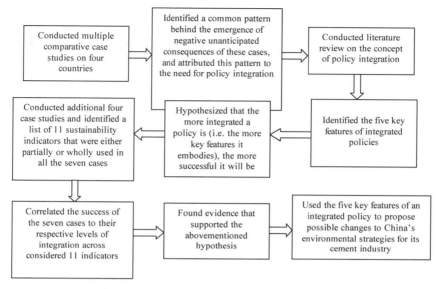

Figure 10.1: Summary of the research methodology

III. The Four Cases

Gravel Tax (GT) of Sweden and Raw Materials Tax (RMT) of Denmark

The goals of the Swedish Gravel Tax (GT), which is collected for extraction and export of domestic raw materials, are clearly stated in the 'Law Concerning a Tax on Natural Materials' (1995: 1667) (Swedish EPA, 2001). One of these goals is to ensure that not more than 30 per cent of the aggregate required nationwide comes from natural gravel, with its substitutes making up the rest of the 70 per cent. The consumption of natural gravel should not be more than 12 million tonnes in the year 2010. As an integral element of the Swedish Environmental Objectives – Interim Targets and Action Strategies (Bill 2000.2001: 130), approved in 2001, it was expected to contribute towards the reduction of the nationwide landfill waste by 50 per cent of the 1994 level by 2005 (Arm, 2003). With regard to reducing life cycle costs, environmentally/ socially-derived costs and market dependence on domestic extraction of gravel, GT was expected to have some indirect effects. However, the government did not set specific targets with regard to reducing industrial energy consumption, air pollution water pollution and toxic waste production.

A similar tax was passed in Denmark. The main goal of the RMT, which is collected for the extraction and import of raw materials, is to enforce the vision of the Danish Raw Materials Act and help ensure that exploitation of raw material deposits is based on the principles of sustainable development. The supply of raw materials to the society shall be ensured in the long term, and the raw materials are to be used according to their quality. Furthermore, waste shall be used to the greatest possible extent as a substitution for natural materials. For this latter goal to be achieved, the Waste Tax was implemented in 1987. The Action Plan for Waste and Recycling (1993–1997) also had the general objectives of increasing recycling as well as reducing the requirement for landfilling. Under this Action Plan, one of the main targets included achieving 54 per cent recycling, 21 per cent landfilling and 25 per cent incineration. It was reported that most of these targets were met in the late 1990s, including the construction and demolition waste category (Green Alliance, 2002). The RMT aims to reduce both life cycle and environmentally/socially-derived costs, while increasing its market share (or to protect its competitiveness in international natural materials trade).

While both the GT and RMT, together with other related policies, resulted in considerable increases in waste recycling rates and reductions in domestic extractions of the concerned raw materials, a glaring negative outcome was the steady increase in the *imports* of raw materials from neighbouring countries. Two years after the implementation of the GT in 1996, the import of crushed stones by Sweden reached its first peak of 250,000 metric tonnes (Swedish EPA, 2000; Eurostat, 2005). This also corresponded with the first peak in the local demand for aggregates after the tax implementation. The second peak occurred in 2001, when Sweden imported 110,000 metric tonnes of gravel and pebbles to cater to a peak in the local demand that occurred around the same time (Eurostat, 2005). Denmark's import of gravel recorded a steady increase beginning 1992. As the demand for aggregate for construction reached its first peak in 1998–1999, the import of gravels also peaked around that time, with a capacity of 2.4 million metric tonnes (Eurostat, 2005). After a reduction between 1999 and 2000, the import has been steadily increasing since 2000. This reflects the demand trend in the local industry. Thus, in both these countries, the increase in imported aggregates very closely reflects the increasing trend of local demands for these materials.

Policies to Regulate the Use of Substitute Fuels for Cement Production in the United States and the United Kingdom

In the United States, the Clean Air Act, the Resource Conservation and Recovery Act (RCRA), the Comprehensive Environmental Response, Compensation and Liability Act (CERCLA), the National Emission Standards for Hazardous Air Pollutants (NESHAP) and the Comparable Fuel Exclusion (CFE) are applied concurrently to ensure that, while cement kilns reduce their reliance on fossil fuels, the air quality in the vicinity of the plants is not compromised. Legislation such as the CFE was implemented to accelerate the fuel approval procedure. In fact, the call for public feedback on the CFE and the disclosure of the life cycle assessment results of plants using recycled liquid fuel as kiln fuel (Chem Systems Ltd, 1999) exemplified versions of public participation and feedback mechanism in the United States. In totality, this set of legislations aims to address a wide range of indicators. In the United Kingdom, the Integrated Pollution Prevention and Control (IPPC) Directive, the Best Available Technique Standard (proposed under IPPC), the Substitute Fuels Protocol and the Waste Incineration Directive play the same role as the above-mentioned policies in the United States.

Although the strategies of these two countries are different, they share a similarity in a certain unanticipated outcome – the numerous disputes that have erupted amongst the different stakeholders regarding safety in the use of certain types of substitute fuels, particularly, discarded waste chemicals. Proponents and opponents of the uses of these fuels were able to present data to support their respective claims. For example, data that showed a reduction of air pollutions in the emission from plants (Chem Systems Ltd, 1999; ERAtech Group LLC, 1999) that use these fuels were pitched against data that showed otherwise (Delles, 1992; Ginns & Gatrell, 1996; Schwartz et al., 1996). The crux of the problem is that the content of emissions is highly dependent on the combustion conditions in the kilns. For the discrepancies in opposing data to be rectified and for the antagonistic stance of the opposing stakeholders to be resolved, an independent system that examines the validity of the data, with respect to the operational conditions of the kilns, is needed.

Reforestation and Afforestation Incentives of Chile

Between 1540 and the mid-1900s, the large numbers of immigrants from Europe who settled in southern Chile brought about the clearance of considerable areas of forests to make way for agriculture, livestock grazing and construction. By 1955, about one-third of Chile's original forests had been destroyed. In 1974, the Forest Development Law (DL701) was promulgated for two purposes: (1) to privatize the forest industry and increase the international competitiveness of the local forest product industry, and (2) to provide subsidies to encourage nationwide reforestation and afforestation (Neira et al., 1992). Under the subsidies, as much as 90 per cent of the cost of planting trees was subsidized.

What were the direct outcomes of the DL701? Between 1975 and 1990, the wood industry of Chile boomed. It became one of the world's leading exporters of wood chips. In fact, currently, 10 per cent of Chile's GDP is accounted for by wood product exports. However, there was also a long list of negative outcomes. It became apparent towards the end of the 1980s that this growth was brought about by the widespread replacement of native forests by commercial, exotic crops. Small-forest owners were also persuaded by big multinational corporations to sell away their native forest lands to the latter, which was accelerated by a drop in the value of native forests with respect to commercial plantations. The current average replacement rate of native forests is around 20,000 hectares per year. Although a national wildlife protection system exists, the region – Region X – which has the second-largest biodiversity, including some species of flora and fauna not found anywhere else in Chile, is the least protected. In their efforts to cut operational costs and increase global competitiveness, many plantations outsource their jobs to non-unionized workers. This is reported to have resulted in plummeting wages and worsening working conditions.

Forest Management and Biodiversity Regulations in the United States

As an integral part of the Climate Change Action Plan under the Clinton Administration, the Northwest Forest Plan (NWFP) was proposed and executed in 1993 to achieve the dual goals of mandating forest management in the Pacific Northwest federal forests and protecting the endangered spotted owls. For the economic ill effects of reducing timber harvest in these forests to be buffered, the Northwest Economic Adjustment Initiative (NWEAI) was proposed. Plans under NWEAI included job training and shifting the overall focus of job creation from the wood/forest industry to the environmental management and civil engineering industries.

Many negative and unanticipated outcomes were observed. While the harvesting of timbers from the federal forests in the Northwest was effectively reduced between 1991 and 1995 – nationwide, the domestic production was actually reduced from 33,500 to 32,000 square metres – the demand in wood for construction alone increased by almost 50 per cent. This resulted in an increase in import from 37 to around 40 cubic metres per capita between 1991 and 1995 (Howard, 2002; FEMAT, 1993; Skog et al., 1998;). In fact, this increase in import set the stage for an ongoing trade dispute between the timber industries of Canada and the United States. Furthermore, the owl protection target was not successful either. The protected forests attracted exotic species into the habitat, including the more aggressive species of owls that either preyed on or competed with the Spotted Owls within the habitat. There were also fewer incentives to manage these protected forests due to the lower projected revenue from the timber products. As a result, the forest management goals were not entirely met. NWEAI's promise was also compromised by several problems in the execution of the initiative. For example, many of the civil engineering projects that were planned to provide retrenched timber workers with jobs required new skills for which timber workers needed to be trained. However, due to many social and psychological reasons, many timber workers did not respond well to the need for retraining. Many of these projects were also technologically-intensive and short term. To exacerbate the situation, many NWEAI local officials were not adequately trained to apply for funding for local developmental projects from the NWEAI.

IV.   Lessons on Integrated Policymaking from Case Studies

Why Negative, Unanticipated Outcomes Occur

These case studies showed that unintended and negative consequences occurred to sustainability indicators when:

1.   they are completely ignored by policymakers; and/or
2.   policymakers fail to identify intrinsic but inconspicuous links between seemingly unrelated indicators.

Although the virgin material taxes of Sweden and Denmark achieved domestic reductions in extraction and material substitution, the likely impact caused by any increased need for transportation of these materials was largely unknown. The shifting of problems associated with aggregate extraction to their trading partners was also not addressed at all in these policies. In regulating the use of substitute fuels in the United States and the United Kingdom, policymakers did not put sufficient attention to bridging the gaps in the data on which the different stakeholders rely for evaluating the costs and benefits of using these fuels. This resulted in public objections to some waste utilization projects, which would have prevented more widespread adoption of such waste reduction strategies. Amongst other shortcomings, the NWFP and NWEAI did not pre-empt how the protection of forests could result in fewer incentives for forest management and how these protected forests would attract predatory species into the habitat of the spotted owl. DL701 accelerated the reforestation process by boosting the local timber and wood products industries, which relied heavily on commercial plantations. The fact that this economic priority subordinated the need to protect local biodiversity caused more native forests to be replaced, and this might even have compromised the chance of further developing the ecotourism and other non-timber products/services sectors.

Based on the lessons learnt, one may postulate that, if policymakers conceptualize policies more broadly, which means considering a wider range of sustainability problems together from the outset and formulating policy strategies to address these problems concurrently, as far as possible, one should also aim to co-address these diverse sustainability indicators with the policy solutions. These concerns had been discussed in the literature under the topic of policy integration.

The Concept of Policy Integration

Current research on policy integration can be divided generally into theoretical/ conceptual and application-based approaches. Even though many of these studies are based on the integration of only a few indicators that are related to only one or two of the three sustainability domains, all acknowledge the fact that sustainability solutions must have an integrated approach in order to be effective. Conceptually, a wide variety of terms related to policy integration are used in the literature. The main ones are policy coherence (Challis *et al.*, 1998), policy consistency, intergovernmental management (Agranoff, 1996), cross-cutting policymaking, holistic government, joined-up government (Ling, 2002), policy collaboration (Gray, 1989; Huxham, 1996) and policy coordination (Mulford & Rogers, 1982; Meijers & Stead, 2004). Additionally, Huxham discussed the importance of interactions between governmental and non-governmental stakeholders. In their discussion on policy coordination, Meijers and Stead also stressed the need to coordinate across different policy instruments that are essential to address different aspects of a sustainability problem.

Many current applications and studies of policy integration focus on the environmental aspects of policymaking. For example, environmental policy integration (EPI) was given its political status by Article 6 of the 1999 Amsterdam Treaty, which states that 'environmental protection requirements must be integrated into the definition and implementation of the European Community policies and activities referred to in Article 3, in particular, with a view of promoting sustainable development' (Lenschow, 2002). This led to various policy developments that are aimed at providing the necessary procedures to materialize the injunction of EPI. Other primarily environmental efforts include those by Coffey and Dom (2004) and Feindt (2004). However, narrowly focusing on EPI overshadows the other domains of sustainable development: employment and economy. There are also other studies that focus on the cultural, employment (e.g. Sprenger, 2004) and economy aspects (e.g. Luken & Hesp, 2004) of development. It is important to note that what works well in the environmental, economic or employment domain separately (or even between two of these domains) may not be as effective when all three domains are considered within a common and integrated framework.

There is relatively little emphasis on how to achieve integration of policies across diverse terms of election or appointment for government officials. That is, the question of how to ensure that sustainability policy objectives remain coordinated and integrated after a government renews itself is seldom touched on in the literature. This is a form of temporal integration, which was mentioned by Briassoulis (2004). Realistic instruments that enhance stakeholder engagement in order to improve policy integration also require more studies, especially those related to all three domains of sustainability.

Based on these current and past researches and the lessons gleaned from the four case studies, a truly integrated policy must embody the following five characteristics:

- Addresses indicators in all three domains of sustainability (i.e. economy, environment and equity);
- Utilizes a range of policy instruments;
- Engages non-governmental stakeholders in the policymaking;
- Engages governmental stakeholders in the policymaking; and
- Promotes interactions amongst governmental and non-governmental stakeholders.

How do we know that, in the real world, integrated policies work better than policies that are not integrated? Before we describe the comparative advantages of policy integration, a brief description of four additional case studies is given in the next section. These additional case studies, together with the four cases described earlier, serve to test the hypothesis that the more integrated a policy is, the more successful it will be.

Advantages of Policy Integration

Additional case studies

In order for us to draw more rigorous conclusions regarding the advantages of policy integration, four additional cases were surveyed. They are described as follows, on the next page.

1. *Santa Monica Sustainable City Plan (SCP)*
   This was the city development plan put forth by the Santa Monica Government. The master plan looked into a total of 62 sustainability indicators. However, two out of the 62 indicators showed conclusively positive outcome. They were the reduction of GHG in the waste sector and the increase in waste diversion from the landfill. By addressing these two indicators in an integrated manner – increasing their curbside recycling program while upgrading their waste-collecting machineries and fleet – the waste-minimization policy was able to further ensure that the increased operation capacity did not increase net GHG emission.

2. *EPA's environment performance track*
   The Environmental Performance Track aimed to create a partnership with firms in which these firms agreed to exceed regulatory requirements and adopt environmental management systems in exchange for concessions, such as reduced priority for inspection. Participating firms needed to commit to improving at least four indicators, including various environmental indicators that were divided into upstream, input, non-product output and downstream stages. In other words, if successful, each participating firm would have addressed a maximum of four indicators under an integrated framework.

3. *Jobs for youth – Boston*
   The aim of this program was to empower the underemployed and unemployed by training them in environmental works, such as Brownsfield remediation. Indicators addressed include toxic waste removal, waste management, environmental costs reduction, increase in the number of jobs, purchasing power increase, health and safety improvement of workers, skills upgrade and Brownsfield utilization for greater economic values. Records show that all eight of these indicators show positive outcomes.

4. *SO_2 cap-and-trade Scheme*
   In order to reduce SO_2 emissions in the context of acid rain reduction under Title IV of the CAA amendments of 1990, this Scheme aimed to balance market force and government regulations to achieve objectives using least-cost methods. Indicators addressed were air pollution, acid rain and water quality, reducing environmentally derived cost, reducing socially derived cost, increasing public health and safety, and increasing economic competitiveness of businesses in trading. The creative use of auctioning procedures, expensive fines for infringement of emission rights and flexibility in SO_2 reduction technologies caused the Scheme to be highly successful. As a result, positive outcomes could be observed in all these eight indicators addressed.

Sustainability indicators

The eight cases studied vary according to the number and nature of sustainability indicators addressed. At one extreme, the Santa Monica SCP addressed 62 indicators. On the other end, Sweden's GT and Denmark's RMT address only four indicators. It is thus useful to formulate a standard list of key sustainability indicators with which these cases and proposed solutions have been assessed. A review of these seven cases yielded the following list of sustainability indicators:

- Energy use (including the fuel mixes involved);
- Air, water pollution and toxic waste production;
- Biodiversity;
- Environmentally derived costs;
- Socially derived costs;
- Market share and economic competitiveness;
- Number of jobs;
- Purchasing power and wages;
- Workers' safety and health;
- Levels and diversity of skills; and
- Opportunity for upgrading and innovation of industry.

This selected set of 11 indicators, to a large extent, encompasses, or is related to, the various other types of indicators included by many of the sustainability models reviewed. A detailed discussion of this observation is beyond the scope of this article.

Evaluating the effectiveness of policy integration

All the seven cases were plotted in the graph in Figure 10.2. The circles in the graph represent all the cases. The x-axis stands for the *number of sustainability indicators explicitly addressed in the policy*, whereas the y-axis is the *number of sustainability indicators integrated into the policy*. Whenever a policy is designed and implemented to address a number of indicators (i.e. it co-addresses these indicators) or considers how addressing one indicator can affect other indicators, that policy can be viewed as having integrated those indicators considered. In our correlation between the level of policy success and the level of policy integration in each of the seven case studies, policy success was defined as the ability of the policy outcome to match its original expectation. In other words, when a policy manages to yield an anticipated outcome, it is considered successful in that specific sense. A few key conclusions could be arrived at from these plots:

- The number of indicators integrated into the policy is a more significant predictor of anticipated positive outcomes than the number of indicators explicitly addressed, as indicated by the fact that cases with a higher level of integration show a higher number of achieved policy goals.

- Policies integrating more indicators also show a lower number of negative unanticipated outcomes.

- Even though a policy considers a very wide range of indicators, as in the case of the Santa Monica SCP, the number of such indicators integrated into a policy may actually be very low. In other words, the policy addresses an indicator in a fragmented manner.

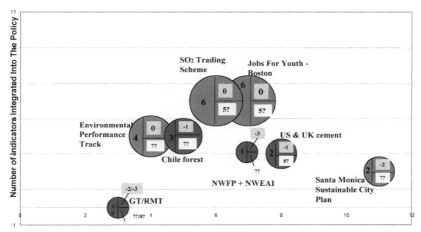

Figure 10.2: Relationship between the number of indicators addressed, the number of indicators integrated into the concerned policy and the policy effectiveness as measured by the number of anticipated positive outcomes. The size of the circles corresponds to the number of positive anticipated outcomes – the more the positive outcome, the bigger the circle. The different segments of the circles are interpreted as follows:

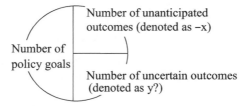

Even the most successful policy, the SO_2 cap and trade, does not look into all 11 key sustainability indicators identified earlier. Henceforth, these seven cases showed, to a great extent, that policy integration will ensure the success of policies in meeting their intended goals. What kind of integrated policies can China implement to address problems with its cement industry? How will these policies be different from existing ones? Existing policies and the policymaking structure of China are summarized in the following sections, before integrated policies are proposed.

V.   Improving China's Strategies for Its Cement Industry

Strategies to Address the Sustainability of the Cement Industry

To reduce the environmental impacts due to cement production and consumption, the Chinese Government has developed strategies to specifically address the improvement of two aspects of the industry: energy efficiency and pollution control. Domestically, the Chinese Government has promulgated the Energy Conservation Law and Environmental Protection Law, which include measures to encourage industries to curb emissions and promote energy efficiency (Wang, 2008). Furthermore, different project-level initiatives have been (and will be) implemented to eliminate outdated production processes, deploy more advanced technologies and establish production lines. For example, newly-developed, highly energy-efficient precalciner kilns with low NO_x emissions are now widely used, and dust collectors and electrostatic precipitators (to eliminate particulates) are already operating in most Chinese cement plants (Wang). In addition, inefficient cement plants with a combined capacity of about 140 million tonnes of production were closed in 2007 in an effort to improve the environmental performance of the industry.

Many cement manufacturers are also using alternative fuels and raw materials with the aim of reducing CO_2 emissions, as well as costs, from the use of fossil fuels. These alternate fuels include industrial by-products, sludge and mining residues. In fact, some cement kilns are being used to process consumer waste, either as a fuel source or as a cement component.

China has been proactively engaging various international initiatives, such as the Asia-Pacific Partnership on Clean Development and Climate, which aim to help China adopt advanced cement production technologies and efficient management skills. The Chinese Government is also working with the World Resources Institute to adopt its GHG Protocol Initiative, which is used internationally as an accounting tool for GHG emissions.

Problems Encountered in Implementing These Policies

Structural issues that may compromise the full potential, and thus shortchange the benefits of the abovementioned policies, exist in the current policymaking system of China. Firstly, as observed by Wang and Lin (2010), China follows a predominantly top-down approach in implementing its environmental policies, a system that contains opportunities of mismatch between the grand overall environmental performance goals and the motivation at the local levels to meet those goals. At the national level, the main agency for environmental management and implementation is the State Environmental Protection Commission. Its main role is to work out general principles for the policies for each sector and set national standards for a certain sector (Wu, 1987). The executive arm of this State Council organization is the National Environmental Protection Agency (NEPA), which oversees all provincial Environmental Protection Bureaus (EPBs) in all administrative regions. Each provincial organization also runs its own network of prefectural and county EPBs, which in turn oversee the city and township protection organizations (i.e. the local agencies). Wang and Lin (2010) observed that, at these local agencies, oftentimes the incentive systems do not reward the meeting of environmental targets, and more emphasis is placed on the ability to meet economic production targets. Hence national-level targets, including those pertaining to the improvement of the cement industry, may not even be incorporated into local-level policy decision-making.

The second structural problem is the fragmented nature of its policymaking process, even at a given geographical and political level. For example, at the national level, soil and water conservation is managed by the Ministry of Water Resources, grassland restoration by the Ministry of Agriculture, forestry by the State Forestry Bureau, and air and water pollution by the Ministry of Environmental Protection (MEP). MEP creates and manages a forum for reconciling conflicts amongst various ministries, coordinating their activities and facilitating enforcement of environmental regulations. Its authority in such matters, however, is limited by the perceived need to seek consensus among governmental representatives from more than 20 ministries. It cannot compel cooperation because it lacks any formal mechanism and jurisdiction to do so. As a result, several more powerful ministries have been known to have ignored the call for cooperation and costly environmental compliance (Edmonds, 1994).

Thirdly, the extensive use of waste materials as alternate fuels for cement kilns in China may give rise to disputes (or even lawsuits) in the future, similar to those observed in the United States and the United Kingdom (i.e. the second case study covered in this work).

Finally, the urban population of China increased to a staggering 629 million people in 2009, compared with 459 million in 2000 (Friedman, 2005). Liu et al. (2008) found that between 1996 and 2008, 8.4 million hectares of cropland were lost due to construction and other causes. In 2008 alone, construction eliminated 313,000 hectares of productive agricultural land. Mandatory closures of the polluting kilns in the rural regions may exacerbate these and related problems because these closures may force an entrenched workforce to move into the cities in search for employment opportunities and thus increase the rate of urbanization. The sharp increase of city labour force (estimated at around 10 million every year) has also placed great pressure on the job market in the urban areas.

Integrated Policy Solutions

Based on the five characteristics of integrated policies defined in section IV on page 228, the following strategic changes are proposed for the China cement industry.

Integrating the policymaking structure and sustainability indicators

Fragmentation of the policymaking process must be minimized. At the national level, MEP can be given more power to exercise decisions to compel different government bodies to comply with environmental standards, as well as initiate cross-ministerial collaborations to meet certain environmental targets. However, this is likely to be unsuccessful if Ministries do not see the importance of environmental protection. Henceforth, in tandem with this aforementioned idea, MEP and all other Ministries must be given new lists of key performance indicators, which will contain economic, environmental and social indicators. In other words, performance targets can be set for a wider range of different indicators, spanning the three bottom lines of sustainable development. Some of these indicators can be created to link economic or social performances with environmental performance. For example, the governments at all levels are required to show how their economic or social performance (measured by certain agreed indicators) has improved due to the improvement in the environmental performance.

This concept can also be applied to, and communicated across, the different hierarchical levels of government in China; that is, *as far as possible*, provincial and city governments are required to set performance targets in selected economic, environmental and social indicators agreed upon by the different levels of governments. In fact, the performance incentive systems at the different levels need to be modified so that meeting of these sustainability targets will be rewarded.

Engaging non-governmental stakeholders and promoting stakeholder interactions

Since the 1990s, environmental non-governmental organizations (ENGOs) have been actively involved in an expanding scope of environmental issues, including resource management and the protection of environmental rights (Zhang, 2011). In the aforementioned proposal, ENGOs can be engaged to suggest indicators in the social and environmental realms that the central government can use as a guide to communicate sustainability targets to its subordinate governments. However, there are examples in which government stakeholders do not participate in initiatives created by ENGOs. For example, the Alxa SEE Ecological Association (SEE), an organization set up by eminent Chinese entrepreneurs, was established in 2004. Most of SEE's work is on the theme of nature protection. In 2010, SEE worked on the carbon trading market with a belief that NGOs must participate in the dialogue to ensure that a fair interest-sharing mechanism can emerge from the initiative. Carbon exchange firms, energy management companies and research institutes were engaged. Unfortunately, due to the lack of government participation and the uncertainty of future policies, the project was aborted in 2010. While a detailed analysis of this failure is beyond the immediate scope of this paper, it underlines the importance of bringing governmental and non-governmental stakeholders to the table on sustainability issues.

How can ENGOs in China be engaged? ENGOs may be engaged by (or even engage) appropriate government bodies to address a specific theme or topic related to the sustainability of the cement industry. For example, an ENGO can form an advisory group to evaluate the safety of using a certain type of waste materials as alternate fuels for cement kilns. This approach, which may be termed 'focused thematic integration', can help create lists of indicators (introduced in section V on page 232) that the central government can use to set sustainability. An advantage of 'focused thematic integration' is that a theme can be used to select and engage relevant stakeholders to focus their sustainability discourse on the said theme. This may be more productive and efficient than engaging them to discuss more general topics on which they may have limited knowledge and interest.

Ensuring safety in the use of alternate fuels with the 'green fuel innovation' credit scheme and a public demonstration and outreach program

The use of alternate fuels for cement production should not be suppressed. The crux of the challenge is in ensuring that their usage is closely regulated and monitored. For this challenge to be addressed, an integrated policy consisting of three main policy components may be conceived. Firstly, a demonstration and public outreach project spearheaded by an agency in charge of coordination across the different agencies may be implemented. At the national level, MEP may play this role. The purpose of this project is to share information and ideas on the advantages and disadvantages of different types of alternate fuels under different thermodynamic conditions. Besides members of the public, cement facilities managers and operators are also the main target audience. These stakeholders can be invited to visit facilities where demonstration projects are conducted – facilities that can be developed into knowledge clearing house on safety in the use of alternate fuels.

Complementing the above policy component, a second policy component known as the 'green fuel innovation' credit scheme may be adopted. Under this scheme, cement companies are given different levels and types of incentives for trying out alternate fuels and reporting emission data from the use of these fuels. Incentives may be in the form of tax rebates, and the level of incentive has to match the extent of emission reduction. Thirdly, cement companies can also earn additional credits under the said scheme if they agree to openly display emissions data of their kilns on the Internet (and verified by approved measurement technology).

It must be stressed that the aforementioned policy components must be executed in an integrated manner. That is, companies that visited the demonstration and public outreach project can be invited to participate in the 'green fuel innovation' credit scheme, or companies that are interested to give the 'green fuel innovation' credit scheme a try must commit to a crash course at the facilities hosting the demonstration and public outreach project.

VI.   Conclusion

The analysis of seven case studies supported the hypothesis that the more integrated a policy is, the more successful it will be. The current situation of China's cement industry – the role it plays in the global market, its significance in contributing to climate change, current policy strategies utilized to regulate the industry and observed problems in these strategies – has pointed us to a few possible solutions. Based on the five characteristics of integrated policies, these proposed solutions are hypothetical at best. To be fair, their eventual usefulness and effectiveness will depend heavily on the actual conditions under which they are implemented by the various stakeholders, especially the different levels of governments. These proposed solutions certainly do not imply that other forms of integrated policy solutions that are not covered in this paper are going to be less effective. The proposed solutions served as guides to trigger further dialogue on how to use the concept of integrated policies to address the many intertwined challenges facing China's cement industry in particular, and the world in general, today.

References

Agranoff R (1996) Managing Intergovernmental Processes. In: Perry JL (ed) *Handbook of Public Administration,* Second Edition. Jossey-Bass, San Francisco.

Arm M (2003) *Mechanical Properties of Residues as Unbound Road Materials – Experimental Tests on MSWI Bottom Ash, Crushed Concrete and Blast Furnace Slag.* KTH Land and Water Resources Engineering, Stockholm.

Augenbroe G, Pearce AR, Kibert CJ (1998) *Sustainable Construction in the United States of America, A Perspective to the Year 2010.* CIB-W82 Representative, Georgia Institute of Technology, Atlanta, GA.

Challis L, Fuller S, Henwood M (1998) *Joint Approaches to Social Policy: Rationality and Practice.* Cambridge University Press, Cambridge.

Chem Systems Ltd (1999) *Substitute Liquid Fuels (SLF) Used in Cement Kilns – Life Cycle Analysis.* Environment Agency Technical Report p. 274. Environment Agency, United Kingdom.

Coffey C, Dom A (2004) Environmental Policy Integration in Europe: Looking Back, Thinking Ahead. *Proceedings of the 2004 Berlin Conference on the Human Dimensions of Global Environmental Change: Greening of Policies – Inter-linkages and Policy Integration.* Berlin, Germany.

Delles J *et al.* (1992) *Trace Metals in Cement and Kiln Dust from North American Cement Plants.* Construction Technology Laboratories for Portland Cement Association, Illinois.

Edmonds RL (1994) *Patterns of China's Lost Harmony: A Survey of the Country's Environmental Degradation and Protection.* Routledge, London.

ERAtech Group LLC. (1999) Alternate inputs of waste materials and cement manufacturing [Article on the Internet]. [Cited 2 October 2011]. Available from: http://www.eratech.com/papers/aiwmcm.htm.

Eurostat (2005) http://epp.eurostat.cec.eu.int/portal/page?_pageid=1090,300706 82,1090_33076576&_dad=portal&_schema=PORTAL.

Feindt PH (2004) Greening European Agricultural Policy – Multi-Level Strategies between Trade, Budget, Agricultural and Environmental Policy. *Proceedings of the 2004 Berlin Conference on the Human Dimensions of Global Environmental Change: Greening of Policies – Inter-linkages and Policy Integration*. Berlin, Germany.

FEMAT (1993) *Outlook for National Forest Products Markets*. In Chapter 6: Economic environment of the options, forest ecosystem management: An economic, environmental and social assessment. Forest Service (United States Department of Agriculture), Bureau of Land Management (United States Department of the Interior), Texas, United States of America.

Friedman J (2005) *China's Urban Transition*. University of Minnesota Press, Minneapolis, MN.

Ginns SE, Gatrell AC (1996) Respiratory health effects of industrial air pollution: A study in East Lancashire, UK. *Journal of Epidemiology and Community Health* **50**, 631–5.

Gray B (1989) *Collaborating Finding a Common Ground for Multiparty Problems*. Jossey-Bass Publishers, San Francisco.

Green Alliance (2002) *Creative Policy Packages for Wastes: Denmark*. United Kingdom.

Howard J (2002) *US Timber Production, Trade, Consumption, and Price Statistics, 1965-2001,* Research Paper FPL-RP-595. Forest Product Laboratory, US Forest Service, Department of Agriculture, Madison, Wisconsin.

Huxham C (ed) (1996) *Creating Collaborative Advantage*. Sage, London.

Hwang BG, Tan JS (2010) Green Building Project Management: Obstacles and Solutions for Sustainable Development, *Sustainable Development*. Online version available at http://onlinelibrary.wiley.com/doi/10.1002/sd.492/ abstract (DOI: 10.1002/sd.492).

Lenschow A (ed) (2002) *Environmental Policy Integration: Greening Sectoral Policies in Europe*. Earthscan, London.

Ling T (2002) Delivering joined-up government in the UK: Dimensions, issues and problems. *Public Administration* **80(4)**, 615–42.

Liu J *et al.* (2008) Ecological and socioeconomic effects of China's policies for ecosystem services. *Proceedings of the National Academy of Science of the United States of America* **105**, 9477–82.

Luken RA, Hesp P (2004) Developing and transition economy efforts to achieve policy integration for sustainable development. *Proceedings of the 2004 Berlin Conference on the Human Dimensions of Global Environmental Change: Greening of Policies – Inter-linkages and Policy Integration*. Berlin, Germany.

Meijers E, Stead D (2004) Policy Integration: What Does It Mean and How Can It Be Achieved? A Multi-disciplinary Review. *Proceedings of the 2004 Berlin Conference on the Human Dimensions of Global Environmental Change: Greening of Policies – Inter-linkages and Policy Integration.* Berlin, Germany.

Mulford CL, Rogers DL (1982) Definitions and models. In: Rogers DL, Whetten DA (eds) *Inter-organizational coordination: Theory, research, and implementation.* Iowa State University Press, Ames.

Neira E, Verscheure H, Revenga C (1992) *Chile's Frontier Forests: Conserving a Global Treasure.* A Global Forest Watch Chile Report. World Resources Institute, Washington D.C., United States of America.

Schwartz S *et al.* (1996) *Domestic Markets for California's Used and Waste Tires.* University of California, CA.

Skog K, Ince P, Haynes RW (1998) Wood Fiber Supply and Demand in the United States. *Proceedings of the Forest Products Study Group Workshop, Forest Products Society Annual Meeting,* June 23, Merida, Yucatan, Mexico. North American Forestry Commission,

Sprenger RU (2004) Integration of Environmental and Employment Policies: Assessment of the EU Experience to Date. *Proceedings of the 2004 Berlin Conference on the Human Dimensions of Global Environmental Change: Greening of Policies – Inter-linkages and Policy Integration.* Berlin, Germany.

Swedish Environmental Protection Agency (EPA) 2000:5077 (2000) *Naturgrusskatten: utvardering av skatteeffekterna.* Naturvardsverket, Stockholm.

Swedish Environmental Protection Agency (EPA) 2001:5155 (2001) *Avgifter, skatter och bidrag med anknytning till miljovard.* Naturvardsverket, Stockholm.

van Oss HG, Padovani AC (2003) Cement manufacture and the environment, Part II: Environmental challenges and opportunities. *Journal of Industrial Ecology* 7(1), 93–126.

Wang L (2008) *The Energy Management and Emission Control with Chinese Cement Industry, Cleaning Up China's Cement Sector.* China Environment Forum, Washington, DC.

Wang C, Lin Z (2010) Environmental policies in China over the past 10 years: Progress, problems and prospects. *Procedia Environmental Sciences* 2, 1701–12.

Worrell E, Price L, Martin N, Hendriks C, Ozawa ML (2001) Carbon Dioxide Emissions from the Global Cement Industry. *Annual Review of Energy and the Environment* 26, 303–29.

Wu Z (1987) The origins of environmental management in China. In: Glaeser B (ed) *Learning from China? Development and Environment in the Third World Countries.* Allen & Unwin, London.

Zhang Z (2011) *The Impact– Environmental NGOs in China's Clean Energy Policy.* Hauser Center China Field Study Report. Harvard University, Cambridge.

CHAPTER ELEVEN

Green Buildings: New Perspectives

George OFORI
School of Design and Environment, National University of Singapore, Singapore

"In Singapore, the regulations, government incentives and market demand have led to progressive action by clients, designers and consultants, and there are indications that providing green features in buildings is becoming the norm in Singapore. However, even here, it is recognized that much more needs to be done. The questions that remain include the extent of provision; and how to ensure the effective operation of the facility."

– George OFORI

Green Buildings: New Perspectives

George OFORI[1]

School of Design and Environment, National University of Singapore, Singapore

Abstract

Several aspects of sustainability are relevant to construction. These include the need to change consumption patterns, the implications of climate change and pollution control. The construction industry plays an important role in the economy and in long-term national development by establishing the physical environment for productive activity and for enhanced quality of life. On the other hand, construction has adverse impacts on the physical environment. This comes from the materials and equipment that are used in the construction process, the activity which creates the constructed items, and from the products of the industry. In these ways, construction manifests the economic, social and environmental pillars of sustainable development.

This chapter considers the sustainability impacts of construction activities. It focuses on climate change, and considers the role of 'green building' in combating this phenomenon. It considers the benefits of green building as its drivers, and then its obstacles and challenges. In this chapter, examples of these main aspects in some national programmes are presented, focusing on Asian countries. The green building programme in Singapore is also discussed. The programmes in some Asian countries are compared and set against those in some industrialized countries. It is observed that approaches differ in various countries and that much progress has been made in most countries. It is recommended that national strategies involving all key stakeholders' strong leadership be adopted in each Asian country and that there is scope for, and merit in, regional co-operation.

Key words: barriers, climate change, drivers, green building, initiatives, Singapore

I.   Sustainable Development: Context of the Chapter

There is a lack of agreement on, inadequate understanding of and controversy over 'sustainable development'. This is discussed in some of the chapters in this book. It is necessary to consider the concept, albeit briefly, in order to provide

---

[1] Department of Building, National University of Singapore, 4 Architecture Drive, Singapore 117566. Email: bdgofori@nus.edu.sg; Tel: 6516 3421.

a frame of reference for discussion in this chapter. The expression 'sustainable development' entered the common vocabulary after it was used by the Bruntland Commission (World Commission on Environment and Development, 1987). It attained even greater currency after the United Nations Conference on Environment and Development (UNCED) in 1992. One of the outcomes of UNCED, *Agenda 21*, called for action in all countries. Several countries prepared their national strategic plans for sustainable development before, or following, the conference. Singapore's national blueprint has been periodically revised (see Ministry of Environment and Water Resources, 2006). Some action has ensued in all countries, but the debate on the adequacy and pace of the action continues. There are questions on whether such actions are necessary at all, what is appropriate, what order of priority should be adopted and who should act.

There are many definitions of 'sustainable development'. A government agency in Hong Kong special administrative region (SAR) defines it as:

> 'Sustainable development in Hong Kong balances social, economic and environmental needs, both for present and future generations, simultaneously achieving a vibrant economy, social progress and better environmental quality, locally, nationally and internationally, through the efforts of the community and the Government'. (Planning Department and Environmental Resources Management, 2000)

The aims of Malaysia's National Policy on the Environment (MOSTE, 2002) are to continue the country's economic, social and cultural progress, and enhance the quality of life of its people through environmentally sound and sustainable development policies. The objectives of the policy are to achieve a clean, safe, healthy and productive environment for the present and future generations of Malaysians; to conserve the country's unique and diverse cultural and natural heritage with effective participation by all sectors of society; and to engender sustainable lifestyles and patterns of consumption and production. Singapore's Inter-Ministerial Committee on Sustainable Development (IMCSD) (2009) noted that 'for Singapore, sustainable development means achieving both a more dynamic economy and a better-quality living environment for Singaporeans now and in the future' (IMCSD, 2009: 13). Thus, various authors highlight different ranges of components of sustainable development. The Hong Kong Government adopted the following 'guiding principles' of sustainability (Planning Department and Environmental Resources Management, 2000): economy, health and hygiene, natural resources, social, biodiversity, cultural vibrancy, environmental quality and mobility.

Of the strands of sustainability, this chapter focuses on the environmental aspects. Climate change is, arguably, the subject of greatest current concern on the sustainable development agenda. Extreme changes in the earth's climate have increased global awareness that urgent action is needed to address the causes of these phenomena, which have significant adverse economic and social implications for many parts of the world. The Fourth Assessment Report of the

Intergovernmental Panel on Climate Change (IPCC) (2007) indicates that the number of heat waves has increased since 1950. Evidence also suggests that there have been substantial increases in the intensity and duration of tropical storms and hurricanes since the 1970s. In 2007, the Meteorological Office, UK, stated that the 11 warmest years on record had all occurred in the previous 13 years.

The International Energy Agency (IEA) (2008) has set a target of 77 per cent (or 48 gigatonnes) reduction in the earth's carbon footprint by 2050. Considering their contribution to this required effort, many countries have set their own targets, formulated policies and launched initiatives in this regard. The Climate Change Act of 2008 commits the UK Government to reducing carbon dioxide (CO_2) emissions by 80 per cent by 2050 over the1990 levels (Department of Energy and Climate Change, 2009a, 2009b). Subsequently, the interim target of reducing emissions by 26 per cent by 2020 was raised to 34 per cent by law (Office of Public Sector Information, 2008). The *Singapore Sustainable Development Blueprint* has set a target of reducing energy use by 35 per cent, from the 2005 levels, by 2030 (IMCSD, 2009). There have also been targets at subnational levels. For example, Seattle, USA, aims to reduce energy use by 20 per cent by 2020 (Koch, 2011).

To attain the targets set at the country, state or city level, each sector of the economy must make a contribution. Singapore's initiatives to respond to climate change have three main thrusts (IMCSD, 2009): (a) promote energy efficiency, developing enablers to make consumers more energy-efficient/fuel-efficient and improving energy management practices of businesses; (b) promote the use of clean energy (such as natural gas); and (c) encourage demonstration projects on renewable energy (such as solar and biomass).

II. Research Objectives

The objectives of this chapter are to:

- consider the (positive and negative) sustainability impacts of construction activity and its products, and use construction to illustrate the main pillars of sustainable development;
- discuss the nature of 'green building' and its role in combating climate change;
- consider the benefits, drivers and challenges of green building;
- discuss green building programmes in some Asian countries and consider the way forward; and
- suggest some ways and means to attain progress in green building programmes in Asia.

This chapter considers the sustainability impacts of construction activities. It discusses sustainable development to provide a contextual background for the chapter. It then explains the nature of the construction industry and its activities.

The relationship between construction and sustainability is explained, focusing on climate change. The concept of green building is then introduced, and the role of green building in combating climate change is discussed. This is followed by consideration of the benefits, drivers, problems and challenges of green building. Examples from many countries are given, with a focus on Asia. It then discusses the actions to attain sustainable building in Singapore, and compares some aspects of green building programmes in some Asian countries against those in industrialized countries. Finally, some recommendations for action in urban Asia are presented.

III. Building and the Environment

The construction industry is defined as the sector of the economy that plans, designs, constructs, repairs, rehabilitates and maintains, and eventually demolishes, buildings and structures of all types (Ofori, 1990). The products of the construction industry can be broadly categorized into 'buildings' and 'infrastructure'. These are the basic components of the 'built environment', which makes up cities, towns, villages and other settlements where humans undertake almost all their activities. Buildings provide the physical facilities for all forms of production (such as in factories, offices, hotels and shopping complexes) and leisure (such as in cinemas, sports halls, theatres and night clubs), and contribute to the enhancement of the quality of life (such as in schools and hospitals) of people. The items of infrastructure provide the utilities for production and other activities (such as in water, power and gas supplies), serve as linkages amongst the production facilities (such as roads, highways, railways, airports and harbours) and adjuncts to the production facilities (such as dams and irrigation systems). Thus, buildings and infrastructure are necessary for economic activity and growth, and for long-term national socioeconomic development.

The construction industry is large. In most countries, construction contributes 5–10 per cent of gross domestic product (GDP). There are several backward and forward linkages between construction and many other sectors of the economy (Hillebrandt, 2000; Ive & Gruneberg, 2000). Thus, construction activity has significant multiplier effects, as it stimulates activities in several other sectors. For this reason, it is suggested that governments use their investments in construction to regulate the economy (see, e.g. Hillebrandt, 2000). Thus, construction activity is indispensable for development. However, the creation of buildings and infrastructure has major implications for the environment.

Ofori (1992) argues that 'the environment' should be the fourth client objective of construction projects (after the traditional criteria of cost, time and quality). What does 'the environment' mean here? Ofori suggests that the scope of 'the environment' in the context of construction is as follows: (a) resource conservation; (b) prevention of all types of pollution; (c) protection and preservation of natural ecosystems; (d) safeguarding the fabric of constructed facilities in changing atmospheric conditions; (e) promotion of the health and well-being of users of the built facilities; and (f) development of environmentally conscious lifestyles of the users. The following can be added: preservation of land, and protection of the health and safety of workers as well as occupants of nearby buildings. Table 11.1 shows the environmental impacts of building activity. The contents show the potential adverse cradle-to-cradle implications of construction activity – from the extraction of the necessary materials; the processing and transportation of materials, components and equipment; the on-site construction; the operation of the completed facility; and the demolition of the building and disposal of wastes.

Examples of the extent of these impacts were highlighted in a study in North America by the Commission for Environmental Cooperation (CEC) (2009), which found that, in Canada, buildings account for 33 per cent of all energy used, 50 per cent of natural resources utilized, 12 per cent of non-industrial water consumed, 25 per cent of landfill-waste generated, 10 per cent of airborne particulates produced and 35 per cent of greenhouse gases (GHGs) emitted. In the USA, buildings are responsible for 40 per cent of energy used, 12 per cent of water consumed, 68 per cent of electricity used, 38 per cent of CO_2 emitted and 60 per cent of non-industrial wastes produced. Finally, in Mexico, buildings account for 17 per cent of all energy consumed, 25 per cent of the total electricity used, 20 per cent of overall CO_2 emissions, 5 per cent of potable water consumed and 20 per cent of waste generated. Of this range of possible impacts of construction, this chapter focuses on energy use and buildings.

According to the Malaysian Energy Centre (2009) (better known by its Malay acronym PTM, which stands for *Pusat Tenaga Malaysia*), the average building energy intensity (BEI) in the country is 250 kilowatt-hours/square metre/year. Only a few buildings in the nation have BEI of less than, or equal to, 150 kilowatt-hours/square metre/year, a figure indicated by PTM as an average, in terms of performance, for buildings at the high end in Malaysia). Amongst these are the Securities Commission Headquarters building (120 kWh/m²/year), the Low Energy Office building (100 kilowatt-hours/square metre/year), the Green Energy Office building[2] (40 kilowatt-hours/square metre/year) and the offices of the Energy Commission building (80 kilowatt-hours/square metre/year).

Many studies show that, as indicated in the data for Canada, Mexico and the USA above, buildings are major users of energy. The global average is about 40 per cent, but there are broad ranges, with industrialized countries showing higher figures than those of developing nations. In Singapore, it is estimated that buildings contributed to about 16 per cent of the total energy use in 2007 (IMCSD, 2009). This excluded household use and consumers' consumption, which amounted to about another 9 per cent. The same report indicated that the total household consumption of electricity in the country increased from 3,794 gigawatt-hours in 1995 to 6,226 gigawatt-hours in 2007. PTM estimates that the

---

[2] This building was initially named the Zero Energy Building (ZEO), but it was renamed after it failed to achieve carbon neutrality. The first building to achieve this in South East Asia is Singapore's Zero-Energy Building.

maximum level of demand for electricity in Malaysia in 2005 was 13.8 gigawatts, compared to an installed capacity of 19.2 gigawatts (Ahmad-Hadri, 2005). The corresponding figures in 2010 would be 20.0 gigawatts and 25.3 gigawatts respectively. Malaysia's total energy consumption per capita has continuously increased over the years. It nearly doubled from 1,307 kilograms of oil equivalent per person in 1990 to 2,229 in 2000 and then to 2,418 in 2005 (IEA, 2008).

Table 11.1: Environmental impacts and considerations of construction activity

What is used	Where it is built	How it is built	What is built
* Where raw materials are obtained	* Location of facility, nature of terrain and ground conditions, alternative uses of the land	* Methods of construction on site	* Planning and design of facility (e.g. potential of daylighting and natural ventilation)
* How raw materials are extracted, how land is restored after extraction (if necessary)	* Immediate physical environment, proximity to water sources and ecosystems	* Construction project management systems (e.g. quality management systems)	* Life cycle economic, quality and maintainability considerations
* How raw materials are processed	* Social disruption (e.g. displacement of site's inhabitants)	* Site control measures (housekeeping)	* The extent of use of energy and other resources in the operation of the building
* Whether, and how renewable raw materials are regenerated	* Economic disruption (e.g. loss of livelihoods of previous inhabitants)	* Welfare of site workers, neighbours and the general public	* The ease of demolition of the building (deconstruction)
* How materials are transported to and stored on site	* Present infrastructure, need for expansion to serve new building, its impact	* Resource management (including waste minimization)	* Recycling and reuse of demolition wastes
* How materials are moved on site	* Impact on local vehicular traffic		

Source: Adapted from Ofori (1999)

As a corollary of their energy use, buildings also make significant contributions to GHG emissions. IPCC (2007) reported that GHG emissions by sector are as follows: energy supply – 25.9 per cent; industry – 19.4 per cent; forestry – 17.4 per cent; agriculture – 13.5 per cent; transportation – 13.1 per cent; residential and commercial buildings – 7.9 per cent; and waste and wastewater – 2.8 per cent. When CO_2 is considered, the building industry is reported in several studies to contribute 9 gigatonnes of CO_2, an average of 33 per cent of the global total, with variations amongst countries. In the European Union (EU), the proportion is 40 per cent, and it rises to about 50 per cent in the UK (see a review of Kievani *et al.*, 2008). It is estimated that energy use in 26.1 million residential buildings in the UK generates about 27 per cent of the nation's CO_2 emissions (Department of Environment, Food and Rural Affairs, 2006). IEA (2008) estimates that the building sector will need to reduce annual emissions by 8.2 gigatonnes below business-as-usual by 2050.

IPCC's (2007) 'Fourth Assessment Report' identified 'building' as having the capacity to reduce emissions by 29 per cent by 2020. This is the sector with the largest single potential in this regard. The panel notes that the building industry offers the greatest low-cost mitigation opportunities. Oreszczyn and Lowe (2010) note that reducing the energy use associated with buildings has long been identified as a key, relatively easy and cost-effective strategy, and as a focus for policy that can deliver a major impact in the next 30 years. However, despite the perception that energy efficiency in buildings is straightforward and requires minimal investment, Oreszczyn and Lowe cite data from several UK reports, and note that empirical evidence and experience suggest that it will not be easy or cheap to reduce energy use in buildings. They note that the technology to reduce heat loss in buildings has been available for over 30 years, and a theoretical understanding of energy flows in buildings has been developed to a high level of sophistication. The government now assumes that radical targets can be achieved quickly, but, where information is available, it indicates that the industry is far from achieving any of the targets the sector requires.

Governments of many countries have set specific targets for various types of buildings, and housing seems to be the focus of such policies in most countries. For example, in the UK, new housing is to be zero carbon by 2016 (Department of Communities and Local Government, 2006); in the Netherlands, 50 per cent of new housing is to be '(net) zero energy' by 2016 (International Council for Research and Innovation in Building and Construction [CIB], 2009); and in California, USA, all new homes are to be zero net energy by 2020 (Marsden Jacobs Associates, 2009). To attain UK's ambitious emissions reduction targets, it plans to retrofit 14 million homes by 2020 (Royal Institution of Chartered Surveyors [RICS], 2011).

Newton and Tucker (2011) note that, to attain a global 2°C limit to temperature rise, Australia's proportionate part would require reductions on 2000-level GHG emissions of 25 per cent by 2020 and 90 per cent by 2050. However, current commitments by the government fall short of this target: a reduction in

CO_2 emissions by 5 per cent below the 2000 levels (rising to 25 per cent contingent on the international agreement). Newton and Tucker observe that Australia's housing sector (which consumes about 12 per cent of the total energy used) has no clearly defined role in the nation's carbon-abatement schemes and has failed to sustain any significant initiatives in carbon reduction since the introduction of the energy-rating scheme for new homes in 2003.

A four-year study by the World Business Council for Sustainable Development (WBCSD) (2009) shows how energy use in buildings can be reduced by 60 per cent by 2050. However, this will require immediate action to transform the building sector. The research covered the residential and commercial building sectors in the USA, EU, Japan, China, India and Brazil, which, together, consume about two-thirds of the current global energy output. A unique feature was that the study was monitored by an assurance group of external experts. Energy use by building type was analyzed for millions of existing and new buildings, and projected to 2050, accounting for differences such as climate and building design. Using computer simulations, researchers were able to show the market response to various combinations of financial, technical, behavioural and policy options, and to identify the optimum mix to achieve transformation for each market studied.

One characteristic of the construction industry is that, in any period, the additional volume of buildings constructed is very small, compared to the total stock of existing buildings. It is estimated that at least 75 per cent of the homes that will be standing in the UK in 2050 have already been built (Gupta & Chandiwala, 2010). In Singapore, some 80 per cent of the population live in public residential apartments. In 2006–2007, some 1,764 units were completed and another 14,212 were under construction, whereas in 2007–2008, 6,247 units were completed and 18,073 were under construction. These figures are very small fractions of the stock of residential units of about 990,000 in 2007–2008 (Housing and Development Board [HDB], 2008). Indeed, Tan (2011) suggests that existing buildings constitute more than 95 per cent of Singapore's building stock. For this reason, the greatest potential for reducing energy usage lies in the refurbishment and rehabilitation of existing buildings.

IV. Sustainable Construction

The expressions, 'sustainable construction' and 'green building', which are often used interchangeably, have become part of the professional and, subsequently, common vocabulary. RICS (2005: 2) offers this definition: 'Green buildings are those that use resources efficiently, reduce waste and produce superior indoor and other qualities'. CEC (2009) notes that 'Green building refers to the use of environmentally preferable practices and materials in the design, location, construction, operation and disposal of buildings. It applies to both renovation and retrofitting of existing buildings, and construction of new buildings, whether residential or commercial public or private'.

The United States Green Building Council (USGBC) Research Committee (2008) notes that:

> 'Green buildings depend on the continuous improvement of building processes, technologies and performance to minimize negative environmental or health impacts and contribute to environmental restoration and sustainable resource management. Objectives of green buildings ... include (a) climate conditions decoupled from human activities; (b) stable, sustainable energy supplies; (c) clean, renewable and sufficient water resources; (d) restorative use of land for the long-term sustainability of habitats; (e) restorative use of materials and assemblies that account for life-cycle impacts; and (f) enhanced human safety, health and productivity in the built environment'.

The Green Building Index (GBI) Sdn Bhd of Malaysia (2009: 2) states that 'a green building focuses on increasing the efficiency of resource use – energy, water and materials – while reducing building impact on human health and the environment during the building's life cycle through better siting, design, construction, operation, maintenance and removal'. This directly reflects the definition of 'green technology' by Malaysia's Ministry of Energy, Green Technology and Water (2009: 2) as the 'development and application of products, equipment and systems used to conserve the natural environment and resources'. The ministry considers the following criteria as necessary features of green technology: (a) it minimizes the degradation of the environment; (b) it has zero or low GHG emissions; (c) it is safe for use, and promotes healthy and improved environment for all forms of life; (d) it conserves the use of energy and natural resources; and (e) it promotes the use of renewable resources.

From the above definitions, *sustainable (or green) construction* involves creating constructed items using best-practice, clean and resource-efficient techniques, from the extraction of raw materials to the demolition and disposal of its components. Appropriate construction can conserve materials, produce energy (such as 'net sero' and 'net positive' buildings), stimulate activity in several parts of the economy and create green and decent jobs (see, e.g. Hendricks & Madrid, 2011). It must be noted that construction activity will always involve some adverse environmental implications although 'sustainable construction' should reduce their extent. In this regard, sustainable construction is an ideal target. CEC (2009) recommends that leaders should make green buildings a foundational driver for environmental, social and economic improvements.

In many industrialized countries, there has been a significant increase in the volume of green buildings, and the market is growing. In the USA, it was estimated that green buildings accounted for 2 per cent of the new non-residential building market. It was expected to grow to 5–10 per cent by 2010 (CEC, 2009). McGraw-Hill (2008) estimated that the total value of green construction in the USA was $10 billion in 2005, and grew to between $36 and $49 billion in 2008. It also estimated that the market could grow to $96–140 billion by

2013. Similarly, RICS (2011) notes that, in the USA, green building is projected to contribute USD554 billion to GDP from 2009 to 2013. In Malaysia, it is estimated that some 3 million square feet of new and commercial buildings will have been green rated by 2012. The growth is evident in specific segments.

There are several examples of large-scale demonstrations of green buildings and infrastructure in eco-cities in Asia. Examples are Masdar in the United Arab Emirates, and the Singapore–China collaboration in the showcased Tianjin Eco-City. Greater progress on both the individual buildings and macro scale is expected. For example, it is predicted that smart grids will have major implications for 'smart cities', their wired residents and global carbon emissions (Karlenzig, 2011). Smart grids will rely heavily on smart buildings and are a critical solution to making renewable energy more scalable through more efficient energy transmission systems. China is testing a 4-square-kilometre smart grid pilot area in Tianjin Eco-City. They will make renewable energy more effective and provide reliable information to people to make decisions about energy use.

V. Main Drivers

The green building wave has many drivers. Pitt *et al.* (2009) ranked the importance of eight drivers of sustainable building as follows: financial incentives, building regulations, client awareness, client demand, planning policy, taxes/ levies, investment and labelling/measurement. The drivers lie at several levels: international, regional, national (government and industry) and organizational.

One of the prime global drivers of change on the environmental front is *Agenda 21*, which acts as a blueprint for action to make development socially, economically and environmentally sustainable in the 21st century. Several other international conventions, protocols and compacts have been concluded, in which governments have committed themselves to meeting certain targets. Thus, another global driver is the Kyoto Protocol, which was adopted in December 1997 and entered into force in February 2005. It has put in place a number of global initiatives and instruments, and set binding targets for 37 industrialized countries and the European Community for reducing GHG emissions. Several United Nations Framework Convention on Climate Change Conferences of the Parties (UNFCCC COP) were held in the past to find an effective international response to climate change for the period after 2012, when the Kyoto Protocol expires. The intention was to replace the protocol at UNFCCC COP15 in Copenhagen, Denmark, in December 2009. However, only non-binding agreements were made. The 17th UNFCCC COP will be held in Durban, South Africa, in December 2011.

Regional and other groupings of governments have also set common targets and proposed initiatives to enhance energy efficiency. Under the 20-20-20 policy, EU will cut GHG emissions to 20 per cent below the 1990 levels by 2020. In 2008, European leaders adopted a climate change and energy package, with proposals for actions and a set of targets, including the following: for power plants and energy-intensive industries, emissions should be cut to 21 per cent below the 2005 levels by 2020; for sectors not covered by the emissions trading scheme, emissions should be cut to 10 per cent below the 2005 levels by 2020, increasing the share of renewable energy in the EU's total energy consumption to 20 per cent; and boost energy efficiency by 20 per cent by 2020 (EU Centre in Singapore, 2009). At the G8 summit in Hokkaido, Japan, in 2008, the participating countries called for a 50 per cent global reduction in GHG emissions by 2050.

Governments of all countries have formulated policies, statutes, codes and standards, and set up regulatory and enforcement agencies. For example, the *Energy Independence and Security Act 2007*, which requires a 30 per cent reduction in energy consumption by 2015 (calculated from a fiscal year 2003 baseline), and the *Clean Energy and Security Act 2009* are among the recent US national statutes. Governments apply a combination of regulations and economic instruments to encourage businesses and individuals to take actions that will protect and enhance the environment. From the perspective of the construction industry, these include amendments to building control legislation to require buildings to attain certain levels of environmental performance; the requirement for certain licences and approvals; offering of subsidies, tax incentives and grants; and certification and labelling of products. The Worldwatch Institute (2008) notes that more than 50 countries currently have standards or labelling programmes, which have resulted in energy savings, but that much more is needed.

Malaysia launched the National Green Technology Policy in 2009 (MEGTW, 2009). The main goals of this policy are: (a) the formation of the National Green Technology Council; (b) the provision of a conducive environment for green technology development; (c) the development of human capital in the sector; (d) the promotion of research and innovation, and the commercialization of green technologies; and (e) the provision of a basis for public awareness on green development issues in general. There are also four pillars that form the core of the National Green Technology Policy: (a) energy – to maintain energy independence and promote efficient utilization; (b) environment – to conserve and minimize the impact on the environment; (c) economy – to enhance national economic development by utilizing technology; and (d) social – to improve the quality of life. Four sectors in which significant progress and major improvements will be sought under the policy are identified as the building, energy, water and waste management, and transportation sectors. The building sector is expected to adopt green technologies in all aspects of its activities, including the construction, management, maintenance and demolition phases.

In Hong Kong, as incentives to developers for introducing environmentally responsible features in their buildings, the following are excluded from the calculation of gross floor area (GFA) for development control purposes: balconies, wider common corridors and lift lobbies, communal sky or podium gardens, sunshades and reflectors, and acoustic fins (Chau *et al.*, 2005). Again, in Hong Kong, the 'initiatives-associated' GFA concessions involve prerequisites such as a high green building rating, compliance with sustainable design guidelines and disclosure of energy estimates (RICS, 2011). In Malaysia, owners of buildings that are GBI certified between 24 October 2009 and 31 December 2014 are given an income tax exemption equivalent to the additional capital cost incurred to make the buildings GBI compliant. Buyers of such buildings are also given an exemption from the stamp duty (see Ofori & Abdul-Rashid, 2010). In Malaysia, some local authorities, including Kuala Lumpur City Hall, provide 'Green Lane Fast Track Approval' to GBI-certified projects, and often grant a higher plot ratio for such projects.

As different kinds of instruments are used to steer sustainable building, including normative regulatory instruments such as building codes, informative regulatory instruments (such as mandatory labelling) and economic- and market-based instruments (such as certification schemes, fiscal instruments and incentives), there is a debate on the most effective approach (Köppel & rge-Vorsatz, 2007). Owing to the fragmented nature of the construction industry, which comprises many actors, regulations are often considered as the only possible way to proceed. However, rigid normative instruments may hinder innovation. Indeed, incentives also have their critics, and it is suggested that not all such interventions have been effective. For example, among the main initiatives in Europe is the installation of solar panels in many houses with heavy government subsidies, which, some researchers suggest, makes little difference in energy saving or production, owing to the extensive cloud cover (RICS, 2011: 23).

Governments also invest directly in exemplar buildings to test products and concepts; demonstrate and communicate the environmental, financial and health benefits of green building; and build confidence. Examples are the Low Energy Office Building and Green Energy Office Building in Malaysia (built in 2001) and the Zero Energy Building @ BCA Academy in Singapore. In 2006, the Government of Canada launched a five-year net zero-energy housing demonstration project, which would involve the completion of 1,500 net zero-energy homes (CEC, 2009). RICS (2011: 19) notes that, in the USA, 'local governments have led efforts to drive green building into the mainstream, enacting creative and wide-ranging policies that strengthen the market for green building and incentivize sustainable development'. Governments also fund research and development (R&D) on green building. The US Department of Energy's 'Building America' programme conducts research with the private sector to produce homes that consume 30–90 per cent less energy than conventional homes do. The goal is to develop, by 2020, zero-energy homes.

In many industrialized countries, building owners are required to report periodically on the usage of energy in their premises. For example, in the USA, all federal buildings must have a completed Energy Management Data Report since 2006, and the state of California introduced such a requirement for all non-residential buildings in 2009. USGBC (2011) reports that green building advancements have been made in all 50 states in the USA in recent years, despite the economic recession. Many laws on the issue were passed in 2010 in spite of the sharp political divide, and green building policies are being implemented in all the states. In 2011, most US states and cities began to require commercial buildings to measure and disclose their energy use (Koch, 2011). The deadline for 16,000 large buildings in New York City (representing half of its interior space) to report how much energy they used in the past years was 2 August 2011. The city will post the data on a public website. Similar requirements took effect in January 2011 in Washington and began in Seattle, San Francisco and Washington, DC, in October 2011; in Austin, Texas, in June 2012; and throughout California in 2012. The US Department of Energy plans to begin testing a voluntary programme in spring 2012 to rate the energy efficiency of commercial buildings, similar to a pilot programme it completed in June 2011 for rating homes (Koch, 2011). In the UK, all commercial and public buildings and homes when bought, sold or rented are required to have Energy Performance Certificates, and, in public buildings, the certificates must be prominently displayed. The purpose of such energy reporting is to allow comparisons with similar buildings, to help owners and operators to manage their energy consumption and costs, to motivate owners to improve the buildings' energy usage, to make a case to financiers on financial investments and to enable energy efficiency to be considered in property valuation (Elliott, 2009).

Professional institutions and trade associations, such as the American Institute of Architects (1996) and RICS (2005), have released blueprints, reports and guides on environmentally conscious practices. One of the earliest works by the Chartered Institute of Building (1989) is entitled *What Are You Doing About the Environment?* RICS (2011) has presented a 'Green List', which comprises 50 policies, initiatives, case studies and solutions from around the world. Professional institutions offer training and advisory programmes for their members. There have also been national awards for projects and practitioners that perform well in this regard.

International organizations, group institutions and individuals with common interests are also driving forces in the effort to attain environmentally responsible building development. CIB (1999) adopted 'Sustainable Construction' as its 'proactive theme' in 1998 for all its activities over a three-year period. It produced *Agenda 21 for Construction*, seeking to create a global framework and terminology to facilitate initiatives at national and subsectoral levels, and outline R&D activities. CIB proposes the following initiatives: re-engineer the building process to increase partnership amongst participants and enhance quality management; adopt new building concepts, incorporating function-integrated systems; increase interdisciplinary education of professionals; enhance public awareness; introduce standards and regulations such as 'green' certification; and intensify R&D. Several organizations, including CIB and RICS, launched an international charter on environmentally conscious buildings and infrastructure at the World Summit on Sustainable Development in Johannesburg in 2002.

At the corporate level, some designers and contractors are responding to changes in regulations and increasingly stringent demands of knowledgeable clients, who would be reacting to government actions and incentives. All major international contractors, such as Bechtel of the USA, Hochtief of Germany, Skanska of Sweden, Shimizu of Japan, and Vinci of France highlight their commitment to environmental responsibility as a key part of their corporate strategies and an element of their competitiveness. BAM Construct UK (BAM) was named one of the best green companies in the UK in 2011. It launched an environmental reporting system called BAM Sustainability and Measurement Reporting Tool, or BAM SMaRT, in January 2010.[3] Whole value-chain initiatives are also evident as contractors put pressure on their subcontractors, suppliers and other business partners to adopt appropriate green methods (Ofori, 2000). Companies make bold claims for their achievements and plans. For example, the US firm GE's Project Frog is trumpeted as being on a mission to revolutionize the way buildings are created by applying technology to overcome the inefficiencies of traditional construction. 'The result is a structure that is measurably greener and significantly smarter; brighter, healthier spaces that inspire better performance from the people who occupy them' (http:// www.projectfrog.com). New business strategies also include alliances and multidisciplinary joint ventures such as the Skanska/Arup Retrofit Partnership, grouping UK design consultants, Arup and Swedish contractor Skanska (RICS, 2011). The initiatives of these corporate leaders are stimulating a ripple of actions within many construction industries.

In some countries, market forces, such as clients' insistence on better environmental performance, have provided the stimulus. There are cases wherein clients give the design and construction teams an incentive to realize a built product with a high environmental performance. An example is City Development Ltd in Singapore, which gives companies bonuses or assists them in making the necessary capital investments to acquire the required equipment. In Malaysia, growth in the volume of green buildings is attributed to private-sector clients responding to a rise in awareness of the merits of and demand for green-rated space, mainly office buildings (Ofori & Abdul-Rashid, 2010).

The final driver of green building considered here is change in technology and increase in knowledge. The integration of green features with intelligent technologies in the so-called 'smart and green' buildings is an example. There are increasingly more sophisticated mechanical and electrical installations and building control systems. An example is offered by BuildingIQ, a green technology firm built around a prototype technology developed in Australia by CSIRO, which regulates energy use in buildings (Courtenay, 2011). The software can update weather information every five minutes. It guides the building's basic management system, telling it when to run harder or slower as circumstances change, and regulating energy output according to weather forecasts, energy prices and occupancy schedules. There are increasingly more sophisticated analytical tools. For example, computer software such as the MARS Facility Cost Forecast System facilitates simulations and analyses, such as life-cycle assessments. In many countries, there are also norms for most of the variables required for

---

[3] Source: http://www.carbonaction2050.com/case-studies/bam-construct-bam-smart.

various analyses, such as the discount rate and the period. Further R&D is continuously conducted to produce more efficient technologies. For example, in lighting, compact fluorescent light (CFL) bulbs, which are more energy-efficient than traditional filament bulbs, are used in many homes. Presently, an increasing number of consumers are using light-emitting diode (LED) lights, which are lightweight and made from thin organic materials. LED contains no hard metals and offers a complete dimming range from 100 to 1 per cent. However, LED lights are more expensive than CFL bulbs.

VI. Benefits

Studies show that environmentally conscious building development offers many benefits. For example, it is suggested that clients and users gain from the following: compliance with regulations and avoidance of liability claims; savings in operating costs of the buildings, including lower consumption of energy and water; improved indoor environmental quality, leading to better health and productivity of occupants; and accordance with clients' and designers' altruism, and social and professional responsibility (Spiegel & Meadows, 1999; RICS, 2005).

Several studies have considered the measurable attributes of green buildings. Energy use is a commonly analyzed parameter. Diamond *et al.* (2006) found the actual energy consumption in 21 Leadership in Energy and Environmental Design (LEED)-rated buildings to be 27 per cent lower than that of a conventional baseline building. Bradshaw *et al.* (2005) showed that electricity and gas consumption decreased by 38 per cent and 35 per cent respectively in 15 out of 16 green residential buildings in the USA. However, the advantages are not always clear cut. Turner and Frankel (2009) analyzed measured energy performance of 121 LEED-certified New Construction buildings. The results showed that certified projects attained better energy performance than non-LEED buildings did. On average, LEED buildings were delivering the anticipated savings. Each of the three views of building performance shows that the average LEED energy use is 25–30 per cent better than the national average. Average savings increase for the higher LEED rating levels. Newsham *et al.* (2009) matched every one of the 121 green buildings studied by Turner and Frankel with a similar conventional one from the Commercial Building Energy Consumption Survey data set. Using an analysis of variance and t-tests, they showed that LEED buildings consumed, on average, 18–39 per cent less energy, but that 28–35 per cent of them consumed more energy than their matched conventional counterparts did. Scofield (2009) also re-analyzed the same data by considering the energy consumed on-site as well as off-site. He also weighed energy intensities based on the sizes of the buildings. Only half of the previous savings (10–17 per cent) were reported when on-site energy was taken into account. When both on- and off-site energies were taken into account, green offices consumed only 4 per cent less energy than other comparable conventional buildings.

Issa *et al.* (2011) have analyzed data on electricity and gas consumption quantities and costs of 10 conventional, 20 energy-retrofitted and 3 green (LEED New Construction-rated) Toronto schools over an eight-year period. The analysis, using the life-cycle costing approach, showed that energy-retrofitted and green schools spent 37 per cent more on electricity than conventional schools did. Nevertheless, green schools spent 56 per cent and 41 per cent less on gas than conventional and energy-retrofitted schools did respectively. Their total energy costs were also 28 per cent lower than those of the conventional and energy-retrofitted schools. However, the authors found that these savings did not always justify their construction cost premiums.

In Oregon, USA, the Earth Advantage Institute found that new homes with a LEED for Homes or Energy Star designation sell for 8 per cent more than non-certified homes do, while existing homes that are retrofitted to meet LEED or Energy Star guidelines sell for up to 30 per cent more (Knott, 2011). The certified homes were also sold faster. A study in Singapore, based on a sample of 23 commercial properties (offices, retail, hotel and a mix of these uses) found that retrofitting commercial buildings could lead to average expected savings in operating expenses of 10 per cent, and could result in an increase in their capital value of about 2 per cent (Building and Construction Authority [BCA], 2011a). It also found that the upfront cost of retrofitting energy-inefficient buildings could be recovered in about four to seven years. Moreover, the effort to achieve Green Mark certification need not be costly. If the retrofit cost is expressed as a percentage of the current market value of property, it is 0.5 per cent for retail and 1 per cent for offices.

There is an ongoing debate on the cost-effectiveness of providing elements to attain energy efficiency in both new and existing buildings. In Singapore, *DLS Dynamics* (2009) presented an analysis of the cost implications of the work that would be required in a building to enable it to qualify for various ratings such as LEED. The basic rating can be met with a slight premium by undertaking work, including external shading, low-energy lighting and energy-efficient heating, ventilation and air conditioning. However, the more radical enhancements that would be required for a platinum rating, such as low emissivity windows, and photovoltaic and solar panels, might increase the capital cost by 5–15 per cent. One of the major problems in the effort to find results in such analyses is that much of the data for analysing maintenance and retrofit projects are not available and must be found by primary research.

Obstacles and Challenges

There are problems and challenges in the effort to attain environmentally responsive buildings. The barriers to sustainable building identified by Pitt *et al.* (2009) were ranked as follows: affordability, the lack of client demand, the lack of client awareness, the lack of proven alternative technologies, the lack of business case understanding, building regulations, planning policies, and the lack of labelling/measurement standards. RICS Foundation and Slough Estates

(2004: 3) identified the key barriers in the UK as 'sector inertia and an apparent lack of market demand'. A challenge identified in a recent UK study is cultural change: changing the mindsets of developers, architects, engineers, builders and facility managers to ensure that designing, producing and operating low- and zero-carbon buildings are routine, rather than an exception (RICS, 2011). It has also been suggested that the biggest hurdle in the USA is education (RICS). In a study in Finland, Hakkinen and Belloni (2011) found that the adoption of sustainable building approaches is not constrained by a lack of technologies and assessment methods, but rather by organizational and procedural difficulties involved in the adoption of new methods. New technologies are resisted because they require process changes entailing risks and unforeseen costs. The barriers these authors found are the following: steering mechanisms; economics (perception of higher costs and increased risks from new technologies); a lack of client understanding; process issues (procurement and tendering, especially difficulties in specifying measuring requirements, timing of decision-making on green aspects, cooperation and networking amongst project team members to attain the optimum solutions); and underpinning knowledge on sustainability, the availability of relevant methods and tools, and the possibility of innovation.

The obstacles identified in India include the lack of awareness of green solutions and technologies, the perception that green buildings are more expensive than conventional ones, the lack of the necessary regulations and an enforcement framework, and the lack of government action to champion green initiatives in buildings. Other challenges are 'the 2–3 per cent additional cost of a certified green building, the lack of locally manufactured green materials and equipment, and the lack of incentives from the government and statutory bodies' (RICS, 2011: 22). In Hong Kong, the largest hurdle is considered to be transforming the market by making low-carbon, high-performance buildings the top priority of buyers and tenants, whereas, in China, the top challenge is 'how to plan and implement low-carbon cities that are conducive to low-carbon lifestyles, efficient land use, environmentally friendly transport, material resource and water efficiency' (RICS, 2011: 20). It is estimated that, in Malaysia, there is a 3 per cent increase of cost for a standard-size building to be GBI rated, while this can be an even higher 15 per cent for a large building (*The Star*, 2009). However, it has also been found that, over the long term, green buildings have a higher investment return of 6.6 per cent than standard buildings do. Another set of estimates indicates that the cost of developing green buildings in Malaysia is relatively higher than in other countries mainly because, generally, the quality of construction is still quite low. We estimate that it will cost between 10 and 15 per cent more for green buildings. However, most of the players in the industry believe that the cost will come down, as both the public and builders address green building, and local authorities and federal governments promote it. Ofori and Abdul-Rashid (2010) found that the barriers to green building in Malaysia are similar to those elsewhere, and include cost and payback period considerations and the lack of expertise. However, the greatest constraint is the energy tariff owing to subsidies, which makes investments in energy efficiency or renewable energy unattractive.

There are difficulties in the assessment of green buildings. Wallhagen and Glaumann (2011) note that tools developed in different contexts focus on different issues. Thus, when there are several tools in the same market making similar claims for ascertaining sustainability but using differing methods and assigning different ratings for the same building, this can cause difficulties for users. For example, in their study, using three benchmarking systems to rate the same buildings, they found that different assessment tools can influence sustainable building design and the realization of different directions, resulting in significantly different buildings owing to variations in the content of the tools, issues assessed, criteria, scores, aggregation and weighting of categories and issues. They conclude that their findings 'confirm that the concept of 'sustainable building' is far from universal' (Wallhagen & Glaumann, 2011: 32).

There is also a lack of access to information in a form that enables it to be readily understood and put to practical use. For example, a study in the USA found that some 78 per cent of respondents did not realize that incandescent light bulbs would be phased out by 2012 (Buranen, 2009).

In developing countries, not much has been achieved under the theme of green building. CIB and United Nations Environment Programme (UNEP) (2002) highlight the following barriers in these nations: the lack of capacity of the construction sector to implement sustainability principles; an uncertain economic environment and weak government finances; poverty, high demographic growth and low urban investment; declining government investment in construction; the lack of accurate data on environmentally responsible materials, equipment and methods; the lack of interest in sustainability issues amongst stakeholders; technological inertia; and the lack of research on relevant subjects. In developing countries, these issues coexist with the symptoms of social and economic underdevelopment, such as weak policies, institutions and enforcement.

Broader Opportunities Offered by the Green Building Drive

The recent upsurge in awareness of, and enthusiasm for, the environment and climate change mitigation offers many opportunities for beneficial actions for the construction industry and, ultimately, for society. Some of these are now considered. First, the construction industry has a poor social image in most countries. It is considered a part of the economy that offers only dirty, demanding and dangerous jobs (Construction 21 Steering Committee, 1999). Employment in construction is usually low-paying and rather precarious, as employers tend to adopt casual employment practices (International Labour Office, 2001). For these reasons, construction is unable to attract high-calibre personnel. For example, in Brunei, Malaysia, Singapore, South Africa and the Middle East, there is an acute shortage of skilled and unskilled local construction workers. It has been observed by many authors that the desire of many youths to do their part to contribute to the efforts to address the negative impacts of human activity can make the industry attractive to them (Ofori & Milford, 2007).

Second, the heightened interest of clients, end purchasers, users and other stakeholders in the performance of the buildings they invest in or occupy will raise the performance in the construction industry. Third, the stiffer competition and rivalry that the green building wave has engendered in the construction industry will lead to greater efforts by construction companies to go beyond compliance with regulations to innovate. In firms, the green aspiration can lead to the development of broader systems that may revolutionize construction. Fourth, there will be a drive towards greater professionalism in the construction industry at individual, corporate and industry levels. Firms and practitioners will seek to mitigate the environmental impact of construction activities by effectively applying their knowledge, skills and experience.

Finally, as construction work is inherently labour intensive, depending on the technologies adopted, construction can provide employment to significant numbers of people. Investments in energy-efficient new buildings or refurbishment will create direct green jobs in the construction industry. Finally, these will stimulate activities and create jobs (indirectly) in many other sectors such as manufacturing (to produce the necessary equipment or installations), commerce (to supply the required green inputs) and finance (to provide the funds and manage the investments).

Examples of Green Buildings

Tables 11.2 and 11.3 present examples of the green features that can be installed in buildings.

Table 10.2: Environment-friendly features in some residential condominiums in Singapore

	Savannah Condo Park	Changi Rise Condominium	Goldenhill Villas
Water taps built in planter boxes for residents to maintain high-rise gardens	X		
Lush landscaping	X	X	
Energy-saving lighting fitted along ventilation walls, boundary fences and other common areas		X	X
Rooftop air turbines that provide energy-efficient cooling and reduce utility bills			X
Energy-saving air-conditioning systems for all houses	X		

	Savannah Condo Park	Changi Rise Condominium	Goldenhill Villas
Eco-friendly architectural design that creates convection wind to cool the residence naturally			X
Specially designed roof gardens that provide rooftop insulation to minimize dependency on air conditioning			X
Solar panels installed in the clubhouse to convert solar energy into usable electricity to be used in selected rooms and water heaters for the clubhouse	X		
An odourless and mechanized pneumatic waste disposal system that removes solid waste from refuse chutes in individual homes to sealed compactors within bin centres or mobile trucks	X	X	
A designated green corner with colour-coded bins placed at strategic locations to encourage residents to recycle	X	X	
The use of environment-friendly building materials, such as recycled wood chips in laminated flooring instead of timber, to help conserve natural resources and the environment	X	X	
An ecological pond that maintains water clarity and controls odour, algae and bacteria growth by recirculating the water through an aerobically active filter bed	X	X	
Provision of bicycle racks to encourage residents to cycle as an alternative to driving	X		

Source: http://www.cdl.com.sg/cdl2.nsf/ws.htm.

Table 11.3: Environmentally-responsible features of non-residential projects that won the Green Mark Platinum award in Singapore in 2010

Project	Environmentally responsible features
Non-residential buildings	
Education Resource Centre, National University of Singapore (NUS)	• Estimated energy savings: 953,044 kilowatt-hours/year; estimated water savings: 4,132 cubic metre/year; envelope thermal transfer value (ETTV): 39.87 watts/square metre; • Chilled ceiling; • A lighting management system; • Natural ventilation design; • Double-vestibule doors.
Lian Beng Building	• Estimated energy savings: 196,342 kilowatt-hours/year; estimated water savings: 3,579 cubic metre/year; ETTV: 38.69 watts/square metres; • An incorporated extensive building outer skin (screen); • A heat-recovery system; • Motion sensors; • The use of solar photovoltaic cells to produce electricity; • Rainwater recycling system; • An extensive use of *Singapore Green Label Scheme* products and items with 30 per cent or more recycled content; • A roof garden.
Solaris	• Estimated energy savings: 2,828,470 kilowatt-hours/year; estimated water savings: 11,785 cubic metres/year; ETTV: 39.92 watts/square metre; • An operable skylight louvre; • A natural ventilated atrium with daylight design; • A solar shaft to enhance natural daylight in the building; • Eco-Cell and 400-cubic-metre rainwater harvesting tank; • Extensive roof gardens and continuous vertical landscaping.

Project	Environmentally responsible features
United World College South East Asia (East Campus)	• Estimated energy savings: 3,081,960 kilowatt-hours/year; estimated water savings: 83,481 cubic metre/year; ETTV: 39.55 watts/square metres; • Air-conditioning plant system efficiency: 0.58 kilowatt/tonne; • Passive design and building layout to minimize heat gain and maximize natural ventilation; • An extensive use of solar thermal system; • Collection of rainwater for landscape irrigation using a rain garden; • A rooftop garden; green walls on façade to reduce ambient temperature.
W Singapore Sentosa Cove	• Estimated energy savings: 3,338,446 kilowatt-hours/year; estimated water savings: 22,200 cubic metres/year; ETTV: 42.25 watts/square metre; • Air-conditioning plant system efficiency: 0.65 kilowatt/tonne; • Passive design and building layout to minimize heat gain; • An activated sensor to turn off air-conditioning system when the sliding door to the balcony is ajar; • A motion sensor for staircase and toilet; • The use of heat pump for hot-water system; • Harvesting of rainwater and collection of condensate from the air-conditioning system.
Woh Hup Building	• Estimated energy savings: 3,338,446 kilowatt-hours/year; estimated water savings: 22,200 cubic metres/year; ETTV: 42.25 watts/square metre; • Air-conditioning plant system efficiency: 0.65 kilowatt/tonne; • Passive design and building layout to minimize heat gain; • An activated sensor to turn off air-conditioning system when the sliding door to the balcony is ajar; • A motion sensor for staircase and toilet; • The use of heat pump for hot-water system; • Rainwater harvesting and collection of condensate from air-conditioning system.

Project	Environmentally responsible features
Special Buildings	
InnoVillage @ SP	• Estimated energy savings: 119,273 kilowatt-hours/year; estimated water savings: 927 cubic metres/year; • More than 80 per cent of construction materials from recycled railings, doors, windows and office containers; • All toilets and common areas designed to utilize natural ventilation and daylight; • A motion sensor, dimmer control and sun pipes installed; • Test-bedding site for clean energy such as bio-fuel and solar power; • A clean-energy competency centre to showcase and display 10 Singapore iconic solar stations; • Solar photovoltaic with 61 kilowatt-peak capacity.
Samwoh Eco Green Building	• Estimated energy savings: 343,088 kilowatt-hours/year; estimated water savings: 2,828 cubic metres/year; ETTV: 42.8 watts/square metre; • Perforated cladding and solar films to improve ETTV; • VRV 3 air-conditioning system, T5 artificial lighting with high-frequency ballast and motion detectors to reduce energy consumption; • Sanitary wares with WELs, water submeters and regulated irrigation system to improve water efficiency; • A rooftop garden, vertical greening and other greenery provisions; • The use of products certified under Singapore Green Label Scheme; • The use of aggregate with high dosage of recycled materials for structural and external works.
Zero Energy Building @ BCA Academy	• Estimated energy savings: 388,720 kilowatt-hours/year; estimated water savings: 3,620 cubic metres/year; ETTV: 43.79 watts/square metre; • Sunshading devices and efficient glazing; • ACMV System (high-performance chillers, displacement ventilation, personalized ventilation, underfloor air distribution system); • Photovoltaic technology of 190 kilowatts peak capacity; • Solar-assisted stack ventilation; • Mirror ducts, light pipes and light shelves; • Sensors and monitoring system for all rooms.

Source: BCA (2010b)

VII. Green Buildings in Singapore

The green building programme in Singapore is government led and spearheaded by BCA. The mission of BCA is: 'We shape a safe, high-quality, sustainable and friendly built environment[4]. Its vision is to have 'the best built environment for Singapore, our distinctive global city'.

Green Mark as Driver

The Green Mark for Buildings Scheme, launched by BCA in 2005, is the main component of Singapore's programme for enhancing the energy efficiency of its buildings. Its aims 'to move Singapore's building and construction industry towards environment-friendly buildings, and help strengthen Singapore's position as a global city committed to balancing its development with care for the environment' (Foo, 2005). The objectives of Green Mark are to (a) promote environmental sustainability in the construction industry and raise the awareness amongst developers, owners and professionals of the environmental impact of their projects; (b) recognize building owners and developers who adopt practices that are environmentally conscious and socially responsible; and (c) identify best practices in the development, design, construction, management and operation of buildings.

Green Mark is applicable to both new and existing buildings. New buildings are assessed under (a) energy efficiency, (b) water efficiency, (c) environmental protection, (d) indoor environmental quality, and (e) innovation. To ensure that buildings given the Green Mark are well maintained during their operation, they are assessed once every three years. There are four levels of ratings in the Green Mark scheme: Platinum, Gold[Plus], Gold and Certified. These ratings correspond to an energy efficiency improvement of more than 30 per cent, about 25–30 per cent, 15–25 per cent or 10–15 per cent respectively. BCA has been using the results of the assessments to further develop the scheme (it is now in its fourth version) and to prepare design guides. By April 2011, there were over 750 Green Mark buildings in Singapore.[5] The tool has been used to assess buildings in China, India, Indonesia, Malaysia, the Middle East and Vietnam. BCA has worked with other agencies, such as the National Parks Board and Land Transport Authority, to develop relevant versions of Green Mark for various types of works. There are now tools for new and existing parks; new and existing office interiors; infrastructure; a rapid transit system; and district.

The Green Mark GFA Incentive Scheme (New Buildings) encourages developers to attain Green Mark Gold[Plus] and Platinum ratings. Additional GFA could reach 1 per cent of the total, or 2,500 square metres (whichever is lower) for Green Mark Gold[Plus] buildings, and up to 2 per cent of the total or 5,000 square metres (whichever is lower) for Green Mark Platinum buildings. A $100-million Green Mark Incentive Scheme (Existing Buildings) encourages the owners of existing large buildings to retrofit them to include more green building features.

---

[4] See http://www.bca.gov.sg.
[5] See http://bca.gov.sg/GreenMark/green_mark_projects.html.

The government incentives have led to progressive actions by clients, designers and consultants.

From April 2008, all new buildings in Singapore must meet the Green Mark Certified rating. New buildings to be built in key development areas, including Marina Bay and the Central Business District, Jurong Gateway, Kallang Riverside and Paya Lebar Central, which are key new growth areas on land purchased from the state, must achieve higher ratings. New public-sector buildings with more than 5,000-square-metre, air-conditioned floor area must attain a platinum rating.

IMCSD (2009) estimated that an overall 5–10 per cent improvement in energy efficiency for the existing building stock can be achieved if 400–600 existing large buildings have green building features. The government will introduce initiatives to encourage 80 per cent of the existing building stock to achieve a Green Mark Certified rating. Existing government buildings with more than 10,000 square metres of air-conditioned floor area must attain Green Mark Gold^{Plus} rating by 2020, as part of their upgrading and replacement cycle. It is estimated that this work in retrofitting the over 210 million square metres of built-up space in Singapore will be worth some S$300 billion (*The Hindu Business Line*, 2011).

In October 2011, a Bill that would lead to an amendment of the Building Control Act to extend minimum Green Mark standards to existing buildings, as and when they are retrofitted, was released for public comment (BCA, 2011b). This will make Singapore one of the first few countries in the world to make such standards for existing buildings mandatory. Owners are required to conduct three-yearly audit on the cooling systems in their buildings. BCA will require utility companies and building owners to submit energy consumption and energy-related building data. It will use these data to monitor energy consumption patterns, measure the effectiveness of various initiatives adopted to improve energy efficiency, and develop national energy benchmarks.

Under a joint pilot scheme of BCA and participating financial institutions, called the Building Retrofit Energy Efficiency Financing building owners and energy services companies can obtain loans to carry out energy retrofits (from October 2011) (Tan, 2011). In September 2011, eight of the largest developers and building owners in Singapore also signed the Green Pledge to show their commitment to attaining higher energy efficiency of their existing buildings through the BCA Green Mark certification by 2020 (BCA, 2011a).

Green Master Plans

The initiatives of the First Green Building Master Plan for Singapore were launched in 2006 (BCA, 2010a) (and how they were implemented) were: (a) spurring the private sector (through the S$20-million Green Mark Incentive Scheme to offer monetary incentives to developers, established in 2006); (b) imposing minimum standards on environmental sustainability for buildings (introduced through an amendment of the Building Control Act to make minimum standards mandatory from April 2008); (c) promoting R&D in environmental sustainability (through the Ministry of National Development's S$50-million R&D Fund); and (d) building up industry capability (through training programmes).

The Second Green Building Master Plan of BCA (2010a) has the following strategic thrusts: (a) public sector taking the lead; (b) spurring the private sector; (c) furthering the development of green building technology; (d) building industry capabilities through training; (e) profiling Singapore and raising awareness; and (f) imposing minimum standards.

Other Initiatives

BCA has published a number of technical guides on green building for practitioners. These include *A Guide on the Use of Recycled Materials* and *Sustainable Construction – Materials for Building*. It also produces the *Sustainable Architecture* newsletter and case studies of sustainable construction. The buildings it has featured include the winners of platinum awards in various years, such as the National Library Building, the Supreme Court building and the Zero Energy Building @ BCA Academy.

Some professional institutions have launched green building blueprints. One of the main planks of a manifesto of the Singapore Institute of Architects (SIA, 2010a), released in 2007, is:

Sustainable and Environment-friendly Island
(a) Singapore should be a model of sustainability and environment-friendliness.
(b) The successful public housing programme should be turned into a model of a sustainable community.
(c) Urgent research on alternative construction methods and materials is required.
(d) End users and the general public should be educated on the importance of sustainable environment.
(e) Singapore should be a nation that appreciates and cares for its environment.

In 2010, SIA (2010b) published its 'position paper' entitled *12 Attributes of a Sustainable Built Environment*.

Singapore Green Building Council (SGBC), a joint initiative by the public and private sectors, with membership from all sections of the construction industry, was formed in 2009. Its mission is 'to propel the Singapore building and construction industry towards environmental sustainability by promoting green building design, practices and technologies, the integration of green building initiatives into mainstream design, construction and operation of buildings as well as building capability and professionalism to support wider adoption of green building development and practices in Singapore'.[6] The council indicates that its work will complement and support the government's efforts to accelerate the greening of buildings in Singapore by 2030. The 'key focus areas' of SGBC are the following: (a) profiling Singapore as a leading Sustainable Hub in the tropics; (b) enhancing professionalism and knowledge in sustainable development; and (c) functioning as a dedicated certification body for green building-related products and services.

HDB, the national public housing agency, states that public housing estates in Singapore are planned, built and maintained with resource considerations in mind. Self-contained townships minimize the need to travel, and optimize the use of land. The building design facilitates cross-ventilation and natural lighting to reduce energy consumption. In future, HDB will build more eco-friendly public housing, test-bed new technologies such as solar power in public housing and improve resource efficiency in public housing maintenance (IMCSD, 2009).

The S$50-million Ministry of National Development Research Fund for the Built Environment has tended to focus on supporting R&D on green building technology. Joint funds such as one set up by BCA, A*Star and the Ministry of National Development have also been established. Other agencies such as the Energy Market Authority and National Research Foundation also provide grants for research on energy use in buildings. The main centre of research on green buildings is at NUS. The work, as it relates to energy efficiency, has been on the following: energy rating of types of buildings, leading to the development of the Energy Star system; green roof and wall systems; indoor air quality and efficient air conditioning; alternative (clean) energy sources; carbon index for benchmarking buildings; passive design to enhance environmental performance; and life-cycle (costing) assessment of key building materials. Researchers at NUS have also assessed the impact of green features on the value of housing units. NUS has collaborated with statutory agencies such as BCA, HDB, the National Environment Agency and the National Parks Board.

Finally, there has been a comprehensive programme to develop the necessary expertise in green buildings. These include short courses and graduate-level programmes such as the M.Sc. (Building Performance and Sustainability) and M.Sc. (Integrated Sustainable Design) programmes at the School of Design and Environment, NUS.

A Comparison

Countries around the world have adopted different approaches in their green building programmes, and the situation in Asia is similar. For example, whereas Singapore's programme has been led by the government through BCA, in Hong Kong and Malaysia, the private sector has taken the lead. In Singapore, legislation

---

[6] Read about SGBC's mission at http://www.sgbe.sg/index.php/green/about/mission.

has been the main driver, and it has progressively been widened and made more stringent. On the other hand, Malaysia has no mandatory requirements for any aspect of green building. In most of the countries, a national building assessment system is in place. The assessment schemes in Asia (in addition to those in Malaysia and Singapore, which are discussed above) include the following:

- Comprehensive Assessment System for Building Environmental Efficiency (CASBEE), Japan;
- Ecology, Energy Saving, Waste Reduction and Health (EEHW), Taiwan;
- Hong Kong Building Environmental Assessment Method (HK BEAM), Hong Kong SAR;
- The Energy and Resource Institute (TERI) Green Rating for Integrated Habitat Assessment (GRISHA), India; and the
- Green Olympic Building Assessment System (GOBAS), China.

However, the extent to which the assessment system has been instrumental in driving the green building programme has varied; it has played a very strong role in Singapore. In most of the Asian countries, the governments offer incentives but the basis differs. Also different is what is offered. In Singapore, it is a mixture of grants, tax incentives and applications of planning regulations. The range is similar in Hong Kong, but narrower in Malaysia.

It is instructive to compare the green building programme in Malaysia with that in Singapore, which has been discussed above. Ofori and Abdul-Rashid (2010) outlined the main points from their study on green building in Malaysia as follows: (a) the benchmarking tool, GBI, has been the main stimulus in getting the building industry to think green; the endorsement of GBI by the government and its adoption as the basis for an incentive scheme have also been beneficial; (b) the main obstacle to initiatives to enhance the energy performance of buildings in Malaysia is the government subsidy on energy prices; it makes some investments, especially those in renewable energy, uneconomical and increases the payback period; (c) the building industry in Malaysia does not have a maintenance culture; and (d) the administrative responsibilities for aspects of green building, and the leadership of the overall programme are not clear, as the Construction Industry Development Board (the equivalent of Singapore's BCA) has only recently adopted it as a strategic objective.

Possible Action

The solutions that have been proposed for addressing the wider impact of construction on the environment include: undertaking environmental impact assessments on sites to find mitigation measures against the potential negative effects; designing for energy efficiency; choosing appropriate materials (Chau *et al.*, 2005); managing on-site to minimize waste; and maintaining water efficiency. Some of the possible solutions are quite simple. For example, in Singapore, Wong and Yu (2005) found that plants within developments can cool the surroundings and generate lower ambient temperature, reducing the heat load on buildings in the tropics and thus conserving energy. Hakkinen and Belloni (2011) suggest that the most important actions to promote sustainable building are as follows: increasing the awareness of clients about its benefits, development and adoption of methods for sustainable building requirement management; mobilizing sustainable building tools; developing designers' competence and team working abilities; and developing new concepts and services. The idea of 'green taping' projects is suggested. This involves 'removing the red tape that can slow innovation and finding ways of de-risking pathfinder projects for funders and developers' (RICS, 2011: 25). There is a challenge of finding 'a single, agreed way of measuring sustainability performance across building types and geographies' (RICS, 2011: 20).

Some of the proposals in studies on energy performance are now outlined. The main recommendations of the WBCSD (2009) study are to strengthen building codes and energy labelling for increased transparency, to use subsidies and price signals to incentivize energy-efficient investments, to encourage integrated design approaches and innovations, to develop and use advanced technology to enable energy-saving behaviour, to develop workforce capacity for energy saving, and to mobilize for an energy-aware culture. Achieving the goal in the report of the National Round Table on the Environment and the Economy (NRTEE), and Economy and Sustainable Development Technology Canada (ESDTC) (2009) requires industry commitment and stringent regulations, together with policies on energy pricing, regulations, subsidies and information programmes. WBCSD recommends policy changes including the following: applying a market-wide price signal on carbon; adopting regulations such as new building codes, minimum performance standards for buildings and equipment, and mandatory energy labelling; targeting subsidies and financial incentives such as accelerated capital cost allowances and technology funds; and utilizing information programmes to drive voluntary actions by building owners and tenants. Newton and Tucker (2011) propose a strategy for a low-carbon housing future for Australia through a new class of hybrid buildings (energy-efficient envelope, energy-efficient plug-in appliances and local energy generation linked to a national grid). They explore a portfolio of technical and policy options for decarbonizing the housing sector.

Recommendations

In each country, the sustainable construction programme should be a multi-pronged strategy. First, there should be a strategy for sustainable construction, drawn from the sustainable development programme that is put together by all key stakeholders in both the public and private sectors. This should provide the background and framework for national policies, programmes and action plans with appropriate measurable targets in terms of building performance. Second, there should be an effective implementation of the plans, with administrative and industry institutions that fit the context. There should also be a partnership of stakeholders including government agencies, professional institutions, trade associations (including those for clients) and labour unions. In some countries, the government will lead; in others, it will be a public–private collaboration; and in some, the private sector is sufficiently strong to lead.

Third, all the participants in the construction industry and its value chain sectors should consider sustainability in all their activities and in the context of the national sustainable construction strategy. These participants should include companies, institutions and associations within the industry, and their business partners including materials manufacturers and suppliers, and financial institutions. Fourth, it is necessary to continue to pursue locally relevant technology development and innovation by carrying out R&D on materials, equipment, methods, processes, and procedures, or adapting and applying new technologies from overseas. The subjects of the R&D, under the broad umbrella of energy efficiency of buildings, would include materials, equipment, management systems, statutes, standards and economic instruments.

Fifth, human resource development should be given attention. This should include both formal tertiary education and continuing professional development, and should cover the technical knowledge and soft skills. Sixth, appropriate practices and procedures that support the implementation of relevant elements of sustainable construction at all stages of the construction process should be adopted. For example, during the key stage of procurement, would be the consideration of environmental experience and achievement during the selection of consultants and contractors for projects. Seventh, an information base that will provide the vital inputs into analyses should be built up. Assessment systems and tools that are suitable for application in the tropical regions, and, in the particular, resource- and expertise-scarce contexts of the countries concerned, should also be formulated. Finally, there should be international partnerships among the neighbouring Asian countries. Most relevant would be networks of developing countries for joint action to address common problems, sharing of experiences. It would also useful to establish mechanisms for knowledge and technology-transfer relationships between developing and industrialized countries.

VIII. Conclusion

The construction industry and its activities demonstrate the various aspects of sustainable development. The industry, its operations and its products are of great economic and social significance. On the other hand, they have negative impacts on the physical environment and can also cause various forms of social disruption. At the same time, construction is seen as having great potential to contribute to the efforts to reduce energy use and carbon emissions and to mitigate against climate change. There is a general trend towards the adoption of green building approaches, and much progress has been made. However, whereas many drivers are providing the impetus for such an approach, several impediments and challenges remain to be addressed. It is likely that progress will be slower and require more effort in the future.

In Asia, progress varies from one country to the next. Main approaches are also different. In Singapore, the regulations, government incentives and market demands have led to progressive actions by clients, designers and consultants, and there are indications that providing green features in buildings is becoming the norm in Singapore. However, even here, it is recognized that much more needs to be done. The questions that remain include the extent of provision and ways of ensuring the effective operation of the facility. There are lessons from Singapore's programme that other countries can adopt. These include the need for strong leadership and championing, and for attention to the development of human resources. Finally, regional collaboration on green building programmes is of great merit.

References

Ahmad-Hadri H (2005) Importance of holistic PV market development approach – lessons learnt from Malaysia. Paper presented at SOLARCON, 21 May, Singapore.

American Institute of Architects (1996) *Environmental Resources Guide*. Wiley, New York.

BCA (2010a) *2nd Green Building Masterplan*. BCA, Singapore.

BCA (2010b) *Green Mark 2010*. BCA, Singapore

BCA (2011a) BCA–NUS study shows that greening existing buildings can increase property value [Article on the Internet]. [Cited 19 Sept 2011]. Available from: http://www.bca.gov.sg/Newsroom/pr16092011_BN.html.

BCA (2011b) National Development Minister of State Tan Chuan-Jin Opens International Green Building Conference 2011 [Article on the Internet] [Cited 19 September 2011]. Available from: http://www.bca.gov.sg/Newsroom/pr14092011_IGBC.html.

Bradshaw W, Connelly E, Fraser Cook M, Goldstein J, Pauly J (2005) *The Costs and Benefits of Green Affordable Housing: Opportunities for Action*. Tellus Institute, Boston, MA.

Buranen M (2009) US Green Building Council meets to help customers see the light [Article from the Internet] [Cited Nov 2009]. Available from: http://www.bizlex.com/LPrintwindow.LASSO?-token.editorialcall.

Chartered Institute of Building (CIOB) (1989) *What Are You Doing About the Environment?* CIOB, Ascot.

Chau KW, Wong SK, Yiu CY (2005) Improving the environment with an initial government subsidy. *Habitat International* **29**, 559–69.

CEC (2009) *Green Building in North America: Opportunities and Challenges*. CEC, Montreal.

Construction 21 Steering Committee (1999) *Re-inventing the Construction Industry.* Ministry of Manpower and Ministry of National Development, Singapore.

Courtenay A (2011) Rise of the smart building. *Sydney Morning Herald* [Article on the Internet] 7 September 2011. [Cited 13 Sept 2011]. Available from: http://www.smh.com.au/business/key-leaders/rise-of-the-smart-building-20110907-1jx6d.html#ixzz1XHDs6YB5.

Department of Communities and Local Government (2006) *Building a Greener Future: Towards Zero Carbon Development.* DCLG, London.

Department of Energy and Climate Change (DECC) (2009a) *Community Energy Saving Programme (CESP): Consultation.* DECC, London.

Department of Energy and Climate Change (DECC) (2009b) *Heat and Energy Saving Strategy: Consultation.* DECC, London.

Department of Environment, Food and Rural Affairs (2006) *Climate Change: The UK Programme 2006.* DEFRA, London.

Diamond R, Opitz M, Hicks T, Vonneida B, Herrera S (2006) *Evaluating the Energy Performance of the First Generation of LEED-Certified Commercial Buildings.* American Council for an Energy-efficient Economy, Washington, DC.

DLS Dynamics (2009) Cost premium of going green: Myth or reality? October, pp. 1–2.

Elliott, G. (2009) Energy reporting: It was only a matter of time [Article on the Internet] [Cited Nov 2009]. Available from: http://www.smart-buildings.com/downloads/energyreportsoct2009.pdf.

EU Centre in Singapore (2009) *Road to Copenhagen 2009: The European Union and Climate Change Action – Executive Summary.* EU Centre in Singapore, Singapore.

Foo C (2005) BCA launches Green Mark for Buildings Scheme announced by Mr Cedric Foo, Minister of State (National Development and Defence) in the opening address at the Construction & Property Prospects Seminar (organised by BCA and REDAS) on 11 January 2005 (Press Release) [Article on the Internet] [Cited October 2011]. Available from: http://www.bca.gov.sg.

Green Building Index Sdn Bhd (2009) Green Building Index, Kuala Lumpur [Cited 19 Feb 2010]. Available from: http://www. greenbuildingindex.org, accessed on 19 February 2010.

Gupta R, Chandiwala S (2010) Understanding occupants: Feedback techniques for large-scale low-carbon domestic refurbishments. *Building Research and Information* **38(5)**, 530–48.

Hakkinen T, Belloni K (2011) Barriers and drivers for sustainable building. *Building Research and Information* **39(3)**, 239–55.

Hendricks B, Madrid J (2011) A star turn for energy efficiency jobs: Energy efficiency must have a starring role in putting America back to work [Article on the Internet] September 2011. [Cited 12 Sept 2011]. Available from: http://www.americanprogress.org/issues/2011/09/pdf/energy_efficiency_jobs.pdf.

Hillebrandt PM (2000) *Economic Theory and the Construction Industry*. Macmillan, Basingstoke.

Housing and Development Board (HDB) (2008) *Annual Report 2007–08*. HDB, Singapore.

Intergovernmental Panel on Climate Change (IPCC) (2007) Background paper 2b, Institutional Efforts for Green Building: Approaches in Canada and the United States. IPCC, 2007, Climate Change 2007: Mitigation of Climate Change. In: Metz B, Davidson OR, Bosch PR, Dave R, Meyer LA (eds.) *Contribution of Working Group III to the Fourth Assessment Report of the Intergovernmental Panel on Climate Change*. Cambridge University Press, New York. Available from: http://www.ipcc.ch/SPM040507.pdf.

Inter-Ministerial Committee on Sustainable Development (IMCSD) (2009) *A Lively and Liveable Singapore: Strategies for Sustainable Growth*. Ministry of National Development, Ministry of Environment and Water Resources, Ministry of Finance, Ministry of Transport and Ministry of Trade and Industry, Singapore.

International Council for Research and Innovation in Building and Construction (CIB) (1999) *Agenda 21 on Sustainable Construction*. CIB Report Publication 237. CIB, Rotterdam.

International Council for Research and Innovation in Building and Construction (CIB) and United Nations Environment Programme (UNEP) (2002) *Agenda 21 for Sustainable Construction in Developing Countries: A Discussion Document*. CSIR Building and Construction Technology, Pretoria.

International Council for Research and Innovation in Building and Construction (CIB) (2009) The Euregion facing a second transition. *CIB News*, December.

International Energy Agency (IEA) (2008) *Energy Technology Perspective 2008*. Organisation for Economic Co-operation and Development, Paris.

International Labour Office (2001) *The Construction Industry in the Twenty-First Century: Its Image, Employment Prospects and Skill Requirements*. ILO, Geneva.

Issa MH, Attalla M, Rankin J, Christian AJ (2011) Energy consumption in conventional, energy-retrofitted and green LEED Toronto schools. *Construction Management and Economics* **29,** 383–95.

Ive GJ, Gruneberg S (2000) *Economics of the Modern Construction Sector*. Macmillan, London.

Karlenzig W (2011) Smart grids will revolutionize cities. *The Energy Collective,* 6 September. [Article on the Internet] [Cited 12 Sept 2011]. Available from: http://theenergycollective.com/warrenkarlenzig/64332/smart-grids-will-revolutionize-cities.

Kievani R, Tah J, Kurul E, Abanda FH (2008) Green Jobs Creation through Sustainable Refurbishment in the Developing Countries – A literature review and analysis conducted for the International Labour Organisation (ILO), Geneva.

Knott J (2011) LEED homes sell faster & for more money [Article on the Internet] 29 August 2011. [Cited 2 Sept 2011]. Available from: http://www.cepro.com/article/leed_homes_sell_faster_for_more_money/.

Koch W (2011) US cities, states require large buildings to cite energy use. *USA Today.* [Article on the Internet] 2 August 2011. [Cited 3 Sept 2011]. Available from: http://www.usatoday.com/news/nation/2011-07-31-rules-require-buildings-disclose-energy-use_n.htm?csp=34news.

Köppel S, Ürge-Vorsatz D (2007) Assessment of policy instruments for reducing greenhouse emissions from buildings. Report for UNEP SBCI. Central European University, Budapest.

Malaysian Energy Centre (2009) *Green Energy Building Features* [Article on the Internet] [Cited 17 Nov 2009]. Available from: http://www.ptm.org.my/PTM_Building/Geo.html.

Marsden Jacobs Associates (2009) Pathways for improving the energy-related performance of residential buildings. Report prepared for the Department of Sustainability and Environment Victoria, Melbourne, VIC.

McGraw-Hill (2008) *2008 Green Construction Outlook.* McGraw-Hill, New York.

Ministry of Environment and Water Resources (2006) *The Singapore Green Plan 2012, 2006 Edition.* MEWR, Singapore.

Ministry of Environment, Green Technology and Water (2009) *National Green Technology Policy.* MEGTW, Kuala Lumpur.

Ministry of Science, Technology and Environment (2002) *National Policy on the Environment.* MOSTE, Kuala Lumpur.

National Round Table on the Environment and the Economy and Economy and Sustainable Development Technology Canada (2009) *Geared for Change – Energy Efficiency in Canada's Commercial Building Sector.* NRTEE and ESTDC, Ottawa.

Newsham G, Mancini S, Birt B (2009) Do LEED-certified buildings save energy? Yes but …. *Energy and Buildings* **41(1),** 897–905.

Newton PW, Tucker SN (2011) Pathways to decarbonizing the housing sector: A scenario analysis. *Building Research and Information* **39(1),** 34–50.

Office of Public Sector Information (2008) *The Climate Change Act 2008 (2020 Target, Credit Limit and Definitions) Order 2009.* OPSI, London.

Ofori G (1990) *The Construction Industry: Aspects of Its Economics and Management.* Singapore University Press, Singapore.

Ofori G (1992) The environment: The fourth construction project objective? *Construction Management and Economics* **10**, 369–95.

Ofori G (1999) Satisfying the customer by changing production patterns to realise sustainable construction. *Proceedings, Joint Triennial Symposium of CIB Commissions W65 and 55, Cape Town, 5–10 September* **1**, 41–56.

Ofori G (2000) Greening the construction supply chain in Singapore. *European Journal of Purchasing and Supply Management* **6(3–4)**, 195–206.

Ofori G, Abdul-Rashid AA (2010) Energy Efficiency and Green Jobs in the Building Industry in Malaysia. Report prepared for the International Labour Office, Geneva.

Ofori G, Milford R (2007) Conclusions and recommendations (May). Proceedings of CIB World Building Congress, Cape Town. Available from: http://www. cibworld.

Oreszczyn T, Lowe R (2010) Challenges for energy and buildings research: Objectives, methods and funding mechanisms. *Building Research and Information* **38(1)**, 107–22.

Pitt M, Tucker M, Riley M, Longden J (2009) Towards sustainable construction: Promotion and best practices. *Construction Innovation* **9**, 201–24.

Planning Department and Environmental Resources Management (2000) *Studies on Sustainable Development for the 21st Century: Final Report.* Government of the Hong Kong Special Administrative Region, Hong Kong.

Royal Institution of Chartered Surveyors (RICS) (2005) *Green Value: Green Buildings, Growing Assets.* RICS, London.

RICS (2011) The green list. *Modus* (Asia Edition) Issue No. 07, pp. 14–29.

RICS Foundation and Slough Estates (2004) *Sustainability and the Built Environment: An Agenda for Action.* Upstream, London.

Scofield JH (2009) Do LEED buildings save energy? Not really… *Energy and Buildings* **41(12)**, 1386–90.

Singapore Institute of Architects (SIA) (2010a) Manifesto 2007. [Article on the Internet] [Cited 15 Oct 2010]. Available from: http://www.sia.org.sg/.

Singapore Institute of Architects (SIA) (2010b) 12 Attributes of a Sustainable Built Environment. [Article on the Internet]. Available from: http://www.sia. org.sg/resources/2010/12AttributesOfGreenArchitecture.pdf.

Spiegel R, Meadows D (1999) *Green Building Materials: A Guide to Product Selection and Specification.* Wiley, New York.

Tan CJ (2011) Speech by BG(NS) Tan Chuan-Jin, Minister of State for National Development and Manpower at the opening of the International Green Building Conference 2011 and BEX Asia 2011 on Wednesday, 14 September 2011 at 9:10 a.m., Suntec City Convention Centre. [Article on the Internet] 14 September 2011. [Cited 19 Sept 2011]. Available from: http://app.mnd. gov.sg/Newsroom/NewsPage.aspx?ID=2921&category=Speech&year=2011 &RA1=&RA2=&RA3=.

The Hindu Business Line. Singapore sees S$300-b opportunity in green buildings programme [Article on the Internet] 13 September 2011. [Cited 19 Sept 2011]. Available from: http://www.thehindubusinessline.com/industry-and-economy/article2450621.ece.

The Star. Green buildings: the new frontier [Article on the Internet] 25 May 2009. [Cited 4 Dec 2009]. Available from: http://dev.kkr.gov.my/en/node/13049, accessed on 4 December 2009.

Turner C, Frankel M (2009) *Energy Performance of LEED® for New Construction Buildings, Final Report.* US Green Building Council, Washington, DC.

United States Green Building Council (USGBC) Research Committee (2007) *Green Building Research Funding: An Assessment of Current Activity in the United States.* USGBC, Washington, DC.

USGBC Research Committee (2008) *A National Green Building Research Agenda.* USGBC, Washington, DC.

USGBC (2011) Advancing Green Building Policy in the States. [Cited 12 Sept 2011]. Available from: http://www.usgbc.org/ShowFile. aspx?DocumentID=10055.

Wallhagen M, Glaumann M (2011) Design consequences of differences in building assessment tools: A case study. *Building Research and Information* **39(1),** 16–33.

Wong NH, Yu C (2005) Study of green areas and urban heat island in a tropical city. *Habitat International* **29,** 547–58.

World Business Council for Sustainable Development (WBCSD) (2009) *Green Jobs: Towards Decent Work in a Sustainable Low Carbon World.* United Nations Environment Programme, International Labour Organisation, International Organisation of Employers and International Trade Union Confederation, Paris.

World Commission on Environment and Development (1987) *Our Common Future.* Oxford University Press, Oxford.

Worldwatch Institute (2008) *Green Jobs: Towards Decent Work in a Sustainable Low-carbon World.* United Nations Environment Programme, Nairobi. [Cited 17 Sept 2011]. Available from: http://unep.org/PDF/UNEPGreenJobs_ report08.pdf.

Environmental Management: Reflections

CHAPTER TWELVE

Urban Governance and Sustainability: Environmental Management Systems for Cities

LYE Lin-Heng
Faculty of Law, National University of Singapore

"On the question of appropriate penalties for breach of the law, this must be carefully assessed. In many developing countries, where the government is less than honest and corruption is a way of life, enhancing the quantum of fines may not be the right solution as it may only serve to enrich the pockets of the enforcers. New approaches are needed, such as economic incentives to induce compliance; or the use of the media to praise and reward industries that meet the standards and to shame those that do not, into doing better."

– LYE Lin-Heng

Urban Governance and Sustainability: Environmental Management Systems for Cities

LYE Lin-Heng[1]

Faculty of Law, National University of Singapore

Abstract

Human activities damage the environment. They deplete natural resources, generate pollution and wastes, accelerate the loss of forests and biological diversity, and threaten the water supply. As populations increase, these problems are exacerbated. Cities bear the brunt of increased human activities on limited land space with limited resources. It is therefore essential that cities adopt a system of environmental governance that will help ensure sustainability. As each city has its own mix of geographical, social, economic, political and environmental problems, it would be simplistic to suggest that there is a formula for sustainability that would fit every city. What is clear is that every city needs an effective environmental management system to manage its many activities and to ensure that development is controlled, environmental damage is minimized, natural areas are preserved and its citizens have an enhanced quality of life. This paper examines the ingredients for sound environmental management in cities, particularly those in the developing world. It submits that a sound environmental management system (EMS) for a city must first start with sound environmental policies and land use planning. It looks at ISO 14001 certification in the context of a city and concludes that environmental management systems in their current context focus largely on resolving problems of pollution. There is a lack of ecological dimensions in environmental management systems, as exemplified by the ISO 14000 series. This paper submits that environmental stewardship and ecological sustainability is at the heart of sustainable development and that the integration of the natural environment within a city has been largely overlooked. It advocates bringing the natural environment back to our cities and the incorporation of this dimension into environmental management systems.

Key words: cities, conservation, EMS, environmental management, governance, land use planning, law, ISO 14001, sustainable development

[1] LLB (Sing), LLM (Lond), LLM (Harv); Advocate & Solicitor. 469G Bukit Timah Road, Eu Tong Sen Building, Singapore 259776. Email: lawlyelh@nus.edu.sg; Tel: (65) 6516 3583; Fax: (65) 6779 0979.

I. Introduction

Sustainable Development and Sustainable Cities

The issue of the sustainability of cities is complex, as few can agree on what 'sustainability' means and how is it measured in the context of a city.[2] Although there is no agreed definition on the terms 'sustainable cities' and 'sustainable human settlements', it is clear that a city encompasses many dimensions, including environmental, economic, social, political, legal, demographic, institutional and cultural.

Fundamentally, cities that strive to be 'sustainable' face the tensions between economic development and environmental stewardship. This issue of environmental stewardship traverses beyond the confines of the city limits, as cities draw on resources beyond their boundaries for sustenance. This is particularly true of wealthy cities in the developed world, which almost invariably depend on well-developed global transportation and communication systems to bring resources from afar for consumption by its citizens, thereby enlarging their ecological footprints far beyond their city boundaries.[3] Research by the Earth Council in 1997 indicated that the 10 most unsustainable nations in ascending order of ecological deficit per capita are as follows: Switzerland, Israel, Japan, Germany, United Kingdom, the USA, the Netherlands, Belgium, Hong Kong and Singapore.[4]

---

[2] David Satterthwaite, 'Sustainable Cities or Cities that Contribute to Sustainable Development?' Chapter 5, The Earthscan Reader in *Sustainable Cities*, David Satterthwaite, Ed. 1999; Peter Hall, *Cities of Tomorrow* (Oxford: Blackwell, 1996, p. 412); Jorge E Hardoy, Diana Mitlin, David Satterthwaite, 'Sustainable Development and Cities', Chapter 6 in *Environmental Problems in Third World Countries, Earthscan Publications, London, 1992;* Genevieve Dubois-Taine, 'Introduction' *Cities of the Pacific Rim – Diversity and Sustainability, Genevieve Dubois-Taine and Christian Henriot, Editors,* Pacific Economic Cooperation Council Sustainable Cities Taskforce; *Habitat Debate –* United Nations Human Settlements Program – UN-Habitat: http://www.unhabitat.org/HD/.

[3] Ecological footprint is the corresponding area of productive land and aquatic ecosystems required to produce the resources used and to assimilate the wastes produced by a defined population at a specified material standard of living, wherever on Earth that land may be located – per William E. Rees, Professor of Community and Regional Planning at the University of British Columbia, 'Ecological footprints and appropriated carrying capacity: What urban economics leaves out'. Environment and Urbanisation 4(2), 121–30. Rees also states that that the 'so-called "advanced" economies are running massive, unaccounted ecological deficits with the rest of the planet... Even if their land area were twice as productive as world averages, many European countries would still run a deficit more than three times larger than domestic natural income'. Rees W (1996) Revisiting carrying capacity: Area-based indicators of sustainability. In: Population and Environment: A Journal of Interdisciplinary Studies 17(2), January 1996. Human Sciences Press Inc.

[4] See 'Ecological Footprint of Nations' web page at http://www.ucl.ac.uk/dpu-projects/drivers_urb_change/urb_environment/pdf_Sustainability/CES_footprint_of_nations.pdf; also, http://www.footprintnetwork.org/en/index.php/GFN/page/basics_introduction/. It should be noted that Singapore has defended its position, as a unique city-state without natural resources, needing, therefore, to rely on its neighbours and beyond for basic necessities such as food, water and energy. The city of Tokyo has an ecological footprint of 1.2 times the total land area of Japan. See http://gdrc.org/uem/mea/case-study-4.html.

The Brundtland Commission's definition of 'sustainable development' is familiar to most as 'development that meets the needs of the present without compromising the ability of future generations to meet their own needs'.[5] The juxtaposition of the word 'development' with 'sustainable' highlights the dilemma that faces all urban environments. As cities are almost invariably the engines of growth that fuel the economy of a nation, they are constantly in the forefront of the myriad challenges that arise from the need to find food, shelter, employment, transportation, energy sources, health care and other essential services for an ever-growing population. Indeed, it has been said that 'the battle for sustainability will be won or lost in cities'.[6] How, then, can a city ensure that its many activities are sustainable?

It should also be noted that the concentration of people, enterprises and motor vehicles in a city, while often viewed as a problem, can also bring certain advantages, such as lower costs per household and per enterprise for the provision of the infrastructure and services that minimize environmental hazards, such as sewage treatment systems and systems for the removal of domestic and industrial wastes.[7] Cities with well-managed transport facilities also reduce the stress on the natural environment, as a good public transport system will minimize the use of motor vehicles. The concentration of industries in a city also brings savings in the enforcement of environmental legislation, reducing the length of journeys required for inspections by the authorities. Indeed, with intelligent planning, the more closely people live and work, the greater the potential for fuel efficiency and the ability to avoid sprawl and wasted resources.[8]

How, then, do we measure the environmental performance of a city? Are the considerations similar between cities in developed and developing economies? Is it a matter of governance? If so, what are the ingredients required in sound environmental governance?

---

[5] *Our Common Future - Report of The World Commission on Environment and Development*, Gro Harlem Brundtland, Chair (1987). That same year, the term 'eco-city' was coined with the book *Ecocity Berkeley: building cities for a healthy future*, by Richard Register. This paper does not attempt to discuss eco-cities, as the few examples today are still in an embryonic stage. See "Sustainable Cities: Oxymoron or the Shape of the Future?", Annissa Alusi, Robert G Eccles, Amy C Edmonson, Tiona Zuzul, Harvard Business School Working Paper 11-062, 2010, http://www.hbs.edu/research/pdf/11-062.pdf . See also the *Centre for Sustainable Asian Cities,* School of Design and Environment, National University of Singapore, http://www.sde.nus.edu.sg/csac/index.htm; and the Singapore government's *Centre for Livable Cities* – http://clc.org.sg/index.php?q=about-centre-liveable-cities. Singapore and China are in collaboration to build two eco-cities - in Tianjin (Sino-Singapore Tianjin Eco-City) and Nanjing (Eco-Hightech Island). See http://www.tianjinecocity.gov.sg/; and http://www.channelnewsasia.com/stories/singaporebusinessnews/view/1075103/1/.html.

[6] Ahmed Djoghlef, Executive Secretary, Convention on Biological Diversity, 2009 – http://www.cbd.int/doc/speech/2009/sp-2009-02-10-cbi-en.pdf.

[7] "Cities as Solutions in an Urbanizing World", UN Centre for Human Settlements, chapter 3, *The Earthscan Reader in Sustainable Cities,* David Satterthwaite, Ed.; reprinted from chapter 13 (pp 417–421) of UNCHS (Habitat), *An Urbanizing World: Global Report on Human Settlements, 1996.*

[8] Thus, according to Columbia University's Earth Institute, New York City is one of the most energy-efficient places in the United States, consuming a quarter of the national average in energy consumption and emitting a quarter of the national average of carbon dioxide. See http://www.columbia.edu/event/new-york-city-sustainable-city-48283.html.

II. Strategies for Urban Environmental Management

The comparison of environmental performance between diverse urban centres is fraught with difficulties.[9] But it may generally be said that a city must ensure that its citizens are provided with a safe, adequate and sustainable supply of food, water, energy and essential services. It must control infectious and parasitic diseases by improvements in water quality, sanitation, drainage and garbage collection. A city must ensure that its energy sources are sustainably managed and must implement processes for clean, affordable and renewable energy. It must have in place a sound system for the disposal of different waste streams (industrial, municipal, toxic, hazardous, etc.). It must provide good roads and an efficient and affordable public transportation system as well as good communication facilities that will, in turn, reduce the need to travel, thereby reducing the impacts on the environment. A city must provide decent and affordable housing to all sectors of the population.[10] It must provide the resources for economic growth, as well as schools and educational institutions to ensure the continuing education of its residents, It must ensure a good quality of life through the provision of parks, open spaces, playgrounds and recreational areas. In this process, a city should strive to minimize ecological destruction. It must protect its natural and cultural heritage through sound land use planning, ensuring the sustenance of existing ecological systems and the preservation of its built as well as historical heritage to sustain its 'soul'. These processes should ideally be transparent and should involve the participation of the people.[11] And, finally, a city should ensure that its consumption of natural resources is sustainable – that the goods and services required to meet the needs of the population are delivered 'without undermining the environmental capital of nations and the world'.[12]

Efforts to manage the environment in any city or country must stem from policies that can only be determined after a thorough examination of its particular problems. Urban environmental problems differ with the stage of evolution of each city, and, therefore, the priorities as well as the solutions will differ.[13] For cities that are poor, the main concern will be public health – securing a system of access to clean water and sanitation facilities, setting up an efficient system for the collection of municipal and industrial wastes, as well as the prevention of organic pollution of water bodies. As cities start to industrialize, the main problems would be air pollution through higher levels of sulfur dioxides (SO_x) and particulates,

---

[9] David Satterthwaite, *Sustainable Cities, op cit* n.1 at. p. 83.

[10] The issue of housing will not be discussed in any detail as it is beyond the scope of this paper.

[11] Public participation is an important component of a city's EMS. It features strongly in Agenda 21 (Earth's Action Plan) see http://www.un.org/esa/dsd/agenda21/.

[12] Rees, *op cit.* note. 2 at p. 84.

[13] See the analysis of the transformation of cities as they move from poverty to rapid industrialization to relative affluence, and the environmental problems faced at each stage – David Sattethwaite, 1997 "'Environmental transformations in cities as they get larger, wealthier and better managed'", The Geographical Journal 163, no.2:216-24. See also the analysis of the stages of evolution of the urban environment in East Asia - Xuemei Bai and Hidefumi Imura, "'A Comparative Study of Urban Environment in East Asia: Stage Model of Urban Environmental Evolution'", International Review for Environmental Strategies, Vol. 1 No. 1 pp. 135-158, 2000, IGES.

water pollution from organic sources as well as from industrial effluent (heavy metals), and land contamination from solid and hazardous wastes. There should also be concerns for the health and safety of workers in these industries. As cities continue to industrialize, with increased affluence and sophistication, there will be increased consumption, increased waste, increased demands on energy, increased motor vehicles and increased numbers of industries; leading to increased levels of pollution, particularly from carbon dioxide emissions, sulphur dioxide, nitrogen dioxide and dioxins. There will also be increased noise levels and loss of natural areas, leading to losses in biodiversity.

The solutions will differ depending on the resources of each city and its special circumstances. Thus, in regard to waste management, one country may decide on incineration instead of landfills (as in the case of Singapore, which has insufficient land for landfills but can afford the substantial cost of incinerators), while another country may vigorously oppose incinerators (as in the case of the Philippines, where the law virtually prohibits all forms of incineration of waste). But both Singapore and the Philippines and, indeed, all cities must strive for waste minimization and resource conservation as well as seek the best solutions for its environmental problems. The challenges for achieving sustainability will also differ for developed and developing, as well as for new and old/decaying cities.

This paper focuses on the problems of cities in the developing world, with illustrations drawn mostly from the cities of South-east Asia. To what extent can cities in the developing world achieve sustainability, where most are facing problems of poverty, mass migrations from rural areas, a poorly educated workforce and political corruption? To what extent can an environmental management system help a city achieve sustainability?

III. EMS for Cities

The Basic Requirements for an EMS

A sound Environmental Management System (EMS) for a city starts with sound environmental management policies. These must then be implemented through institutional, administrative, legal and physical infrastructures. A sound EMS for a city should comprise the following:

1. Sound institutional and administrative structures;
2. Comprehensive land use planning laws and policies;
3. Effective environmental laws and enforcement;
4. Physical infrastructure for the provision of essential services such as clean water, electricity, transport and communications; and
5. Physical infrastructure for pollution control, including facilities for the treatment of sewage, collection of garbage, management of hazardous substances and control of air emissions.

It should be emphasized that there must be coherence in the various policies and in their implementation amongst the various institutions, and this must be integrated into local, regional and national policies and legal frameworks.

(i) Institutional and administrative structures

A ministry for the environment?

The administrative structure for each country and each city will differ, and not all will have a dedicated ministry that focuses entirely on the environment. Is it essential that there be a dedicated environmental authority here?

Prior to the Stockholm Summit on the Human Environment in 1972, a dedicated Ministry for the Environment was quite unheard of for most countries. Environmental issues were hitherto regarded as health matters and were the province of the Ministry of Health. It was only after 1972 that countries started to institute a separate Environment Ministry. Even so, not all countries today have a separate ministry for the environment. For many years, Thailand and Vietnam subsumed the environmental portfolio under their Ministry of Science, Technology and the Environment (MOSTE). It was only in 2002 that Thailand established its new Ministry for Natural Resources and the Environment (MNRE). Power previously exercised by agencies such as the Ministry of Agriculture and Fisheries are now transferred to the new MNRE, but there is great uncertainty about how extensive the powers of MNRE will be, as against the existing sectoral ministries. There are also concerns as to how the new MNRE can assert its influence against the National Environment Board, a high-level body constituted by the Prime Minister.[14]

Similarly, in Vietnam, a new Ministry of Natural Resources and Environment was created in 2002 to perform the functions previously exercised by the National Environment Agency (NEA), which was constituted under the Ministry of Science, Technology and Environment (MOSTE). This new Ministry will initially take charge only of terrestrial environmental issues. Other natural resources such as fisheries, agriculture, oil and gas continue to be managed by the Ministry of Fisheries, the Ministry of Agriculture and Rural Development, and the Vietnam Oil and Gas Corporation respectively. It is expected that power will gradually be devolved to the new MNRE, but there are concerns that the new ministry will remain poorly funded and relatively weak as compared to other more powerful agencies such as the Ministry of Planning and Investment (MPI) and the Ministry of Agriculture and Rural Development(MARD).[15] These changes are translated to the provincial governments where functions previously exercised by the Departments of Science, Technology and Environment (DOSTE, the local branch of MOSTE) are transferred to the Departments of Natural Resources and Environment (DNRE).

---

[14] Alan K J Tan, "'Recent Institutional Developments on the Environment in South-east Asia – A Report Card on the Region'", [2002] 2 Singapore Journal of International and Comparative Law.
[15] Alan K J Tan, op cit, n. 9.

In contrast, the Philippines has a comprehensive body (the Department of Environment and Natural Resources) that is responsible for all facets of the environment. In Singapore, the Ministry of the Environment and Water Resources (MEWR) only deals with pollution,[16] while nature conservation is the domain of the Ministry of National Development's National Parks Board.

Which of these countries has the most effective system? What organizational structure achieves optimal effects? This must also be examined in the context of a city, which may have its own administrative structure.

In evaluating whether a city has an effective EMS, it is essential to first examine the effectiveness of the roles of the Environment Ministry and other supporting government ministries, such as the Ministry of Agriculture, Ministry of Industry and the Ministry of Development, to see if there are synergies in policies and administration or portfolio conflicts and to ascertain whether the Environment Ministry has adequate powers to decide on policies as well as to implement them.

It is also essential to examine the effectiveness of the authorities that are in charge of the city's economic activities and trade, as well as the authorities on land use planning and development, transport and communication, as well as manpower/labour. Each sector has important roles to play in the development of the city. An effective EMS requires communication between the various authorities and third parties such as the private sector, non-governmental organizations (NGOs) and citizens. This helps to ensure that, before any project is approved or a new industry is allowed to set up its operations, its environmental impacts are carefully evaluated and measures are adopted to ensure that its operations will not have an adverse impact on the environment. Indeed, this process, if properly planned, will ensure that industries that are highly pollutive are denied a licence to operate. If allowed into the city, they must be appropriately located where they do not pose unmanageable health and safety hazards and pollution problems. A good EMS will also ensure that the industry is sited where there can be synergies with other industries (e.g. the waste from one factory can be used as the raw materials for another, an example of 'industrial ecology').[17]

In most developing countries, however, it is not uncommon for highly pollutive industries to be allowed in without consultation with the Environment Ministry or with the planning authorities. There may not be any controls on where

---

[16] Singapore established the National Environment Agency (NEA) under the Ministry of the Environment (now the Ministry of Environment and Water Resources) on 1 July 2002 to focus on the implementation of environmental policies. Under the NEA, the divisions of Environmental Protection, Environmental Public Health, and Meteorological Services work together. See Code of Practice on Pollution Control, http://app2.nea.gov.sg/codeofpractice.aspx.

[17] Industrial Ecology, T.E. Graedel and B.R. Allenby, second edition (Upper Saddle River, NJ: Prentice Hall, 2003); see also "'An Indigo Paper on Industrial Ecology'" http://www.indigodev.com/DefineIE.html.

they may be located. They may be sited near residential or commercial sites, which lack the necessary infrastructure to deal with the wastes and other forms of pollution that ensue. Such cities lack an EMS. A sound EMS acknowledges that environmental problems can be anticipated and prevented through policies on land use planning and the imposition of appropriate controls. An integrated approach is essential and is part of good environmental governance.

The city-state of Singapore is an example of a city with an effective EMS in place. It may be said that Singapore's EMS starts with the identification of the types of industries that can be allowed into the city-state. The Ministry of Environment and Water Resources (MEWR)[18] houses the National Environment Agency (NEA) as well as the Public Utilities Board (which is responsible for the supply of water). MEWR works closely with the Economic Development Board (EDB) and the Ministry of Trade and Industry (MTI), to decide what kinds of industries are needed for the economic development of the city-state. As the EDB identifies the kinds of industries that Singapore would like to attract, discussions are held with other ministries and organizations to ascertain if there are problems in accommodating these industries.[19] In particular, discussions are held with NEA to see if the pollution that ensues can be controlled, with urban planners from the Ministry of National Development (MND) to establish the possible sites for these industries; with the Jurong Town Corporation (JTC), the largest landlord of industrial premises, to discuss the physical logistics of location for these factories; and with the Ministry of Manpower to ascertain the impacts on the workforce.

NEA adopts an integrated approach in the planning control of new developments, so as to ensure that environmental considerations are incorporated into the land use planning, development control and building control stages. While, in the past, highly pollutive low-technology and 'sweatshop' industries were allowed, today, the emphasis is on high-tech industries, such as electronics, bio-technology, chemicals, petro-chemicals and pharmaceuticals. Many of these are highly pollutive, but planners from the Ministry of National Development's Urban Redevelopment Authority (URA) check with the Pollution Control Department (PCD) of NEA on the siting requirements for these new development projects and their compatibility with the surrounding land use.[20] This is part of the planning process, to ensure that industries are located in specially designated areas such as 'eco-industrial parks' with measures to control, manage and minimize pollution

---

[18] This Ministry was re-named the Ministry of Environment and Water Resources as from 1 September 2004.

[19] The EDB was established soon after Singapore's independence, to spearhead industrialization, promote investments, develop and manage industrial estates and provide medium and long-term industrial financing. See *Heart Work – Stories of How EDB Steered the Singapore Economy from 1961 to the 21st Century,* 2002, Singapore EDB and EDB Society; and *Heart Work 2 - EDB & Partners: New Frontiers for the Singapore Economy,* Straits Times Press and EDB, 2011.

[20] Today, these industries are located in Jurong Island, an amalgamation of several smaller islands just off the main island, forming a petro-chemical complex with a well planned emergency system involving the participation of all parties. See http://www.jurongisland.com/

as well as to maximize industrial and technological synergies.[21] In particular, PCD will examine measures to control air, water and noise pollution, the management of hazardous substances and the treatment and disposal of toxic wastes.

To guide land use planning and help industrialists in the selection of suitable industrial premises, industries are classified into four categories, namely clean, light, general and special industries, based on the impact of residual emissions of fumes, dust and noise on the surrounding land. Buffer distances are imposed between pollutive industries and the nearest residential areas. Industrial premises that are located close to residential areas and within water catchment areas may only be occupied by clean or light industries.[22]

Before a proposed development can be constructed under the Building Control Act, the developer is required to submit building plans of the proposed works to the Building Plan and Management Division of the Building and Construction Authority (BCA) for approval. These building plans must also be submitted to and approved by various authorities, including the Fire Safety Bureau, the National Parks Board and NEA. Within NEA, the Central Building Plan Unit (CBPU) of PCD will examine all building plans to ensure they comply with sewerage, drainage, environmental health and pollution control requirements. In particular, CBPU will screen prospective industries to ensure that they:

- are sited in designated industrial estates and compatible with the surrounding land use (industries are classified into four categories depending on their capacity to pollute);[23]
- adopt clean technology to minimize the use of hazardous chemicals and the generation of wastes;
- adopt processes that facilitate the recycling, reuse and recovery of wastes;
- do not pose unmanageable health and safety hazards and pollution problems; and
- install pollution-control equipment to comply with discharge or emission standards.

When the factory building is completed, inspectors from CBPU will inspect the premises to check if the structure has been built in compliance with pollution-control requirements. Only when this is approved will the factory be given a licence to operate.

---

[21] Lye Lin Heng, "Singapore: Long Term Environmental Policies" in *Cities of the Pacific Rim – Diversity & Sustainability*, PECC Sustainable Cities Taskforce, Genevieve Dubois-Taine, Christian Henriot, Eds., 2001 PUCA, p. 155-168. See also Lye Lin Heng, "A Fine City in A Garden – Environment Law and Governance in Singapore" [2008] SJLS 68–117.

[22] See NEA's Code of Practice on Pollution Control (2009) and other Codes -http://app2.nea.gov.sg/codeofpractice.aspx; as well as Guidebooks and Handbooks - http://app2.nea.gov.sg/guidebooks.aspx

[23] See the Code of Practice on Pollution Control, revised in 2009 – http://app2.nea.gov.sg/data/cmsresource/20090312534898283541.pdf.

Coordination amongst government authorities

It is important that government agencies work with each other and that the administration of foreign aid be properly coordinated so as to achieve the greatest efficacy. One example is the drafting of new environmental laws by international legal consultants. The new laws should take into account the technical capacity of the country in implementation. The passing of new laws prescribing trade effluent standards should, therefore, be coordinated with the building of the necessary physical infrastructure to test effluent samples. This coordination amongst different institutional agencies is, however, often lacking in developing countries, resulting in laws that are much stricter than industries can comply with and which cannot be enforced.

Building technical capacity

It is also important for developing countries to build technical capacity. Many international agencies and institutions today help build institutional capacity by providing short-term training programmes for government personnel. The World Bank and the Asian Development Bank (ADB), as well as aid agencies like the Canadian International Aid Agency (CIDA), the Swedish International Development Agency (SIDA), AusAid (the Australian government's overseas aid program) and the Danish International Development Agency (DANIDA), run training programmes for developing countries. Governments of developed countries also conduct special environmental and technical training programmes.[24] These programmes should be properly coordinated and the right persons must be selected for training. Persons trained should be made to return to their home countries to train, in turn, their fellowmen. Again, this requires a system to ensure that the best and brightest are sent abroad for higher education and that they are contractually bound to return home to share what they have learnt. The system should be transparent so that the right persons, not those who are politically connected, are sent for training.

Public participation

Chapter 23.2 of Agenda 21 (Earth's Action Plan), conceived at the United Nations Conference on Environment and Development in 1992 at Rio, states:

'One of the fundamental prerequisites for the achievement of sustainable development is broad public participation in decision-making. Furthermore, in the more specific context of environment and development, the need for new forms of participation has emerged. This includes the need of individuals, groups and

---

[24] The MEWR's Singapore Environment Institute regularly conducts training programmes on various aspects of environmental management for government officers from neighborbouring countries. The IUCN Commission on Environmental Law, the ADB, UNU (United Nations University) and the Asia-Pacific Centre for Environmental Law (APCEL) ran two intensive capacity-training programmes on environmental law for law professors from the Asia-Pacific region in 1997 and 1998. APCEL with the Ministry of Foreign Affairs (MFA) also runs a yearly programme on Urban Environmental Management for government officers from developing countries. See http://law.nus.edu.sg/apcel/.

organizations to participate in environmental impact assessment procedures and to know about and participate in decisions, particularly those which potentially affect the communities in which they live and work. Individuals, groups and organizations should have access to information relevant to environment and development held by national authorities, including information on products and activities that have or are likely to have a significant impact on the environment, and information on environmental protection measures.'[25]

In particular, Agenda 21 calls for the participation of non-government organisations (NGOs) as 'partners for sustainable development', as well as the participation of women, workers and trade unions, businesses and industries, and the scientific and technical communities in ensuring a sustainable environment.

(ii) Land use planning laws and policies

Planning theorists have defined urban sustainability as comprising economic, social and environmental dimensions, and emphasized that these must be integrated, interlinked and guided by the proposals contained in Agenda 21, which specifies concrete planning strategies.[26]

The conflict between development and conservation is particularly marked in an urban environment. Thus, it is especially important that environmental considerations are incorporated in the early stages of development planning so that appropriate measures can be undertaken to balance the pressures for land development with the need to protect the natural environment. Environmentally sensitive land use planning provides the opportunity to institute proper measures and controls at an early stage in the development process. Chapter 10 of Agenda 21 emphasizes the importance of an integrated approach in the planning and management of natural resources.

The purpose of urban planning is to reconcile competing claims for the use of limited land, so as to provide a consistent, balanced and orderly management of land use to provide a good physical environment for the promotion of a healthy life and to provide the physical basis for a better urban community life.[27] While concern for human well-being is a major focus in planning, the conservation of

---

[25] http://www.un.org/esa/dsd/agenda21/res_agenda21_23.shtml.

[26] See Kahn M (1995). Concepts, definitions and key issues in sustainability development: The outlook for the future. *Proceedings of the 1995 International Sustainable Development Research Conference, Manchester, England, 27–28 March 1995.* Keynote paper, 2–13, discussed by Basiago AD (1999) in 'Economic, social and environmental sustainability in development theory and urban planning practice. *The Environmentalist* 19, 145–61, p. 150.

[27] Foley D (1960) 'British Town Planning: One Idealogy or Three?', British Journal of Sociology 2. One planner sees the issues as a triangle where planners have to reconcile 'not two, but at least three conflicting interests: to 'grow' the economy, to distribute this growth fairly, and in the process, not degrade the ecosystem. Thus, the triangle points are the Economy, the Environment and Equity. Also Campbell S (1996) 'Planning: Green cities, growing cities, just cities? Urban Planning and the Contradictions of Sustainable Development'. Journal of the American Planning Association 62(3), pp. 296–312.

the earth's natural capital (including both natural resources and ecosystems) is also of primary importance, although it may not always be a major concern in practice. Strategic planning is important, as it enables the identification of areas of key environmental concern at the outset and pre-empts potential difficulties.

It is clear that a city that is well-planned will not have as many environmental problems. Its industries will be located in areas where facilities that can deal with the wastes generated by manufacturing processes exist. Highly pollutive industries will be located away from residential and commercial areas. And where economic conditions permit, proper land use planning in a city will ensure that it is provided with:

- efficient public transport facilities, which will reduce the need for individual commutes via motorized vehicles;
- efficient telecommunications, which will reduce the need to travel to work;
- piped water supplies at an affordable rate for the poor;
- drainage and sanitation facilities;
- a system for the collection, disposal and recycle of various wastes;
- adequate housing for the people; and
- schools, commercial and recreational areas and nearby amenities.

All of these require careful strategic land use planning,[28] and is part of a good environmental management system.

In the development of a city, land use plans must anticipate the city's future needs within a specific time frame. In particular, it should have regard to the following:

- Identification of developmental constraints and major land uses that affect the environment, such as areas for highly pollutive and hazardous industries, airport zones, storage of military supplies, live-firing areas for military training (in the context of a city-state like Singapore);
- Projection of land needs for environmental infrastructure, such as sewage treatment plants and refuse disposal facilities (dump sites and incinerators);
- Projection of land needs for transportation and communication (such as for future airport expansions, new railway lines, mass rapid transport systems, expressways, satellite receiving and transmitting stations);
- Identification of possible areas for major utility installations and infrastructural needs that are potentially hazardous or pollution prone, such as power stations, gas works, storage facilities for explosives and other hazardous materials;

---

[28]   Malone-Lee L C "Environmental Planning" in Capacity Building for Environmental Law in the Asian and Pacific Region – Approaches and Resources, Vols. I and II, Asian Development Bank, Donna Craig, Nicolas A Robinson, Koh Kheng Lian, Eds.2002. p. 606. See also Lye Lin Heng, "Land Use Planning, Environmental Management and the Garden City as an Urban Development Approach", Land Use for Sustainable Development, ed. John Nolon, Nathalie Chalifour, Lye Lin Heng, Patricia Kameri Mbote (2007). New York: Cambridge University Press; Lye Lin Heng, "Landscape Protection Laws in the Evolution of Modern Singapore". Landscape, Nature and Law, ed. Antonio Herman Benjamin (2005): 119 134. Sao Paolo, Brazil: Laws for a Green Planet Institute.

- Identification of ecologically sensitive areas for nature conservation;
- Protection of water catchment areas; and
- Identification of new areas to be opened up for major developments, such as new industrial estates, business or science parks, and new housing estates or towns.

It should be emphasized that public participation is an important component in this process. Part III of Agenda 21 details the participation of various sectors of the population and the emergence of new forms of participation. These include the need of individuals, groups and organizations to participate in environmental impact assessment procedures and to know about and participate in decision-making, particularly those that potentially affect the communities in which they live and work. Individuals, groups and organizations should have access to information relevant to the environment and development held by national authorities, including information on products and activities that have or are likely to have a significant impact on the environment, and information on environmental protection measures.[29]

(iii) Legal infrastructure

Another requirement for a sound EMS for cities is the establishment of sound legal institutions: judicial, prosecutorial, inspection and enforcement agencies; reliable administrative offices with sound laws and respect for the rule of law. Caring for the Earth, the successor to the World Conservation Strategy, spelt out the minimum content of environmental law for sustainability. These can be applied to cities. At a minimum, governments should ensure that their nations are provided with comprehensive systems of environmental law that encompass:

- land use planning and development control;
- the sustainable use of renewable resources and the careful use of non-renewable resources;
- the prevention of pollution through the imposition of standards for emissions, environmental quality, process and product standards designed to safeguard human health and ecosystems;
- the efficient use of energy through the establishment of energy-efficient standards for processes, buildings, vehicles and other energy-consuming products;
- control of hazardous substances, including measures to prevent accidents during transportation;
- waste disposal, including standards for minimization of waste and measures to promote recycling; and

---

[29] "Strengthening the Role of Major Groups", Chapter 23, Agenda 21. See the Convention on Access to Information, Public Participation in Decision-making and Access to Justice in Environmental Matters, Aarhus, Denmark, 1998. http://www.unece.org/fileadmin/DAM/env/pp/documents/cep43e. pdf. Many developed countries now have passed "right to know" laws regarding environmental hazards; and have also passed statutory laws for Freedom of Information. Closer to home, Thailand passed a Freedom of Information Act in 1997 (see http://www.worldlii.org/int/journals/PLBIN/2000/29. html). India passed its Right to Information Act in 2005.

- the conservation of species and ecosystems through land use management, specific measures to safeguard vulnerable species and the establishment of a comprehensive network of protected areas.[30]

The national legal system for governance of cities should provide for:

- the application of the precautionary principle;
- the use of the best available technology for the setting of pollution standards;
- the use of economic incentives and disincentives based on appropriate taxes, charges and other instruments;
- effective and appropriate penalties for non-compliance;
- environmental impact assessments of all new developments and projects;
- periodic environmental audits of industries, government departments and agencies;
- effective monitoring so as to detect infringements and effect changes in the laws where necessary;
- public access to information including EIAs, environmental audit data and the results of monitoring, as well as information relating to the production, use and disposal of hazardous substances; and
- the facilitation of public participation and citizen access to courts.

An environmental management system will require that these laws be enacted, that they are continuously monitored for compliance, enforcement and relevance, and that they are updated periodically, as the needs of the cities change.

On the question of appropriate penalties for breach of the law, this must be carefully assessed. In many developing countries, where the government is less than honest and corruption is a way of life, enhancing the quantum of fines may not be the right solution, as it may only serve to enrich the enforcers. New approaches are needed, such as economic incentives to induce compliance or the use of the media to praise and reward industries that meet the standards and to shame those that do not into doing better. One example of an innovative approach is the PROPER Prokasih (or Clean Rivers) programme introduced in Indonesia in 1989, targeted at biochemical oxygen demand (BOD) emissions from major industrial polluting plants.

This programme started with the identification of major industrial water polluters along the most polluted rivers. Once identified, vice-governors invited the polluter to sign a voluntary, non-legally binding pollution reduction agreement with the vice-governor and the state environment agency. Thereafter, samples were taken of the plant's effluent to establish a pollution baseline and to assess the degree of compliance with the terms of the agreement. By 1994, 1,400 manufacturing plants had participated. The mean reduction in BOD loads in these plants was 44 per cent. The BOD load per unit of output fell by 55 per cent.

---

[30] *Caring for the Earth: A Strategy for Sustainable Living,* 1991, IUCN, UNEP and WWF, at p. 68.

Despite this success, BAPEDAL, (the State enforcement agency) faced considerable difficulties, including the lack of capacity to reliably monitor air and water quality and the lack of authority to inspect and enforce the laws.[31] It also lacked the authority to issue permits detailing emissions requirements, and the courts refused to grant legal standing to either the new emission standards or the results of monitoring. Thus, other ways had to be found to resolve the pressing industrial pollution problems. BAPEDAL decided to build on the success of the Prokasih programme by turning it into an environmental rating and public disclosure system.

The PROPER Prokasih programme succeeded in increasing the compliance rates substantially within six months.[32] It used a system of ratings based on colours (gold, green, blue, red and black in descending order of compliance, where gold signified compliance with international best-practice standards). No plant received a gold rating, but five received a green rating. These were publicly named, but plants that received lesser ratings were not. Instead, these plants were consulted to ensure that the ratings were correct. Then they were given six months to improve their performance, after which BAPEDAL made it clear that it would rate the plant again and announce the results. Six months later, there was a substantial improvement, with the compliance rate increasing from 36 to 41 per cent.

Six reasons have been advanced for the success of this programme:

1. BAPEDAL, the Indonesian management agency for EIAs, had obtained high-level political support for the programme, including support from the President.
2. It had reached out to manufacturing facilities, environmental NGOs, community leaders and the media to demonstrate how the ratings worked, thus inspiring confidence in the programme amongst all these players.
3. The programme was focused on widely acknowledged and accepted national BOD emission standards.
4. The new PROPER Prokasih ratings programme was grafted on to the well-known and successful Prokasih programme.
5. The participants built on what they learnt in the first Prokasih programme.
6. They kept the rating system 'simple, credible, transparent and honest'.[33]

---

[31] This was because only the Ministry of Industry and the local police had the authority to enter factories to obtain emission samples.

[32] Michael T Rock, "Searching for Creative Solution to Pollution in Indonesia", Chapter 4, *Pollution Control in East Asia – Lessons from Newly Industrializing Economies*, 2002, pp. 65-81; Shakeb Asfah and Jeffery R Vincent, "Putting Pressure on Polluters: Indonesia's PROPER Programme"; chapter 8 in *Asia's Clean Revolution – Industry, Growth and the Environment David P Angel and Michael T Rock, contributing eds*, 2000. See also "Going public on polluters in Indonesia: Bapedal's Proper Prokasih Program", David Wheeler and Shakeb Asfar, http://siteresources.worldbank.org/NIPRINT/Resources/GoingPubliconPollutersinIndonesia.pdf.

[33] Above, note 32 at p. 74.

The success of the PROPER Prokasih programme in Indonesia is an example of the workings of an environmental management system for water pollution. This included:

- the setting up of an appropriate institution (BAPEDAL);
- the identification of the problems (major polluters, limited resources) by BAPEDAL; and
- the identification of possible solutions, which included:
  (a) the formation of a local Prokasih team headed by the vice-governors and the enforcement agency (State Ministry for Population and the Environment) as well as local public officials and representatives from the environmental study centres at Indonesian universities and testing laboratories;
  (b) the identification of the most polluted rivers or the most polluted section of rivers within a province, and the major industrial water polluters along those rivers;
  (c) invitations by vice-governors to the major polluters to sign the voluntary but non-legally binding agreements;
  (d) repeated sampling of a plant's effluent to establish a pollution baseline and to assess the degree to which the plant was meeting the terms of the agreement;
  (e) the identification of funding for the project (funded by the implementing agency and the provincial governments);
  (f) following the creation of BAPEDAL in 1990, the setting of effluent and emission standards for water by BAPEDAL for 14 industries identified as major pollutant sources (these standards were set after discussions with industry associations, gathering evidence on best-practice technologies used in Asia and reaching an agreement with these associations and other government ministries);
  (g) increasing the staff members of the implementing agency, BAPEDAL; and
  (h) coordinating with international aid agencies to assist in the purchase of pollution control equipment,[34] assistance in gathering and analyzing pollution data, drafting of laws, and exploration of the improvement in institutional capacity, such as the feasibility of establishing local (provincial) environmental impact agencies (BAPEDALDA).

However, these developments did not last due to economic and political changes, leading one writer to conclude that 'lasting, sustainable industrial pollution control programmes ... most assuredly depend on creating tough command-and-control environmental agencies that can weather political change'.[35] In particular, decentralization policies exacerbated these problems, as the governments in Jakarta

---

[34] The Japanese provided soft loans for the purchase of pollution control equipment; the World Bank financed several capacity-building projects in BAPEDAL.
[35] Ibid, n. 14 at p. 78-79. See also the papers and discussion on Jakarta's Water Supply in *Sustainable Urban Services – HongKong Seminar*, PECC Sustainable Cities Taskforce, 23–24 November 2001. Genevieve Dubois-Taine, Ed., pp. 21–72.

could not control the increased contamination of the water sources upstream, which were the province of the local governments. This led to a shortage of water in the city, as the new management of water resources raised charges for water that cities downstream were unable to afford.[36] There is no easy solution here, as the increased contamination raised the costs for water treatment, making it difficult even for private operators of water supplies.[37]

A typical example of a non-functioning system in a developing country is that of the city of Bangalore in India at the turn of the millennium,[38] where the rule of law was found wanting due to:

- corruption;
- improper exercise of bureaucratic discretion, leading to failure to enforce laws;
- interference of politics in bureaucratic decision-making;[39]
- inadequate laws as well as woefully inadequate fines and penalties;
- ignorance of the law;
- inadequate allocation of jurisdiction between government institutions – actions taken were fragmented as each activity was the responsibility of a separate agency;
- the lack of power and resources;
- the lack of qualified lawyers;
- an inefficient legal system, leading to long delays and high costs; and
- power imbalance between the government and the individual, leading to difficulties in obtaining information in proceedings against the government.

In contrast, the Indian state of Bihar, once India's poorest and most lawless state, has now become 'a model of reform for the entire country', achieving double-digit growth in just six years under its Chief Minister, Nitish Kumar.[40] It is fitting that Kumar emphasized the importance of the rule of law and set out to eradicate corruption and fight crime. As security improved, growth followed.[41]

(iv) Physical infrastructure

The building of the physical infrastructure for environmental management is crucial for a sustainable city. The best laws will not work if there is no piped

---

[36] Lye Lin Heng, "The Environmental Dimension of Urban Services", in *Sustainable Urban Services, PECC Sustainable Cities Taskforce*, Santiago de Chile Seminar, 10-11 July 2002 p. 165–170

[37] See the 25-year contract with Ondeo to supply water to West Jakarta, in *Sustainable Urban Services, ECC Sustainable Cities Taskforce*, HongKong Seminar 23–24 November 2001, Report on Jakarta's Water Supply, p. 21-72.

[38] Amanda Perry "Sustainable Legal Mechanisms for the Protection of the Urban Environment", chapter 3, *Sustainable Cities in Developing Countries*, Cedric Pugh, Ed. Earthscan, 2000.

[39] "One judge said that "the order of decisions is all wrong. Politicians decide on projects and then scientists simply concur. It should be the other way around, but no one has the guts to make an independent decision", per Amanda Perry *op. cit.* at p. 58.

[40] See TIME, November 7, 2011"Breaking free - How one man turned Bihar, once India's poorest and most lawless state, into a model of reform for the entire country", pp. 26–29.

[41] "I didn't say much," he tells TIME. "I said rule of law will be established", p. 28.

water supply to the urban population, if there are no facilities to deal with sewage and trade effluents, and if there is no system to deal with the different kinds of wastes generated. There is a need for sewage treatment facilities, trade effluent treatment plants, properly constructed dumpsites for wastes, as well as incinerators for hazardous and bio-hazardous wastes. Each city has to carefully consider the most viable solutions to these problems in the context of its political, social and economic situations. Some basic factors remain. The best laws prescribing trade effluent standards, for example, will not work if there are no laboratories to test effluent samples from particular factories or to monitor the water quality from rivers, lakes and seas. And even with the building of the physical infrastructure, there must be a system in place to ensure the collection and the separation of different types of wastes for recycling or for special treatment.

How can cities in developing countries find the financial resources to build the environmental infrastructure? Certainly, a comprehensive redesign to eliminate waste at the source will reduce the size of financing required and may even be self-financing. Where central governments lack the financial resources to fund such projects, loans may be obtained from multilateral agencies, such as the World Bank or the Asian Development Bank, which have accrued significant experiences in a range of sectors including electricity, water and sanitation, and telecommunication.[42] This may also be facilitated through partnerships with the private sector. Public–Private Partnerships (or PPPs) work in a variety of ways, such as where the private operator builds, operates and owns (BOO), or builds, operates and transfers (BOT) possession of the facility to the city or state.[43] Funding can come from the private sector or through the multilateral agencies. However, serious issues of governance can hinder such partnerships and need to be resolved by the state or the city itself. In the case of the water sector, the following have been identified:

- The apparent low priority that central governments give to these issues;
- Confusion of social, environmental and commercial aims;
- Political interference;
- Poor management structures and imprecise objectives;
- Inadequate legal frameworks – weak, absent or inconsistent regulations;
- The lack of transparency in the awarding of contracts;
- Non-existent or weak and inexperienced regulators; and
- Resistance to cost-recovery tariffs.[44]

There are also issues of foreign exchange risks, contractual risks (projects of long duration entered into with poor initial formation) and country risks.

---

[42] Plummer, Janelle (2002) *Focusing Partnerships: A Sourcebook for Municipal Capacity Building in Public-Private Partnerships*, London: Earthscan Publications; Cambessus M and Winpenny J (2003) *Report of the World Panel on Financing Water Infrastructure: Financing Water For All*, World Water Council, 3rd World Water Forum, Global Water Partnership.

[43] See "Public-Private Partnerships for the Urban Environment", UNDP http://www.pppue.org.np/; and for service delivery, see http://www.undp.org/pppsd/.

[44] Report of the World Panel on Financing Water Infrastructure, *op. cit.*, p. 176.

This includes the creditworthiness of the country, as well as its respect for the law and its reputation for honouring contracts made. The private sector will have to do its own due diligence assessments and in particular, work on the risk factors. What is clear is that a country or city with good governance and respect for the rule of law will find it much easier to obtain financing and assistance. Thus, the building of a sound legal framework is fundamental.

Some successful models of PPP can be found in the water sector, such as in Santiago, Chile, which privatized its water supply, sanitation and water treatment facilities.[45] Other examples include the reforms to improve water and basic sanitation services in Cartagena de Indias, Colombia,[46] and the innovative water management system of Adelaide, Australia, which outsourced water and waste water services to United Water, a consortium of Vivendi Water, Thames Water and Kinhill in January 1996, with the assets remaining the property of the State through SA Water.[47] More recent examples are the ventures by successful Singapore corporations, Hyflux, Keppel Seaghers and Sembcorp.[48] However, PPP efforts at privatizing the water supply in Jakarta have not met with success, despite the passing of new laws for Water which legitimized privatization (the Water Resources Law No. 7, 2004) for reasons attributed to the lack of transparency as well as corruption in the tender process.[49]

---

[45] See "Santiago de Chile, Water Supply, Sanitation and Water Treatment in Santiago de Chile: A fully privatized system", *Sustainable Urban Services, Santiago de Chile Seminar Report*, Pacific Economic Cooperation Council Sustainable Cities Taskforce, July 2002, Genevieve Dubois-Taine, Ed.; pp. 161–228

[46] Above, at pp. 39–100.

[47] This partnership has "brought together the public and private sector, combining world class expertise in water and waste water management with local knowledge and appropriate contractual arrangements to achieve the effective delivery of sustainable water services." Per Philippe Laval, Manager Director, United Water "The Adelaide Contract: The Contribution of Outsourcing to Sustainability", *Sustainable Urban Services – Shanghai Seminar, April 2003* Report, Pacific Economic Cooperation Council Sustainable Cities Taskforce, Genevieve Dubois-Taine, Ed., pp. 33–68.

[48] Hyflux has built huge desalination plants in Algeria, China and Singapore (http://www.hyflux.com/biz_trackrecord.html); the Magtaa Desalination Plant in Algeria is the world's largest, at 500,000 m³ per day. See also the projects by Keppel Seaghers in Qatar to develop the largest greenfield wastewater treatment and reuse facility and the first integrated solid waste treatment facility in the Middle East (http://www.kepcorp.com/en/content.aspx?sid=95); and Sembcorp's many projects providing environmental infrastructure worldwide, some of which are entirely privatized (http://www.sembcorp.com/en/global-presence.aspx?type=all).

[49] Renalia I, "Ten Years of Public-Private Partnership in Jakarta Drinking Water Service", see http://researcharchive.vuw.ac.nz/handle/10063/978. The analysis concludes that: Public Private Partnership (PPP) in Eastern Jakarta does not bring improvement to the region's drinking water service. Thames PAM Jaya (TPJ) had failed in fulfilling targets set in the Cooperation Agreement. Lack of transparency and public tendering in the process of forming the public private partnership may have contributed to this poor performance because the proper search for a competent partner was short circuited. Political interference in the bidding process is a form of corruption in which the company granted the contract was clearly complicit. The water tariff in Jakarta is not only the highest in Indonesia, but it is also the highest in the Southeast Asia region. The quality of its service, however, is still of poor quality..."

IV. ISO 14001

ISO 14000 is a series of voluntary international standards setting out the requirements for an environmental management system, developed by the International Organization for Standardization (ISO) based in Geneva, Switzerland. These international standards serve as tools to manage environmental programmes and provide an internationally recognized framework to measure, evaluate and audit these programmes.

At its core is ISO 14001, which provides a framework for the development of an environmental management system, encompassing environmental auditing, environmental labelling and environmental performance evaluation as well as life cycle assessment. Compliance with the series is certified by a third party. Other standards in the series relate to:

- Guidance on the development and implementation of an EMS (ISO 14004);
- General principles of environmental auditing (ISO 14010, now superceded by ISO 19011);
- Specific principles on environmental auditing (ISO 14011, now replaced by ISO 19011);
- Qualification criteria for environmental auditors and lead auditors (ISO 14012, now superseded by ISO 19011);
- Audit programme review and assessment materials (ISO 14013/5);
- Environmental labelling (ISO 14020+);
- Performance targets and monitoring within an EMS (ISO 14030+);
- Life cycle assessment (ISO 14040);
- ISO 14015, Environmental assessment of sites and organizations;
- ISO 14020 series (14020 to 14025), Environmental labels and declarations;
- ISO 14030, which discusses post-production environmental assessment;
- ISO 14031, Environmental performance evaluation – Guidelines;
- ISO 14050, Terms and definitions;
- ISO 14062, which discusses making improvements to environmental impact goals;
- ISO 14063, Environmental communication – guidelines and examples;
- ISO 14063, Measuring, quantifying, and reducing greenhouse gas emissions; and
- ISO 19011, which specifies one audit protocol for both 14000 and 9000 series standards together (replacing ISO 14011 meta-evaluation).

These standards provide organizations with a structured management system which assists them in protecting the environment while carrying out their activities. The implementation of ISO 14001 first requires the development of a clear and comprehensive environmental policy. For its implementation, it requires a planning process, a procedure for audits and checks to ensure compliance, corrective action where necessary, emergency planning procedures and regular reviews from the management. ISO 14000 sets out voluntary standards to be implemented by the organization, which are then subject to external verification and evaluation.

The system brings many benefits to the organization, including:

- compliance with environmental laws;
- a defense of due diligence should there be an accident or incident;
- cost savings through cleaner production, resource conservation and waste minimization;
- cost savings through reduced insurance premiums;
- improvement of industry–government relations;
- enhancement of public image; and
- opening of more markets that look for 'green' production.

In other words, an EMS will help a company improve its cost performance by reducing wastage, reducing its environmental risks and potential liabilities and finally, by enhancing its public image. The ISO 14000 series was initially targeted at corporations in the production chain, but it can apply to any organization, including service organizations and even schools and institutions. In recent years, ISO 14001 has been applied to local governments and cities, particularly to Japanese cities.[50] It must be emphasized that the series is entirely voluntary and that much depends on the goals or objectives of the organization, whether it is an industrial manufacturer, a service provider, a local government or a city.

The Implementation of ISO 14001 in Cities

A number of cities and prefectures in Japan have obtained ISO 14001 certification. They include Shirai city, Itabashi ward of Tokyo, the Tokyo Metropolitan Government and Gifu Prefecture. The benefits of certification in this context have been identified as comprising internal and external components. Internally, it creates a structured management as well as information system from which a cycle of continuous improvements can be established, leading to savings in resources as well as costs. Externally, it demonstrates a city's 'green face' to its citizens, helps to emphasize the need for greater action on the part of urban stakeholders and serves as a model for emulation by other urban governments as well as other stakeholders, particularly private-sector businesses and industries.[51]

Beyond ISO 14001

While obtaining ISO 14001 certification is an indication that a city has an EMS in place, it is submitted that ISO 14001 is lacking in that it is pollution centred and does not take a holistic look at the environment, particularly the ecological processes and systems. In the context of sustainable cities, while the management

---

[50] "ISO and Japanese Cities – The ISO 14001 Initiatives of Public Authorities in Japan", which examines four case studies – Shiroi Town, Tokyo's Itabashi ward, Tokyo Metropolitan Government and Gifu Prefecture, see http://gdrc.org/uem/observatory/iso-cover.html.

[51] Hari Srinivas and Makiko Yashiro, "Cities, Environmental Management Systems and ISO 14001: A View from Japan", http://www.gdrc.org/uem/observatory/seoul-iso14001.PDF.

of pollution is of great importance, so, too, is the need to sustain the natural components of the environment, as well as its historical and cultural heritage. The preservation of nature improves the quality of life for the people, reducing the harshness of the urban environment and providing a refuge from the stresses of urban living. It also assists in the preservation of natural resources and plays an important role in achieving sustainability. The preservation of its historical and cultural heritage sustains the soul of the city.

The development needs of the city must, therefore, be balanced with the needs of the natural as well as the man-made environment. As far as possible, nature should be preserved within a city. This can take many forms, such as the preservation of nature reserves and of natural open spaces such as mangroves and wooded areas; the reservation of green areas as parks and gardens, as well as the provision of green corridors connecting parks, gardens and recreational areas and providing a refuge for wildlife in an urban environment. It should extend to the planting of trees and shrubs in the city, along roads, on balconies of houses, and even on rooftops. It should also extend to the waterways. The city should be landscaped, and this should extend to industrial, commercial and residential premises. The right species of trees and shrubs should be planted, species that are indigenous to the country and which will provide food for the birds, insects, animals and other life forms. This should be facilitated by the planning process and by the passing of laws that should make the cutting down of particular trees and plants an offence, and make the restoration of the natural environment mandatory. [52]At the same time, there should be laws to ensure the preservation of historic, religious and cultural sites and artefacts, as well as the building of museums and centres for the arts to enrich the lives of the residents.

V. The Association of Southeast Asian Nations (ASEAN) Framework for Environmentally Sustainable Cities

The cities of ASEAN (Association of Southeast Asian Nations)[53] have sought to develop a framework for sustainable cities. At a 2003 workshop for 'Environmentally Sustainable Cities in ASEAN' held in Singapore,[54] all 10 nations of ASEAN identified three major environmental problems as priority areas. These were:

1. air pollution from industries and motor vehicles;
2. the lack of infrastructure for sewerage and drainage; and
3. the lack of management for solid waste.

---

[52] Koh Kheng Lian, "Fashioning Landscape for the 'Garden City'", *Landscape Conservation Law – Present Trends and Perspectives in International and Comparative Law*, IUCN Environmental Policy and Law Paper No. 39, (2000), p. 39–47. See also Lye Lin Heng, "Landscape Protection Laws in the Evolution of Modern Singapore". *Landscape, Nature and Law*, ed. Antonio Herman Benjamin (2005): 119 134. Sao Paolo, Brazil: Laws for a Green Planet Institute.

[53] The Association of Southeast Asian Nations (ASEAN) comprises Brunei, Cambodia, Indonesia, Laos, Malaysia, Myanmar, Philippines, Singapore, Thailand and Vietnam.

[54] *Framework for Environmentally Sustainable Cities in ASEAN*, ASEAN Working Group on Environmentally Sustainable Cities in ASEAN, December 2003, Singapore.

These translate to the challenges of providing clean air, clean water, and clean land. The workshop drew up a strategy for specific achievements within these three goals, taking the view that these 'brown' issues must first be addressed and that the 'green' and 'blue' issues can be addressed later.[55] It is submitted that, in the case of ISO 14001, the environment should be tackled as a whole and that ecological considerations should be integrated into the environmental management plans at the outset. Additionally, these efforts should be integrated, with a national sustainable development strategy aimed at eradicating poverty, and linked with an ecosystem approach.[56]

VI. Conclusion

While the issue of the sustainability of cities is highly complex and dependent on many factors, it is submitted that it can be evaluated and the criteria for sustainable cities emerge. This paper has attempted to lay down the basic requirements for an Environmental Management System (EMS) in the context of a city. Whether an EMS can be effectively implemented depends on many factors, particularly the political, social and economic contexts of that city. An effective EMS is a step – indeed, a very significant step – towards sustainability. It is part of good governance. Good governance requires an honest and capable government, with political will. How can this be found, in developing economies which are mired in poverty with a poorly educated populace? Are particular forms of government more effective in managing the environment? These issues, though pertinent, are beyond the scope of this paper. What is clear is that all agencies involved in the management of a city, as well as each individual corporation and inhabitant, should be educated and informed about the environmental challenges that are present as well as forthcoming. An integrated, holistic approach should be taken, with responsibilities assumed by each person. Here, the aspirations contained in the Earth Charter are worth repeating and should be adopted. The Earth Charter was drafted over five years (1995–2000) by a broad spectrum of people 'to promote the transition to sustainable ways of living and a global society founded on a shared ethical framework that includes respect and care for the community of life, ecological integrity, universal human rights, respect for diversity, economic justice, democracy and a culture of peace'.[57] It lays down strong ethical foundations of environmental stewardship and is an inspiration and a guide towards sustainability.[58]

---

[55] Paragraph 5, NEA Workshop Information Paper.

[56] Anantha K Duraiappah Exploring the Links – *Human Well-being, Poverty and Eco-system Services*, UNEP's Institute for Sustainable Development, www.iisd.org/publications.

[57] See the Earth Charter Initiative http://www.earthcharterinaction.org/content/.

[58] The Earth Charter is "a declaration of fundamental principles for building a just, sustainable, and peaceful global society in the 21st century. It seeks to inspire in all peoples a new sense of global interdependence and shared responsibility for the well-being of the human family and the larger living world. It is an expression of hope and a call to help create a global partnership at a critical juncture in history." http://www.earthcharter.org/files/charter/charter.pdf.

'We stand at a critical moment in Earth's history, a time when humanity must choose its future. As the world becomes increasingly interdependent and fragile, the future at once holds great peril and great promise. To move forward, we must recognize that, in the midst of a magnificent diversity of cultures and life forms, we are one human family and one Earth community with a common destiny. We must join together to bring forth a sustainable global society founded on respect for nature, universal human rights, economic justice and a culture of peace. Towards this end, it is imperative that we, the peoples of Earth, declare our responsibility to one another, to the greater community of life, and to future generations.

We urgently need a shared vision of basic values to provide an ethical foundation for the emerging world community. Therefore, together in hope, we affirm the following interdependent principles for a sustainable way of life as a common standard by which the conduct of all individuals, organizations, businesses, governments and transnational institutions is to be guided and assessed.

Principle I – Respect and Care for the Community of Life

1. Respect Earth and life in all its diversity.
2. Care for the community of life with understanding, compassion and love.
3. Build democratic societies that are just, participatory, sustainable and peaceful.
4. Secure Earth's bounty and beauty for present and future generations.

Principle II – Ecological Integrity

5. Protect and restore the integrity of Earth's ecological systems, with special concern for biological diversity and the natural processes that sustain life.
6. Prevent harm as the best method of environmental protection and, when knowledge is limited, apply a precautionary approach.
7. Adopt patterns of production, consumption, and reproduction that safeguard Earth's regenerative capacities, human rights, and community well-being.
8. Advance the study of ecological sustainability and promote the open exchange and wide application of the knowledge acquired.

Principle III – Social and Economic Justice

9. Eradicate poverty as an ethical, social and environmental imperative.
10. Ensure that economic activities and institutions at all levels promote human development in an equitable and sustainable manner.
11. Affirm gender equality and equity as prerequisites to sustainable development and ensure universal access to education, health care, and economic opportunity.
12. Uphold the right of all, without discrimination, to a natural and social environment supportive of human dignity, bodily health and spiritual well-being, with special attention to the rights of indigenous peoples and minorities.

Principle IV – Democracy, Non-violence and Peace

13. Strengthen democratic institutions at all levels and provide transparency and accountability in governance, inclusive participation in decision-making, and access to justice.
14. Integrate into formal education and life-long learning the knowledge, values and skills needed for a sustainable way of life.
15. Treat all living beings with respect and consideration.
16. Promote a culture of tolerance, nonviolence and peace.

Finally, it is fitting that the Charter declares that 'Every individual, family, organization and community has a vital role to play. The arts, sciences, religions, educational institutions, media, businesses, non-governmental organizations and governments are all called to offer creative leadership. The partnership of government, civil society and business is essential for effective governance'.

References

Alusi A, Eccles RG, Edmonson AC, Zuzul T (2010/2011) "Sustainable cities: Oxymoron or the shape of the future?" Harvard Business School Working Paper 11–062.

Angel DP, Rock MT (eds) (2000) *Asia's Clean Revolution – Industry, Growth and the Environment.* Greenleaf Publishing.

Bai X, Imura H (2000) A Comparative study of urban environment in East Asia: Stage model of urban environmental evolution, (IGES). International Review for Environmental Strategies 1(1), 135–8.

Basiago AD (1999) Economic, social and environmental sustainability in development theory and urban planning practice. The Environmentalist 19, 145–61.

Brundtland GH (1987) *Our Common Future – Report of The World Commission on Environment and Development.*

Cambessus M, Winpenny J (2003) *Report of the World Panel on Financing Water Infrastructure: Financing Water For All,* World Water Council, 3rd World Water Forum, Global Water Partnership.

Campbell S (1996) Planning: Green cities, growing cities, just cities? Urban planning and the contradictions of sustainable development. Journal of the American Planning Association 62(3), 296–312.

Cooper R, Evans G, Byoko C (2009) *Designing Sustainable Cities.* Blackwell Publishers.

Craig D, Robinson NA, Lian KK (eds) (2002) *Capacity Building for Environmental Law in the Asian and Pacific Region – Approaches and Resources,* Vols. I and II. Asian Development Bank.

Dubois-Taine G (1992) 'Introduction' Cities of the Pacific Rim – Diversity and Sustainability. In: Dubois-Taine G, Henriot C (eds) *Environmental Problems in Third World Countries*. Pacific Economic Cooperation Council Sustainable Cities Taskforce, Earthscan Publications, London.

Dubois-Taine G (2002–4) Pacific Economic Cooperation Council Sustainable Cities Taskforce Coordinator, Reports from Seminars on Sustainable Urban Services held in Bangkok, Hong Kong, Santiago (Chile), Noumea, Shanghai and Adelaide (2002–4).

Duraiappah AK *Exploring the Links – Human Well-being, Poverty and Eco-system Services*. UNEP's Institute for Sustainable Development. Available from: www.iisd.org/publications.

Evans B, Joas M, Sundback S, Theobald K (2009) *Governing Sustainable Cities*. Earthscan Publications, London.

Fernandes E (ed) (1998) *Environmental Strategies for Sustainable Development in Urban Areas – Lessons from Africa and Latin America*. Studies in Green Research.

Foley D (1960) British town planning: One idealogy or three? British Journal of Sociology 2.

Fook LL, Gang C (2010) *Towards a Livable and Sustainable Urban Environment: Eco-cities in East Asia*. World Scientific.

Graedel TE, Allenby BR (2003) *Industrial Ecology*, second edition. Prentice Hall, Upper Saddle River, NJ.

Hardoy JE, Mitlin D, Satterthwaite D (1992) *Environmental Problems in Third World Cities*. Earthscan Publications, London.

Hari Srinivas and Makiko Yashiro, "Cities, Environmental Management Systems and ISO 14001: A View from Japan", http://www.gdrc.org/uem/observatory/seoul-iso14001.PDF.

Heart Work 2 – EDB & Partners: New Frontiers for the Singapore Economy (2011) Straits Times Press and EDB.

Lye LH (2007) Land-use planning, environmental management and the garden city as an urban development approach. In Nolon J, Chalifour N, Lye LH, Kameri-Mbote LL (eds) *Land Use for Sustainable Development*. Cambridge University Press, New York.

IGES (2000) International Review for Environmental Strategies – Inaugural Issue. IGES.

Inoguchi T, Newman E, Paoletto G (eds) (1999) *Cities and the Environment: New Approaches for Eco-societies*. United Nations University Press.

IUCN, UNEP, WWF (1991) Caring for the Earth: A Strategy for Sustainable Living.

Janssens M (2009) *Sustainable Cities: Diversity, Economic Growth and Social Cohesion*. Edward Elgar, Great Britain.

Koh KL (2000) Fashioning Landscape for the `Garden City. In: *Landscape Conservation Law – Present Trends and Perspectives in International and Comparative Law*, IUCN Environmental Policy and Law Paper No. 39, pp. 39–47.

Ooi GL (ed) (1999) *Model Cities – Urban Best Practices*, Vols. I and II.

Lye LH (2005) Landscape protection laws in the evolution of modern Singapore. In: Benjamin AH (ed) *Landscape, Nature and Law*, pp. 119–34. Laws for a Green Planet Institute, Sao Paolo, Brazil.

Malone-Lee LC (2002) Environmental planning. In: Craig D, Robinson NA, Lian KK (eds) *Capacity Building for Environmental Law in the Asian and Pacific Region – Approaches and Resources*, Vols. I and II, p. 606. Asian Development Bank.

Mega VP (2010) *Sustainable Cities for the Third Millennium: The Odyssey of Urban Experience*, Springer Science and Business Media LLC.

Mega V (2011) *The Desirable Future of Innovative Cities: Cities in Harmony with Nature, People and Society.* Lambert Academic Publishing.

Mitlin D, Satterthwaite D (eds) (2004) *Empowering Squatter Citizen – Local Government, Civil Society and Urban Poverty Reduction.* Earthscan Publications, London.

OECD (2001) *Strategies Towards Sustainable Development.* OECD. Available from: http://www.oecd.org/dataoecd/34/10/2669958.pdf.

Plummer J (2002) *Focusing Partnerships: A Sourcebook for Municipal Capacity Building in Public-Private Partnership.* Earthscan Publications, London.

Pugh C (ed) (2000) *Sustainable Cities in Developing Countries.* Earthscan Publications, London.

Rees W (1996) Revisiting carrying capacity: Area-based indicators of sustainability. Population and Environment: A Journal of Interdisciplinary Studies 17(2).

Rogers R (1997) *Cities for a Small Planet.* Faber & Faber, London.

Satterthwaite D (ed) (1999) *The Earthscan Reader in Sustainable Cities.* Earthscan Publications, London.

Singapore EDB, EDB Society (2002) *Heart Work – Stories of How EDB Steered the Singapore Economy from 1961 to the 21st Century.*

Tan AKJ (2002) Recent institutional developments on the environment in Southeast Asia – A report card on the region. Singapore Journal of International and Comparative Law 2.

Transport and Communications for Urban Development – Report of the Habitat II Global Workshop, 1995 UNCHS (Habitat), 1997.

UNEP (1997) UNEP Environmental Law Training Manual. UNEP.

Walter B, Arkin L, Crenshaw R (eds) (1992) Sustainable Cities: Concepts and Strategies for Eco-city Development. Eco-Home Media, Los Angeles.

TIME (2011) "Breaking free – How one man turned Bihar, once India's poorest and most lawless state, into a model of reform for the entire country. Time, pp. 26–9, 7 November 2011.

Planning Sustainable Cities: Global Report on Human Settlements, 2009 (UN Human Settlements Programme). Earthscan Publications, London.

Sustainable Cities, Parliament of the Commonwealth of Australia, August 2005. Canberra.

CHAPTER THIRTEEN

Environmental Management: Challenges of the Next Decade

Nicholas A ROBINSON
University Professor, Pace University, New York, USA

"How worldwide muddling toward the green economy will evolve is unclear, but what is clear is that the knowledge and skills of the environmental manager will be essential to the success stories. As the importance of environmental management becomes better understood, there will emerge a demand for more graduates in this field. While lawyers and MBAs will come to work with environmental managers, they will not supplant this managerial force. Indeed, one can imagine in this century that schools of environmental management or sustainability will eclipse the still powerful magnet of the business school. The contemporary business school model is so embedded in "business as usual" that it resists taking on the profound reforms that sustainability requires."

–Nicholas A ROBINSON

Environmental Management:
Challenges of the Next Decade

Nicholas A ROBINSON[1]

University Professor, Pace University, New York, USA

Abstract

The skills and knowledge acquired though the still new field of environmental management are fundamental for attaining sustainable development in all sectors. The MEM discipline around the world extends beyond the focus on environmental management systems (EMS) or environmental impact assessment procedures (EIA). These are core elements, to be sure, as was made clear in the Declaration on Environment and Development at the 1992 Rio de Janeiro Earth Summit. But more is evident. The MEM studies produce environmental leadership in all sectors, building toward the shift to the "green economy" that Earth must have as we add two billion more people to the planet in the coming 50–75 years. For these reasons, the next decade is particularly important. Societies will need to move environmental management from a sub-field of management to a central point of authority. Events are demanding adaptation to the effects of climate change, and environmental management has fashioned the analytic tools, proven reform programs, and long-term vision to enable societies to sustain the resilience needed in order to weather the coming changes.

Key words: sustainability, environmental management, green economy, resilience, waste

I. Introduction

Today, the National University of Singapore (NUS) Masters of Environmental Management is a leader in an international academic effort to prepare a new generation of specialists who will guide the transition to a sustainable society. Universities address societal sustainability from different perspectives, but they all embrace the same core knowledge, methodologies and skill sets, synthesized at a *Meta* level from usually the same contributing disciplines. The NUS Masters in Environmental Management shares much in common with comparable degrees, such as those at Yale University's School of Forestry & Environmental Studies; with Columbia University's Masters in Sustainability Management; with Duke University's Master of Environmental Management, with Natural Resource & Environment Masters at the University of Michigan (Ann Arbor); and with

---

[1] School of Law, Pace University. 78 North Broadway, White Plains, New York 10603, USA. Email: nrobinson@law.pace.edu; Tel: (+1) 914-422-4244.

the Masters of Environmental Science and Management at the University of California (Santa Barbara); the Masters of Public Policy and Urban Planning at Harvard University (offered by the Kennedy School of Government and the School of Design), the forthcoming Applied Environmental Management Leadership Masters of Pace University, or the sustainability management degree programmes at the Ashridge School of Management in England.

What all these degree programmes have in common is that they each have embarked upon a rigorous exploration of how to give substantive definition to the objectives that governments and civil societies globally have espoused in calling for 'sustainable development'. Sustainable development was first advanced in 'Caring for the Earth,' a programme of the International Union for the Conservation of Nature & Natural Resources and taken up by the United Nations Commission on Environment and Development, in its report entitled *Our Common Future* of 1987 and thereafter by the United Nations General Assembly in *Agenda 21* in 1992 (Robinson, 1993).[2] Today, three decades into seeking to establish sustainability practices, it is plainly not sufficient to simply declare that sustainability means ensuring that future generations have the same capacity for well-being that the present generations enjoy. One must also demonstrate how to do so and then how to maintain and sustain the systems doing so. This is what all participants in the NUS environmental management programmes address. It is the shared focus of all the various university graduate programmes in environmental sustainability.

In reflecting on what the study of environmental management has become and what challenges it will face in the coming decade, five topics may be examined: (a) coping with the growing prominence of environmental management; (b) articulating the core methodologies that belong more to the field of environmental management, with their underlying knowledge, than to other departments of study; (c) examining how the global press to the 'green economy' will call for greater use of environmental managers; (d) elaborating new curricular components that environmental management programmes will be called upon to address and, in particular, the need to add studies in sustainability finance to the environmental management programmes; and (e) elaborating further the guiding principles for meeting the new challenges that sustainability programmes inevitably will address.

II.   Environmental Management as a Matured Field

Sustainability management integrates knowledge and skills across sectors, from biology to engineering, from social sciences to law, from finance to energy systems. Today, this new synthesis is popularly urged as the 'green economy' (United Nations Environment Programme, 2011), or at the cutting edge, is seen in

---

[2] Approved at the UN Conference on Environment and Development (UNCED) in 1992, under the effective leadership of the Conference Chairman, Ambassador and Professor Tommy Koh, and thereafter adopted by the UN General Assembly.

proposals for a more profoundly transformative 'blue economy' (Pauli, 2010). The environmental manager has a doubly difficult mission: to apply and put in practice what environmental sustainability requires and to educate all other managers about these requisites. The environmental manager is both a doer and a teacher, in short, a leader, a capacity builder. A 2009 survey by the Massachusetts Institute of Technology and Boston Consulting Group asked: What will organizations need to be good at in order to thrive in the emerging sustainability economy? The responses: integrate sustainability in to strategy, understand integrated systems, collaborate with stakeholders, embrace and employ long-term measurements and reporting systems (MIT Sloan Management Review). These are the management systems that progressive multinational companies have fashioned and shared through the World Environment Center and the World Business Council for Sustainable Development. Their roots go back to Joe Ling and the 3M Company's insightful 3P programme 'Pollution Prevention Always Pays'. Properly perceived, the environmental manager is a profit centre, not a cost to a company.

Today, companies routinely issue sustainability reports, and the use of these reports is mandated in the European Union. Their metrics are as important to the company's bottom line as are the fiscal audits and financial reports. They have established, amply, that it is a falsehood to suggest that there is a trade-off between environment and development, or between the environment and the economy. Those who continue to repeat facile arguments, that we cannot afford environmental protection, deny society the benefits of sustainability and labour in the self-induced fog of ignorance. The challenging task of the environmental manager is to dispel this fog and expose the error of such ideological arguments. The environmental manager is not destined to be partisan in a debate about philosophical questions. Rather, he leads by making and applying the empirical case, amassing the facts, sharing the data and case studies as well as establishing and measuring sustainability reforms that confirm the value and effectiveness of sustainability systems.

Cohen (2011: 147), Director of the Masters of Sustainability Management at Columbia University in New York, has ably summed up the attributes of the environmental manager as follows:

> 'We will also need to develop a new kind of management capacity in modern organizations, which will be led by people trained in the field of sustainability management. Most people who lead large and complex organizations are either the lawyers or MBAs. The lawyers are not trained to manage organizations, and the MBAs learn management and finance but have not training in the physical dimension of sustainability. They do not know the science of environmental impacts or the engineering of energy or resource efficiency. Sustainability managers must know enough of each of these fields to draw on and translate the work of others. They must also be able to build staff capacity to analyze and implement organizational changes designed to reduce resource use while building production.

The organizational capacity for managing sustainability requires an ability to understand and analyze new technology, analyze and apply public policy rules and incentives, analyze financial costs and benefits, and management organization change. It requires constant learning, analysis, and teaching. It requires an understanding of a variety of fields and appreciating how to draw in and elicit work from many types of experts. At the global level, it also requires the ability to interact with partners from all over the world and benchmark operational lessons from distant locations'.

As university programmes in environmental management expand globally, NUS will need to reflect on its own growth and benchmark its own capacity and quality to differentiate itself from others while ensuring always the highest academic standards. What should be the niche for the NUS masters of environmental management 10 years from now? Ultimately, NUS will need to test itself by the sustainability concepts and knowledge that it studies and teaches.

III. Core Knowledge and Methodologies for Sustainability Management

University programmes in environmental management today are producing graduates with the capabilities that Cohen (2011) described. The graduates are being employed, and there is an unfilled market demand for more such highly qualified individuals. These graduate programmes are also establishing research and teaching about the management systems that such practitioners of sustainability employ. Not every programme gives the same weight to the different methodologies, but there are core subjects that are increasingly necessary for managers to know if they are to be effective in implementing sustainability objectives. When environmental management programmes are compared, it is important to examine how the following bodies of knowledge and methodologies are taught and what research is being undertaken to critique and refine them.

1. **The Evolution of Concepts of Sustainability** – From the awareness of George Perkins Marsh and the emergency of doctrine and study of the conservation of nature and natural resources to the later concepts of environmental pollution controls. These cannot be divorced from the understanding of the emergence of ecology and the environmental sciences, and of environmental engineering.

2. **Industrial Ecology** – The study of how governmental and commercial and social systems can be understood as analogous to ecosystems by designs that manage the flows of energy and materials, using and reusing assets, and avoiding wasteful inefficiencies.

3. **Environmental Management Systems (EMS)** – As employed by many companies and governments, an enterprise's EMS integrates analysis and

compliance with all environmental stewardship tasks and obligations. EMS is useful in non-profit organizations, in universities, in military departments, in agencies of government, as well as in companies. EMS is increasingly being reshaped as a 'Sustainability Management System' and is always adapted to the special characteristics of the entity employing the system.

4. **International Standards Organization (ISO) 14000 Series** – The ISO system of auditing compliance with environmental and other obligations is a well-used system but essentially measures the routine compliance with environmental rules in place when a facility is located. As such, ISO 14000 tests minimal expectations, and needs to be married with broader concepts of sustainability. Many have adapted the essence of ISO 14000 to a deeper or wider analytic audit.

5. **Annual Sustainability Reporting** – The metrics and the information gathering and assessing systems are essential to an understanding of environmental sustainability. This necessarily includes understanding how to maintain the independence of the reporting and analysis and understanding the importance of designing a system of multi-year measurements, which can mesh with and guide the financial reporting. Analytic approaches for identifying obsolete but still used metrics that falsely report the well-being of an enterprise or agency are important.

6. **Supply Chain Management** – Establishing the standards and contractual obligations and capacity building measures to ensure that all links in the supply systems have their own EMS and quality assurance. Procurement standards for shifting to ever more sustainable designs and products are important.

7. **Environmental Impact Assessment (EIA)** – Procedures for undertaking, communicating and constantly learning from and improving EIA are fundamental to sustainability, and need to be tools not just used by governmental authorities. Private-sector EIA needs to be advanced; at the same time, greater rigour and transparency are advanced within governmental agency EIA. Comparative EIA is important to learn how others have improved their EIA systems and adapted the reforms. EIA needs to be compared with the narrower body of lore and practice associated with 'cost/benefit' analysis.

8. **Technology Assessment** – The more specialized techniques for evaluating proposals for new technologies and their applications are significant components of building sustainability. How technological innovations are evaluated over time and how their use is adjusted in light of unintended consequences are important. This early work of the Congressional Office of Technology Assessment needs to be re-evaluated. Management techniques for using practical assessments give realistic effects to the 'Precautionary Principle' to get beyond unsubstantiated fears about new technologies and, at the same time, avert unintended consequences.

9. **Public Participation in Environmental Decision-Making** – The rationale for a means by which all stakeholders are invited to participate in decisions affecting the public and associated interests. This includes access to environmental information and how such information is disclosed and accessed.

10. **Life Cycle Analysis** – The criteria and methods of evaluating the complete life cycle of a product or of a service are essential tools for the environmental manager. Some specialized systems have emerged for given sectors, such as the Green Buildings Council's LEED programme for energy-efficient new construction. Feedback systems facilitating continual updates of standards, so as to take in to account new technological innovations or understanding, are important.

11. **Environmental Law Norms and Procedures** – Nations continue to elaborate systems for environmental law. From 1970 to 1990, the USA provided innovative models, and the declaration of the 1990s saw developing nations take up the innovative challenge. Since 2000, it has been the European Union that has pioneered refinements in environmental laws. Since the environmental manager works across all national borders, there is a need for him to learn and keep abreast of environmental law developments, since the law is the vehicle by which society defines and establishes sustainability norms.

12. **Ethical and Legal Foundations** – Knowledge of the discipline of environmental ethics, and experience with evaluating ethical challenges in practice are essential for the environmental management. An understanding of environmental law, both national and the system of multinational environmental agreements, is as important as the understanding of environmental justice and the trends in developing more rigorous environmental norms.

Each of these curricular themes will change over time; some may break into subfields or be merged into a high order of integration. Since sustainability is contextual, there also will be other subjects appropriate to the location or characteristics of the entity employing the environmental manager. Therefore, techniques such as these will be supplemented by other subjects. Indeed, the environmental manager needs to have an aptitude for innovations and be able to fashion new sustainability techniques as the need arises. Case studies in how to innovate for sustainability are usefully a part of all these masters programmes. Experiential learning and field trips to study sustainability systems in practice are also key components for learning environmental management.

Columbia's Cohen (2011: 26) identifies one case study that illustrates how many of the above environmental management methodologies are employed. Cohen restates the case study as follows: 'In 2005, Saijo (2005) of Japan for Sustainability presented an excellent case study of the effort of Fuji Xerox Company to redesign the office copier. Its term for green manufacturing is what they call 'inverse manufacturing'. According to Saijo, this is 'a manufacturing system that considers the entire product life (planning and designing, manufacturing, use and disposal) at the design phase and incorporates into the design consideration of the process of collecting used products, dismantling, separating and reusing them in the form of components and materials... Inverse manufacturing minimizes the consumption of resources, energy and volume of waste while also creating value... Inverse manufacturing has been propounded as a model for sustainable manufacturing over the past dozen years or so, but practising it is not easy and only a limited number of companies have succeeded in adopting it as a business model'.

IV. Transitions to the Green Economy

The environmental manager must be equally conversant with the causes of The Great Recession of 2008 (Robinson, 2009) and the fourth Assessment Report of the Intergovernmental Panel on Climate Change or the Millennium Ecosystem Assessment. Economy and ecology are both facets of the same environmental management conundrum. As the human population grows in number from seven billion currently to the United Nation's projection of nine billion in the coming two or more decades, the financial and environmental stresses on Earth will require pervasive changes in human socioeconomic systems. There are myriad calls for a transition to convert current unsustainable systems into a 'green economy', which eliminates waste, maximizes recycling, requires restoration of degraded natural systems and requires energy self-sufficiency from renewable sources. Embedded investments and commitments to unsustainable economic models will make this transition halting and problematic. The environmental manager will be essential in guiding the steps to the green economy and to coping with the resistance to such steps.

The outline of sustainability, popularized by *Our Common Future* and *Agenda 21*, has been overtaken by events. The concepts and reforms espoused in these seminal works are not obsolete, but rather the failure to attain sustainable systems at the end of the 20th century means that human society must design and undertake pervasive new measures in the 21st century. At the same time, new technologies have become available to facilitate these undertakings, such as the Internet and ever more advanced computer modelling capacity, and communications systems such as social networking. The elaboration of the system of multinational environmental agreements allows for a concerted harmonization and integration of national environmental laws and policies to move towards more sustainable systems.

Yet the steps to build a green economy are not likely to come from above. Too many incremental decisions must be made at all levels, and local authorities will take the lead. The transition will be disaggregated and be from the bottom up. This will be facilitated by rapid sharing of information about successful measures in some places, and their adaptation and emulation in other places. Already, the mayors of local authorities are collaborating to mitigate emission of greenhouse gases and institute comparable adaption measure cope with new ambient conditions appearing in the wake of climate change.

To be sure, the use of green technologies will be fostered by the International Renewable Energy Agency, whose headquarters is in Abu Dhabi, and cities will emulate the innovative design of Masdar to meet habitation needs with less energy and impacts. Green walls and roofs on buildings and vast afforestation measures on landscapes are emerging locally, without waiting for any international agreements. Local measures for local generation of electricity, from solar or kinetic to wind energy, diminish the need for fossil fuels or electrical distribution systems. The move to the green economy will be incremental, local and of uneven speed. As nations follow the European Union's concept of 'subsidiarity', they will favour and foster such innovative measures locally and authorize them. At the same time, decentralization in developing nations must be accompanied by efforts to strengthen the rule of law, lest local authorities repeat patterns of corruption learned from above, as nations such as Indonesia or the Philippines have learned. The establishment of environmental courts and tribunals is one way to build the culture of the rule of law.[3] Without the rule of law, sustainability norms are corrupted also.

How worldwide muddling towards the green economy will evolve is unclear, but what is clear is that the knowledge and skills of the environmental manager will be essential to the success stories. As the importance of environmental management becomes better understood, there will emerge a demand for more graduates in this field. While lawyers and MBAs will come to work with environmental managers, they will not supplant this managerial force. Indeed, one can imagine in this century that schools of environmental management or sustainability will eclipse the still powerful magnet of the business school. The contemporary business school model is so embedded in the 'business as usual' concept that resists taking on the profound reforms that sustainability requires.

In short, the transition to the green economy poses pressures that existing environmental management educational programmes will need to address in the coming decade. First, they will find a need to refine, render more effectively, and expand universally the use of the existing core EMS outlined above. It must be admitted that many of these systems today operate at the margins of decision-making in all socio-economic sectors. As these sectors seek more sustainable models, the inadequacy of contemporary environmental management tools will become more evident. Reforms and enhanced rigour will become a priority in research and teaching of today's integrative sustainable energy/environment management programmes. Second, this work will be difficult in itself, but these time-consuming tasks unavoidably take place at the same time the same programmes innovate to confront new and transformative physical conditions emerging as a result of anthropogenic climate change, human migration and demographics, and materials shortages.

---

[3] See the Journal of Court Innovation (NYS Judicial Institute and Pace University) and the conference papers on environmental adjudication (April 2011) assembled at www.law/pace/edu.

In short, the academic establishment of sustainability studies and the emergence of the environmental manager are not just illustrations of new wine in old bottles. The green environment is not just the adoption of a few new technological systems, although many may regard it as such. The environmental manager's knowledge and skills do not indulge such shallow doubts. The environmental managers will be worked overtime to build and refine the green economy.

V. Meeting Fundamental New Challenges

If university environmental management programmes only had to refine and establish more widely the use of sustainability metrics and methodologies, one can imagine the growth of new faculty departments or schools, and formal academic recognition of Environmental Sustainability, as another historic example of how universities have reinvented themselves for each era since their emergence in the Middle Ages in Europe. Berry (2000) contemplated this environmental mission for universities before the full dimensions of the anthropogenic transformation of Earth had become as evident as they are today.

The Antarctic ice core reveals 420,000 years of climate conditions. The current levels of carbon dioxide in the atmosphere are higher than those at any time in this record. The warmer atmosphere is accompanied by the melting of the cryosphere and the consequent rise in sea levels, perhaps 1–2 metres in the coming century. Already, the Dutch Government and Rhine Delta Commission have begun implementing a 200-year plan to protect the Netherlands from the effects of sea level rise and the deeper fluctuations of floods, droughts and temperatures. Environmental managers will need to study the Dutch innovations and adapt them elsewhere. At the same time, China is completing the development of consciously designed 'eco-cities', capable of accommodating large numbers of residents without the sort of environmental impacts that existing cities entail. Environmental managers will need to study the Chinese innovations and adapt them elsewhere. Most human populations live along the coasts of the oceans, and there will be a migration from the coasts, as sea levels rise. Environmental managers will need to guide the infrastructure relocations and engage in a new field that may come to be known as 'coastal morphology', or the human reshaping of the coasts ahead of sea level rise in order to accommodate planting new wetlands, provide habitats and places for biological systems to function and make provisions for human settlements as well as all the traditional uses for commerce, recreation and other ends. Environmental managers will need to study the innovations and experience of Singapore, which has decades of experience in creatively reshaping its coasts.

The entire role of cities, as innovators of new environmental management programmes, will become increasingly important for the environmental manager. How can community gardens or vertical farms become suppliers of locally produced fresh food, with positive environmental benefits and new adverse environmental impacts? Experiments in Brooklyn, New York, may provide answers. Cities are already initiating artificial experiments in sustainability. How can they harvest rainwater, reuse water locally and become more self-sufficient in water supply? Since cities are the home to most university academic environmental management programmes, how can partnerships in experiential learning be forged with environmental sustainability studies programmes? Cities, by necessity, are eco-innovators, and need environmental managers. The converse is also true. Environmental managers need to learn from cities and share their analyses in academic and other media for others elsewhere to study. This approach goes beyond traditional urban studies, which still have their role. Rather, a sustainability engagement with cities allows the invention of new concepts in human habitats.

The managerial consequences of climate change play out in other environmental management sectors. The role of insurance will grow in importance. EMS will need to teach about casualty insurance to cope with losses associated with weather events, and also index insurance to cope with crop failures as a result of droughts or floods, or provision of micro-insurance so that poor communities in developing countries can also have assets to adapt and rebuild after weather-induced losses. The insurance sectors need to be expanded globally, and appropriate insurance administrative and legal regimes need to be established to ensure the integrity, fiscal propriety and effectiveness of insurance regimes. Insurance works only with well-regulated governmental supervision.

The coming two billion inhabitants of the planet will occupy less land. They will be unevenly concentrated. Some vast areas, such as Russia, have declining population numbers, but all regions will find that their coast populations are relocating inland, and form new dense centres, or cluster into existing centres, making them denser. Managing this new urban shift affords opportunities to implement new sustainability systems. Rigorous use of EIA can permit innovations with every action studied. Resystematizing EMS can permit new or denser human settlements to employ continuous sustainability learning processes. Design for no-waste regimes, with local generation of electricity and water stewardship, can remake supply chains and adopt industrial ecology models as operational standards.

This is the promise of positive change in the coming era of adapting to climatic change. The change in the environment is happening. Whether humans will adapt using the knowledge and skills of sustainability remains a question. The environmental manager knows the answer, if not explicitly, rather than just the knowledge of the factors and pathways to be followed to find the answers. Just as large enterprises such as Johnson & Johnson have a Vice President for Sustainability reporting directly to the Chief Executive Officer, so someday soon, Prime Ministers will have a Deputy Prime Minister for Sustainability, and every agency and local authority with have teams of environmental managers.

To be fully effective in this coming era of environmental managers, however, the environmental manager needs knowledge and skills that most EMS do not currently provide. This topic is *finance*. The masters of environmental management programmes need research about how sustainable financing techniques can provide for the coming transition to a sustainable and 'green' economy. Environmental managers need to know how to make the case to show that the transition can be self-financed and need not be costly. A new kind of finance is needed, not one driven by annual budgets and debt service for sunk costs. Tools to identify and retire perverse subsidies for unsustainable programmes need to be part of the environmental manager's skill set. The savings from eliminating wasteful governmental programmes are not unlike the profit centres that companies established to realize the promise that 'pollution prevention always pays'.

Finance studies in environmental management programmes should identify externalities and identify their costs (Richard *et al.*, 1984). They should also identify and measure the costs of maintaining ambient environmental quality, for clean air or for wetlands, which provide absorption of floodwaters and recharge aquifers. There is a need for user fees and taxes to pay for the true costs of maintaining these environmental services. Just as pollution controls and abatement are needed to eliminate externalities to ensure that harm is avoided by requiring the responsible persons to bear the legitimate economic costs of preventing the harmful externality, so there is also a need to ensure that all share in the modest cost of sustaining the flow of ecosystem services and other environmental benefits. The continued availability of such diffuse environmental benefits cannot be ensured without public support. The public has long paid for establishing and sustaining shared parklands. What is needed is for the environmental manager to help identify other environmental services and explain what must be done to invest in sustaining their continued availability. New York City pays property owners near the city's Catskill reservoirs to avert waste discharges that could enter the reservoir. This approach will need to be expanded to all environmental dimensions. Maintaining upland open-space and wetland areas for retaining and slowly releasing floodwaters is a more cost-effective way to avert flood damage than building dikes and levees downstream. Plus, there are biodiversity values associated with land preservation.

Another illustration is that creative financing addresses the need to retrofit old buildings that waste energy with new energy-saving technologies or with alternative sources of renewable electricity. In keeping with the European Union's idea of subsidiarity, local authorities and local property owners can act to find the funding to make these changes. Metrics exist for an energy audit to determine how much a building retrofit would save in demand for electricity. One can compute the cost of the energy-saving technology installed and how long it takes annually for the savings to cover the costs of the new technology retrofit. A programme entitled 'Property Assessed Clean Energy' (PACE) then allows the property owner to borrow the cost of the new technology and install it. The owner continues to pay the same fee for energy as before, but, since the building uses less energy, the differential between the fees with and without generates a sum of money that is used to pay off the loan. If the owner sells the house, the next owner continues to pay the energy fees at the standard rate until the loan is paid off. At that point, the building owner sees the cost of energy drop for the building, but, all the while, society has had less emissions associated with the generation of the electricity. More than 22 states in the USA have authorized their local governments to provide for PACE.

Most waste systems can be 'mined' for money to pay for the transition to a sustainable economic model. It is wasteful, several times over, to keep paying for wastes. The Chinese model of the 'circular economy' needs to be made operational. Means exist to do so, besides eliminating waste by redesigning or recycling. For example, revolving funds can be established to pay for upgrades to infrastructure, with user fees paying for the upgrades and replenishment of the funds over time. Since 1984, this approach has been used in New York to enable local authorities to restore and maintain their ageing sewage collection pipes, without new taxes.

For sustainable financing to become realistic and widespread, new metrics will be needed. Annual budgets must defer to life cycle budgets for ecosystem services. Long-term planning will need to measure the incremental expenses for sustaining systems, and then user fees or taxes will need to be set up to offset those costs. Systems employing short-term approaches and unrealistic discount rates need to be restructured. The environmental manager needs to be able to make the financial case for sustaining environmental services over the long term, with incremental payments to do so. Similarly, the environmental manager needs to address both the science and the economics of trade-offs, such as sequestration of carbon through photosynthesis in trees, along with the costs for sustaining biological diversity in the same forest. Trade-offs under the UN Framework Convention on Climate Change for Reducing Emissions from Deforestation and Degradation (REDD+) need close study by environmental managers if these new systems are to have integrity and effectiveness. Traders often want short-term transaction profits from negotiating trade-offs without investing the time or effort to ensure a proper biological assessment, and sustainable financing model to pay for the life cycle costs of the trade-off in sustaining the forest systems. The environmental manager needs to know how to quantify the costs on a full-life-cycle basis and to structure the financing to cover those costs, with an adequate margin for error and adequate insurance.

Financing is thus a key component of environmental management and needs to be incorporated into the course of study for preparing environmental managers. We need to mine the profligacy of the waste-dependent business, as a usual model to help pay for the transition to a more sustainable model step by step. The time may be at hand for public sovereign debt to be invested into the transition in each country, rather than in investing in other nation's public debt. The models of economic development adopted during the Second World War are increasingly no longer viable, yet the World Bank's business-as-usual model persists in promoting that model. Since reforms are unlikely from the global level, it is from the local level that 'bottom up' reforms will appear and accumulate.

Local innovations in finance and self-sufficient programmes need to be undertaken and encouraged. Environmental management programmes need to study what works and what does not. They need to evaluate efficiencies of scale, leverage a local success regionally or replicate it locally elsewhere. There is a need to apply a life cycle analysis to scarce natural resources and identify appropriate substitute substances of designs.

VI. Principles to Promote Sustainability

None of these reforms to progressively extend the capacity of environmental management to build sustainability throughout socio-economic systems will be easily realized. They will come haltingly and, in some places, not at all. The inertia of old ways is a powerful force. The vested interests that still derive wealth from unsustainable exploitation of resources will not accommodate shifts to sustainable regimes without a fight.

What then can be done to promote a principle that environmental managers, and all of us, can use to muddle through the coming transitions? The existing principles, as in the Declaration of Rio de Janeiro for Environment & Development, or in the UN General Assembly's World Charter for Nature, of course, need to be reaffirmed. We shall need to stress an anti-backsliding concept, or a principle of non-regression, which obliges everyone not to retreat from the past accomplishment and legal norms for environmental protection and sustainable development. But new global environmental conditions now require the recognition of new principles.

In all the methodologies of environmental management, there is a goal to sustain or restore and maintain the resilience of the natural systems in which human life exists. It is time to acknowledge the Principle of Resilience as a general principle of law. The Principle of Resilience states that every decision must be assessed and be taken in ways that will always strengthen resilience as fully as possible, as appropriate to the context of the decision. The Principle shall be applied to all governmental decision-making. When society takes risks without a fallback position or insurance, it is not respecting the need for resilience. Brittle and exhausted systems lack resilience. Before an act is carried out, an analysis is needed about how to 'bounce back' in case something goes wrong. EIA can and should build resilience thinking into its analysis. All decisions should allow a buffer zone, or a regional margin for error, and not try to 'game' the system with a narrow or non-existent protective measures.

If existing national and local biological diversity plans are examined through the analytic filter of Resilience, it will become evident where and how they meet this sort of 'resilience stress test', or where and how they may fail to do so. The analogy to human health is suggestive; annual medical physical examinations are a way to identify resilience problems in human health. Ecologists can study factors that enhance or prevent resilience in ecosystems (Salt & Walker, 2006). The Stockholm Resilience Center (2011) is engaged in studies of socio-economic applications of Resilience. It is time for environmental managers to explicitly study and teach Resilience theory and practice. They can, in turn, convey to all decision-makers the sort of factors that they need to consider to ensure that resilience is sustained, or restored, and enhanced.

With Resilience as a fundamental legal principle, society can build in the margin of safety that will let it recover from unforeseen disruptions associated with climate change and all the other environmental and non-environmental problems of the coming era. Resilience thinking entails mindfulness of our human deliberations. It disciplines us to be careful, and to avoid the hubris of assuming that we can cope with all that we encounter. The environmental manager will see more clearly than most the value of a Resilience Principle. As the NUS Masters of Environmental Management degree completes its first decade, it has shown a measure of Resilience. For it to expand its research and teaching to meet the coming needs, it will need to consciously apply the Principle of Resilience to its future design and coming growth.

VII. Conclusion

Environmental management has proven its worth as a foundation for attaining sustainable practices. Governments have embraced environmental management in many sectors, for example in the decision of the US National Forest Service to use ISO 14000 as the basis for its regulatory EMS in 2008. The European Commission developed its Eco-Management and Audit Scheme in 1993. Most multinational companies have established internal EMS. The demand for specialists educated to develop and apply these sustainability regimes is present and growing. It is likely that the demand will expand rapidly as governments and enterprises see the need to adapt to the impacts of climate change. For example, Reuters (2011) has reported a study concluding that 'countries and megacities in Africa and Asia are among the most vulnerable to the impacts of climate change over the coming years, a global survey shows, underscoring the risk from floods, rising sea levels, droughts and storms. With populations in many developing nations growing quickly, particularly in megacities with 10 million or more people, already creaking infrastructure could be overwhelmed by an increase in deadly disasters'.[4] In the coming years, universities will find that they need to expand their support for programmes in environmental management or create such programmes where they do not yet exist. If academic research and teaching comes to embed in environmental management teaching a role for resilience planning and practice, our MEM graduates will have the knowledge and skills to cope with the challenges of maintaining sustainable societies and economies amidst the uncertainties of climate change.

---

[4] For example, amongst the Asian cities facing extreme risk are Dhaka, Chittagong, Kolkata and Jakarta. Cities with high risk include Lahore, Delhi and Guangzhou. Cities with medium risk include Shanghai and Beijing.

References

Berry T (2000) *The Great Work*. Harmony Books, New York.

Cohen S (2011) *Sustainability Management: Lessons from and for New York City, America and the Planet*. Columbia University Press, New York.

MIT Sloan Management Review **51**, 1. Maurice B, Townsend A, Khyat Z, *et al.* Survey: The Business of Sustainability Now. [Article on the Internet] Fall 2009. Available from: http://sloanreview.mit.edu/the-magaizine/2009-fall/511808/.

Ottinger RL, Wooly RD, Robinson NAS, Hodas DR, Babb SE (1984) *The Environmental Costs of Electricity*. Oceana Publications, New York.

Pauli G (2010) *The Blue Economy – 10 Years, 100 Innovations, 100 Million jobs.*

Reuters. Factbox: Global climate change risk rating. [Article on the Internet] 25 October 2011. [Cited 27 Oct 2011]. Available from: www.reuters.com/assets/porint?aid=79P00A201110256.

Robinson NA (ed) (1993) *AGENDA 21: Earth's Action Plan*. Oceana Publications, New York.

Robinson NA (2009) Hedging against wider collapse: Lessons from the 'meltdowns'. In: Lye LH, *et al.* (eds) *Critical Issues in Environmental Taxation – International and Comparative Perspectives*, Vol. VII. Oxford University Press.

Saijo E. (2005) Redesigning the office copier – One manufacturer's efforts to conserve resources. [Article on the Internet] Available from: www.japanfs.org/en/mailmagazine/newsletter/pages/027881.html.

Salt D, Walker B (2006) *Resilience Thinking – Sustaining Ecosystems and People in a Changing World*. Island Press.

Stockholm Resilience Center. See research for governance of socio-ecological systems available at www.stockholmresilience.org. Stockholm University. [Article on the Internet] Cited 2011. Available from: www.stockholmresilience.org/.

UN World Commission on Environment and Development (1987) *Our Common Future*. The Brundtlund Commission Report. Oxford University Press.

United Nations Environment Programme. Paper for the UN's 2012 'Rio+20' Conference in Brazil. Submitted in April, 2011.